D0860478

Bioremediation Technologies for Polycyclic Aromatic Hydrocarbon Compounds

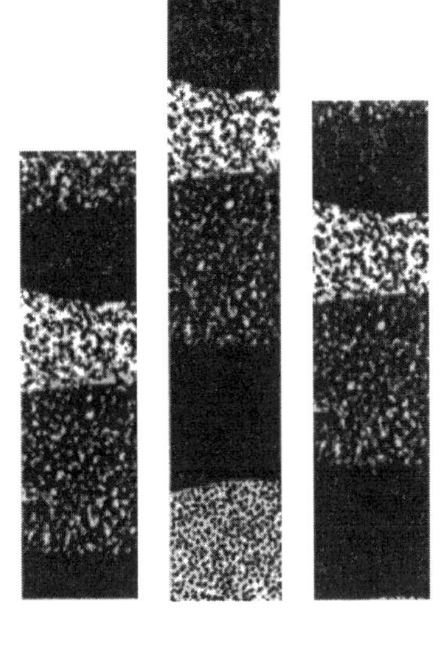

Editors

Andrea Leeson
and Bruce C. Alleman
Battelle

The Fifth International In Situ and
On-Site Bioremediation Symposium

San Diego, California, April 19–22, 1999

BATTELLE PRESS
Columbus • Richland

Library of Congress Cataloging-in-Publication Data

Bioremediation technologies for polycyclic aromatic hydrocarbon compounds / editors, Andrea Leeson and Bruce C. Alleman
p. cm.
Proceedings from the Fifth International In Situ and On-Site Bioremediation Symposium, held April 19–22, 1999, in San Diego, California.
Includes bibliographical references and index.
ISBN 1-57477-081-0 (hardcover : alk. paper)
1. Phytoremediation Congresses.
I. Leeson, Andrea, 1962– . II. Alleman, Bruce C., 1957–
III. International Symposium on In Situ and On-Site Bioremediation (5th : 1999 : San Diego, Calif.)
TD196.A75B566 1999
628.5'2--dc21 99-23398
CIP

Printed in the United States of America

Battelle Press
505 King Avenue
Columbus, Ohio 43201, USA
614-424-6393 or 1-800-451-3543
Fax: 1-614-424-3819
Internet: press@battelle.org
Website: www.battelle.org/bookstore

For information on future environmental conferences, write to:
Battelle
Environmental Restoration Department, Room 10-123B
505 King Avenue
Columbus, Ohio 43201-2693
Phone: 614-424-7604
Fax: 614-424-3667
Website: www.battelle.org/conferences

CONTENTS

FOREWORD

The Fifth International In Situ and On-Site Bioremediation Symposium was held in San Diego, California, April 19–22, 1999. The program included approximately 600 platform and poster presentations, encompassing laboratory, bench-scale, and full-scale field studies being conducted worldwide on a variety of bioremediation and supporting technologies used for a wide range of contaminants.

The author of each presentation accepted for the program was invited to prepare a six-page paper, formatted according to specifications provided by the Symposium Organizing Committee. Approximately 400 such technical notes were received. The editors conducted a review of all papers. Ultimately, 389 papers were accepted for publication and assembled into the following eight volumes:

Natural Attenuation of Chlorinated Solvents, Petroleum Hydrocarbons, and Other Organic Compounds – Volume 5(1)
Engineered Approaches for In Situ Bioremediation of Chlorinated Solvent Contamination – Volume 5(2)
In Situ Bioremediation of Petroleum Hydrocarbon and Other Organic Compounds – Volume 5(3)
Bioremediation of Metals and Inorganic Compounds – Volume 5(4)
Bioreactor and Ex Situ Biological Treatment Technologies – Volume 5(5)
Phytoremediation and Innovative Strategies for Specialized Remedial Applications – Volume 5(6)
Bioremediation of Nitroaromatic and Haloaromatic Compounds – Volume 5(7)
Bioremediation Technologies for Polycyclic Aromatic Hydrocarbon Compounds – Volume 5(8)

Each volume contains comprehensive keyword and author indices to the entire set.

This volume presents the latest research in polycyclic aromatic compound (PAH) remediation theory and practice. PAHs are common and challenging contaminants that affect soil and sediments. Methods for treating PAHs undergo continuing change and refinement. The papers in this volume cover topics ranging from the remediation of manufactured gas plant (MGP) sites to the remediation of sediments. The papers present laboratory and field studies, characterization studies, comparison studies, and descriptions of technologies ranging from composting to thermally enhanced bioremediation to fungal technologies and other innovative approaches.

We would like to thank the Battelle staff who assembled the eight volumes and prepared them for printing. Carol Young, Lori Helsel, Loretta Bahn, Gina Melaragno, Timothy Lundgren, Tom Wilk, and Lynn Copley-Graves spent many hours on production tasks—developing the detailed format specifications sent to each author; examining each technical note to ensure that it met basic page layout requirements and making adjustments when necessary; assembling the volumes; applying headers and page numbers; compiling the tables of contents and author and keyword indices, and performing a final check of the pages before submitting them to the publisher. Joseph Sheldrick, manager of Battelle Press, provided valuable production-planning advice and coordinated with the printer; he and Gar Dingess designed the covers.

The Bioremediation Symposium is sponsored and organized by Battelle Memorial Institute, with the assistance of a number of environmental remediation organizations. In 1999, the following co-sponsors made financial contributions toward the Symposium:

Celtic Technologies
Gas Research Institute (GRI)
IT Group, Inc.
Parsons Engineering Science, Inc.
U.S. Microbics, Inc.
U.S. Naval Facilities Engineering Command
Waste Management, Inc.

Additional participating organizations assisted with distribution of information about the Symposium:

Ajou University, College of Engineering
American Petroleum Institute
Asian Institute of Technology
Conor Pacific Environmental Technologies, Inc.
Mitsubishi Corporation
National Center for Integrated Bioremediation Research & Development (University of Michigan)
U.S. Air Force Center for Environmental Excellence
U.S. Air Force Research Laboratory Air Base and Environmental Technology Division
U.S. Environmental Protection Agency
Western Region Hazardous Substance Research Center (Stanford University and Oregon State University)

The materials in these volumes represent the authors' results and interpretations. The support of the Symposium provided by Battelle, the co-sponsors, and the participating organizations should not be construed as their endorsement of the content of these volumes.

Andrea Leeson and Bruce Alleman, Battelle
1999 Bioremediation Symposium Co-Chairs

NEW PLUG FLOW SLURRY BIOREACTOR FOR POLYCYCLIC AROMATIC HYDROCARBON DEGRADATION

Samia Gamati, Christian Gosselin, Eric Bergeron, Martin Chenier, Tri Vu Truong, Sodexen Group, Quebec, Canada
Jean-Guy Bisaillon, INRS-Institut Armand-Frappier, Quebec University, Canada

ABSTRACT : A pilot-scale plug-flow (PFR) reactor has been developed to efficiently treat polycylic aromatic hydrocarbon (PAH) contaminated soils and sediments. Innovations of the bioslurry reactor include specifically-designed Venturi jet aerators for optimal mixing and oxygen distribution, as well as the development of mixed bacterial cultures selectively adapted to high molecular weight PAHs. Rapid biodegradation is obtained because of enhanced mass transfer rates and optimal microorganism/contaminant contact. Based on microcosm degradation results, specific bioenhancing agents were added to the inoculated slurry to optimize bacterial activity and increase substrate availability. Recirculation of up to 50 % (w/v) slurry was achieved by the use of centrifugal pumps, in conjunction with the jet aerators and water nozzles. Volatile organic compounds were treated by biofiltration. Removal efficiencies ranged from 63 to 90 % depending on PAH molecular weights. Residence time was about 10 days, with longer periods needed for 4- to 6-ring PAHs. This pilot study provides performance information as basis for the up-coming 4 t/day demonstration-scale test, which will be conducted on a selected platform provided by the Montreal Center of Excellence in Brownfields Rehabilitation.

INTRODUCTION

Development of a new slurry-phase bioreactor is largely motivated by the need for more effective and cost-efficient remediation processes. Other technologies have shown their viability; however they are time-consuming and ineffective for reaching the desired level of decontamination in case of recalcitrant contaminants such as PAHs.

Slurry-based bioremediation processes offer an attractive option for recalcitrant organic-contaminated soil. They provide favourable conditions for enhanced mass transfer rates and optimal soil/hydrocarbon/microorganisms contact. Several reports described bioslurry reactors for PAH soil biodegradation (Lewis, 1993; Jerger, 1993; Brown et al, 1995; Puskas et al, 1995; Woodhull and Jerger, 1995; Zappi et al. 1996; Banergee et al. 1997; U.S.EPA, 1997; Mulder et al. 1998; Tabak et al. 1998). These are mostly based on continuous-stirred tank reactor (CSTR) design equipped with mechanical agitation alone or in conjunction with air diffusers. High energy requirements are associated with such mechanical devices. Moreover, although some success has been achieved in these slurry treatments, optimization of certain key operating conditions, such as increased slurry solid concentration, reduced energy and maintenance, or reduced residence time, still needs detailed investigations.

On the other hand, jet loop bioreactors have long been used for biological wastewater treatment and were used to operate continuous high cell density fermentation (Garcia-Salas and Flores-Cotera, 1995). However, information on the efficiency and limitation of high slurry solid content in a jet aerated plug-flow reactor (PFR) for PAH-soil treatment and subsequent scale-up has not been studied.

OBJECTIVE. The objective of this study is to evaluate and optimize a pilot-scale plug flow reactor for enhanced PAH-contaminated soil biodegradation. One main goal is to investigate the feasibility of operating a 50 % (w/v) soil:water ratio with minimum energy consumption and maximum oxygenation for optimal removal of the EPA priority PAHs. Another focus of the study is to investigate the effect of specific bioenhancing agents on the biodegradation activity of the developed bacterial consortium. Pilot study results will provide necessary data for the upcoming 4 t/day demonstration phase test, that will be conducted on a selected site provided by the Montreal Center of Excellence in Brownfields Rehabilitation.

MATERIALS AND METHODS

Soil. The soil used for this study came from a contaminated site in the Montreal region. This site is known to have received industrial wastes between 1966 and 1968. Soil sample was composed of 51 % sand, 27 % silt and 22 % clay. The organic content was 2 % by weight. Total PAHs and total petroleum hydrocarbons concentrations were 746 mg/kg and 3 581 mg/kg respectively.

Bacterial consortium. Adapted mixed bacterial cultures were obtained by enrichment from a PAH-contaminated site. Cultures were maintained on Bushnell-Haas minimal salt medium (BH), supplemented with 100 mg/L of phenanthrene, pyrene, benzo(a)pyrene and 0.025 % yeast extract (Dagher et al, 1997).

Microcosm studies. 50 % (w/v) slurry were prepared in 500 mL Erlenmeyer flasks containing a 200 mL BH medium. Inoculated (10 % v/v) microcosms, were agitated at 180 rpm and incubated at 25 ^{0}C in the dark. Each reactor received Tergitol 0.01 % (v/v), organic compost 0.02 % (w/v), salycilate 110 mg/L and succinate 15 mg/L. Control microcosms, containing sterile soil, were included for abiotic loss determination.

Pilot scale reactor . Cylindrical pilot-scale PFR 200 L is equipped with a series of Venturi jet aerators placed at equal distances along the reactor. A schematic diagram of the design is shown in Fig.1. Mixing and aeration is provided by submerged jet aerators supplied with specific diffusers and water nozzles. Bacterial consortium (10 % v/v), amendments (Tergitol 0.01 % (v/v), salycilate 110 mg/L and succinate 15 mg/L) and water are added together with the screened soil to prepare 50 % (w/v) slurry. The slurry is mixed and recirculated by the use of a centrifugal pump and the jet aerators. Temperature is controlled by a heat exchanger. Collected volatile organic compounds are treated by a biofilter.

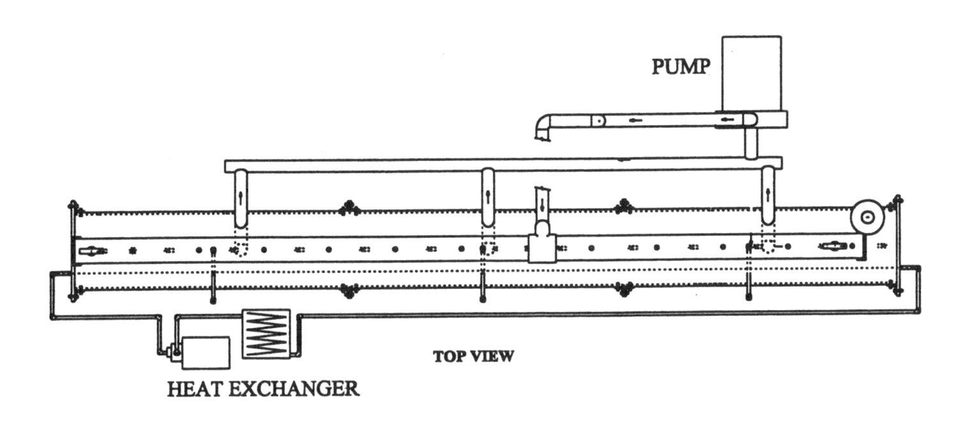

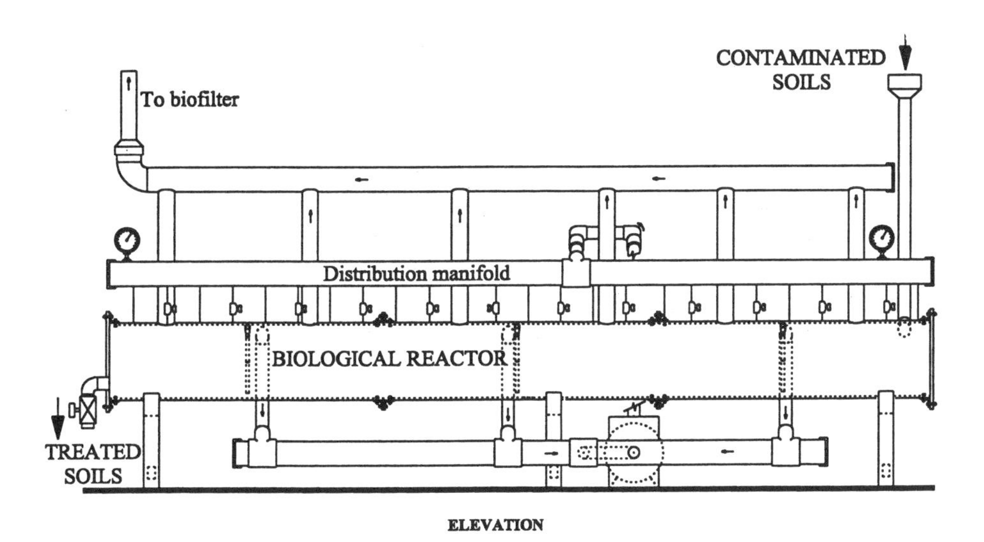

FIG.1: SCHEMATIC DIAGRAM OF THE PILOT PLUG FLOW REACTOR (PFR) FOR SOIL REMEDIATION

Biodegradation Procedures. The PFR was continuously fed with 50 % (w/v) slurry and operated at room temperature. Inoculated mixed cultures contained total heterotrophic bacteria at a concentration of 9.8 x 10^6 CFU/ mL. Nutrients (ammoniacal-N and inorganic phosphate) were adjusted according to a C/N/P ratio of 110/10/1. Salicylate, a known inducer of naphthalene catabolic operon on NAH plasmids (Gamati and Greer, 1993), and succinate, a by-product of naphthalene metabolism, were added at respective concentrations of 110 mg/L and 15 mg/L to enhance PAH degradation. Reactor pH, temperature and dissolved oxygen were monitored daily, while ammoniacal nitrogen, nitrate nitrogen, total inorganic phosphate, residual PAH concentrations and total solids were analyzed every 3 days for a 20-day operation period. Samples were centrifuged to monitor solid and aqueous phases. Total heterotrophs and PAH degraders were enumerated by spread plate technique at similar intervals. PAHs were analyzed using GC/MS according to a modified EPA Method 8270. Calculated PAH removal rates were adjusted for abiotic losses.

RESULTS AND DISCUSSION

Results obtained demonstrated the feasibility of operating the continuously-fed plug-flow reactor at high slurry solid content to treat PAH-contaminated soil. The bioslurry reactor allowed adequate operational conditions and mass transfer rates. A previously conducted study has shown the superiority of a PFR over a CSTR to efficiently treat hydrocarbon-contaminated soil (Bergeron et al. 1997).

Initial operation tests showed that increasing the slurry solid content above 40 % required certain design modifications to move settled material. These included the number, size, position and distribution of the aerators. Internal shape of the reactor was also modified in order to obtain adequate mixing and solid suspension.

A major goal of this pilot study was to demonstrate that oxygenation and mixing of a 50 % (w/v) solid content was provided by the developed aeration system. Results showed that modifications incorporated in the nozzles, the jet aerator diameter size and internal reactor shape, were key factors for vigorous mixing and optimal aeration. During process operation, the reactor temperature was maintained at 25 ^{0}C ± 5 ^{0}C, the pH at 7.1- 7.7 and the dissolved oxygen concentrations from 3.5-7.8 mg/L throughout the reactor. These oxygen levels demonstrated the efficiency of the aeration system to ensure oxygenation levels needed for organic oxidation. PAH-utilizing bacteria inoculated at a concentration of 2.7×10^5 CFU / mL, were maintained at 3.8×10^7 - 1.1×10^8 CFU / mL.

Results, presented in Figure 2, indicate a rapid biodegradation (80-90 %) of the more readily-available 2- and 3- ring PAHs after a residence time of about 10 days. Higher phenanthrene degradation (79.3 %) over anthracene (60.5 %) was probably due to its higher solubility. Similar observations were reported by Banergee et al. (1995). Overall biodegradation rates of more sorbed 4- to 6-ring PAHs were low, suggesting the need for longer residence times, improved bioavailability and better operation control during treatment.

Parallel microcosm biodegradation studies, revealed that pyrene and benzo(a)pyrene removal rates average was 60 % following a 21-day incubation. Under these conditions, the bacterial consortium, enriched and maintained on a mixture of PAHs containing phenanthrene, pyrene and benzo(a)pyrene for more than a year, has developed the potential to readily metabolize 4- and 5-ring PAHs.

Moreover, the nutrients found in the solid organic matrix of the compost, including nitrogen and phosphorous, contributed to enhance PAH degradation. Consistently, Kästner and Mahro (1996) reported a stimulatory effect on soil PAH bidegradation upon organic compost addition. They attributed this effect to increased bioavailability and activation of cometabolic degradation mechanisms of high molecular weight PAHs.

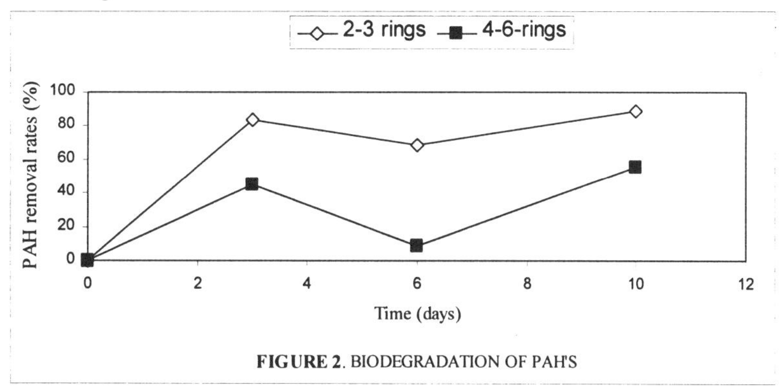

FIGURE 2. BIODEGRADATION OF PAH'S

Optimization of the pilot PFR operation and high molecular weight PAH degradation is under way using organic compost and an emulsifying agent to increase their bioavailability. Additional performance evaluation of the PFR will be obtained during the upcoming 4 t/day demonstration phase that will be conducted on a site provided by the Montreal Center of Excellence in Brownfields Rehabilitation.

ACKNOWLEDGEMENTS : The authors gratefully acknowledge the National Research Council of Canada, Environment Canada, Economic Development of Canada and the Centre Québécois de Valorisation des Biomasses et des Biotechnologies for financial participation in this project.

REFERENCES

Banerjee D.K., Gray M.R., Dudas M.J. and Pickard M.A. (1997). « Protocol to enhance the extent of biodegradation of contamination in soil ». *In Situ On-Site Biorem., 4th Symp. 5: 163-168. Battelle Press, Columbus, Ohio.*

Bergeron E., Gosselin C., Gamati S., Truong TV. and Bisaillon J.G. (1997). « Economic study on the feasibility of treating hydrocarbon-contaminated soils in a new slurry phase bioreactor ». *GASReP/AMERICANA Symposium, March, Montreal, QC*

Brown K.L., Davila B., Sanseverino J., Thomas M., Lang C., Hague K. and Smith T. (1995). « Chemical and biological oxidation of slurry-phase polycyclic aromatic hydrocarbons ». *Biological unit processes for hazardous waste treatment.* 9 :113-115

Dagher F., Deziel E., Lirette P., Paquette G., Bisaillon J.G. and Villemur R. (1997). « Comparative study of five polycyclic aromatic hydrocarbon degrading bacterial strains isolated from contaminated soils ». *Can. J. Microbiol.* 43 :368-377

Gamati S. and Greer W.C. (1993). «Enhanced bioremediation of naphthalene-contaminated soil using non-radioactive gene-probe enriched indigenous bacteria ». *GASReP/BIOQUAL Symposium, Sept, Quebec, QC*

Garcia-Salas S. and Flores-Cotera L.B. (1995). « Influence of operating variables on liquid circulation in a 10.5-m^3 jet loop bioreactor ». *Biotechnology and Bioengineering.* 46 :408-414

Jerger D.E. (1993). « Slurry phase treatment of wood preserving wastes : A practical success story ». *Intertech Conferences, Applied Bioremediation 93, Fairfield, New Jersey*

Kästner M. and Mahro B. (1996) « Microbial degradation of polycyclic aromatic hydrocarbons in soils affected by the organic matrix of compost ». *Appl. Microbiol. Biotechnol.* 44 :668-675

Lewis R.F. (1993). « SITE demonstration of slurry-phase biodegradation of PAH contaminated soil ». *AIR and WASTE.* 43 :503-508

Mulder H., Breure A.M., Van Andel J.G., Grotenhuis J.T.C and Rulkens W.H. (1998). « Influence of hydrodynamic conditions on naphthalene dissolution and subsequent biodegradation ». *Biotechnol. Bioeng.* 57(2) :145-154

Tabak H.H., Govind R., Gao C. and Fu C. (1998). « Protocol for evaluating biokinetics and attainable end-points of polycyclic aromatic hydrocarbons (PAHs) in soil treatment ». *J. Environmental Science and Health, Part A : Toxic/Hazardous Substances & Environmental Engineering.*33(8) :1533-1567

U.S. Environmental Protection Agency Report (1997). *Innovative methods for bioslurry treatment.* SITE, EPA/540/SR-96/505

Woodhull, P.M. and Jerger D.E. (1995). « Temperature effects on kinetics and economics of slurry-phase biological treatment ». *Microbial Processes for Bioremediation* 8 :289

Zappi M.E., Rogers B.A., Teeter C.L., Gunnison D. and Bajpai R. (1996). « Bioslurry treatment of a soil contaminated with low concentrations of total petroleum hydrocarbons ». *J. Hazard. Mater.* 46(1):1-12

TREATMENT OF TOWN-GAS SOIL POLLUTANTS WITH BIOSLURRY CONDITIONS

John A. Glaser, Paul T. McCauley, and Ronald Herrmann (USEPA, National Risk Management Research Laboratory Cincinnati, OH), and Majid A. Dosani (IT Corporation, Cincinnati OH)

Abstract: Soil, contaminated with town-gas pollutants, was excavated from the Calhoun Park property located in Charleston, SC. Total polynuclear aromatic hydrocarbon (PAHs) encountered in the starting blended soil ranged from 2100-3400 mg/kg in the soil. In past slurry research, we treated nutrient composition as an unimportant component for consideration. Recent studies suggest that ther may be some significance to nutrient composition (Tabak, 1998). A series of five nutrient mixtures, supplemental carbon sources to assist biomass growth, and process controls were evaluated to determine the effect of each experimental condition on the biodegradation of polynuclear aromatic hydrocarbons in the slurry. The conditions showing greatest removal of total PAHs consisted of nutrients, and 0.1 vol % molasses. Treatment results will be discussed along with an unanticipated problem of extensive pH drift encountered during treatment.

Introduction

Town gas properties are often found among those referred to as "brownfields". Generally, they are proximate to population centers since their products were important to commerce about a century ago. The thermally based technology extracted, from coal and other carbonaceous materials, volatiles and gases that could be transformed into an early form of combustible gas called "producer gas" which is lower in heat producing content than natural gas. The process residues were tars and some low molecular weight organic aromatic hydrocarbons, cyanide and sulfur where the coal was of high sulfur content.

The treatment of PAH contamination derived form these sources has received considerable attention in the past ten years (Pradhan et al, 1994; Srivastava et al., 1994; Ginn et al. 1994). A survey of the literature shows that significant difficulties have been encountered with all treatment technologies pointing to need for the properly deigned application of bioremediation (Cutright & Lee, 1994; Luthy et al., 1994).

The soil for this study was obtained from the Calhoun Park site in Charleston, South Carolina. Based on historical land use information, the contamination is believed to have originated from a manufactured gas plant (including coal-gas and water-gas). A portion of the site operated as a coal-gas plant from 1855 to 1910 and as a water-gas plant from 1910 to 1957. Coal tar is produced as a by-product of these processes. The manufactured gas portion of the site also operated as a propane distribution facility (1957 to 1979) and as an electrical substation (1979 to the present). Other portions of the site have operated as a saw mill, lumber yard, chemical company (creosote production, paint manufacturing, and charcoal production), junk yard, steel storage yard, Naval store yard (rosin yard), playground, dog pound, and housing project (USEPA, 1998).

Soil Classification and Characterization. Twenty-two 55-gallon drums of contaminated soil were excavated from the site to provide sufficient material for evaluations of bioslurry, composting and land treatment techniques. The raw soil was size classified by passing it through a two-stage vibrating one-inch screen to remove large rocks and debris. The undersized material was blended in a cement mixer and mixed for 15 minutes to obtain a mixed soil composition for research investigations. One drum of the homogeneous soil was passed through a 1/4-inch screen for use in the bioslurry study.

Experimental Design. Past bioslurry treatment investigations conducted by the authors exhibited little response to changes of nutrient composition. Small scale studies conducted by coworkers showed a significant change in the extent of treatment based on nutrient composition (Tabak, 1998). The current study was designed to investigate the importance of these findings at a larger scale of operation. We chose to investigate two nutrient compositions to determine their effect on the extent of bioslurry treatment for representative contaminated soils.

Five experimental conditions were tested in duplicate using 10 bench-scale and 2 pilot-scale bioslurry reactors. The tests were designed to determine the effect of several variables on the biodegradation of PAHs. The experimental conditions are shown in Table 1. Each of the test conditions utilized a 20 weight percent solids slurry. The first condition used inorganic salts (ammonium sulfate and dibasic ammonium phosphate) as nutrients for the microbes. The second condition used these same nutrients and 0.1 volume percent molasses as a putative cometabolite. The third condition used a complex set of inorganic nutrients and a vitamin solution developed by the Organization for Economic Cooperation Development (OECD). The fourth condition used the OECD nutrients and 0.1 volume percent molasses. The fifth condition used 2 weight percent formalin as a "kill control" reactor to evaluate abiotic removal of PAHs. The third experimental condition (slurry + OECD) used in the bench-scale reactors were duplicated in the two pilot-scale bioslurry reactors. The pilot-scale reactors were primarily used to obtain a sufficient amount of post-treatment sample at the end of the study to perform toxicological analysis. Nutrient levels for all of the experimental conditions were adjusted to provide a 100:10:1 carbon:nitrogen:phosphorus (C:N:P) ratio. Because the recipe for OECD contains a large amount of phosphorus, additional ammonium chloride was added to achieve the 100:10:1 C:N:P ratio.

Table 1. Experimental Design.

Reactor	Treatment Condition
1 & 2	Standard Nutrients[b]
3 & 4	Standard Nutrients[b] & 0.1 vol.% Molasses
5 & 6	OECD Nutrients[b]
7 & 8	OECD Nutrients[b] & 0.1 vol.% Molasses
9 & 10	2 wt.% Formalin

[a] Each test condition was evaluated in duplicate, the first 10 reactors are 8L bench reactors and reactors 11 & 12 are 70L pilot reactors, at 20 wt % solids..

[b] Nutrient levels were adjusted to provide a 100:10:1 carbon:nitrogen:phosphorus ratio.

RESULTS AND DISCUSSION

The presence of sulfur and its significance to treatment results was not recognized at the start of the investigation. Fortunately, daily pH was part of the operation protocol since significant pH drift was observed and adjusted by caustic solution addition. This daily vigilance kept the pH fluctuations to a range of 5 and 8 except for one occasion where pH 4.0 was observed. The lowering of the reactor pH was attributed to the presence of sulfur-oxidizing microorganisms but was not confirmed by microbiological assay. The two best treatment involved the use of molasses as a supplemental carbon source.

Each treatment condition exhibited comparable capability (84 to 90% removal) to degrade 2 & 3 ring PAHs. The large (4-6) ring PAHs were found to be treated at a average of 37% and a range of 27-54% for all conditions except those including molasses. The molasses amended treatment conditions showed and average of 57% removal for the large (4-6) ring PAHs and a range of 41-69%. These results are under further scrutiny to understand the true factors contributing to this enhance treatment.

Reductions of 31-82% total PAH concentrations were observed for the range of treatments in the experimental design. Two different treatments exhibited comparable removal of total PAHs of 78 & 84%, small PAHs 90 & 90% and large (4-6 ring) PAHs of 60 & 69% (Figure 1). The Figures 1.A & B correspond to the treatment observed in reactors 3 and 8 respectively of the experimental design. Statistical analysis of the PAH analyte concentrations for each treatment condition shows them to be indistinguishable. Control conditions (reactor 9 & 10) were designed to be formaldehyde killed treatment conditions. Previous research using a one-time addition of formaldehyde failed to maintain the intended sterility throughout the entire treatment period. The failure was confirmed by the regrowth of large (10^8 - 10^9) populations of heterotrophs. We expect that the formaldehyde was simply air-stripped after which regrowth ensued. In this study reactors 9 & 10 were dosed weekly with formaldehyde without significant effect. The continued

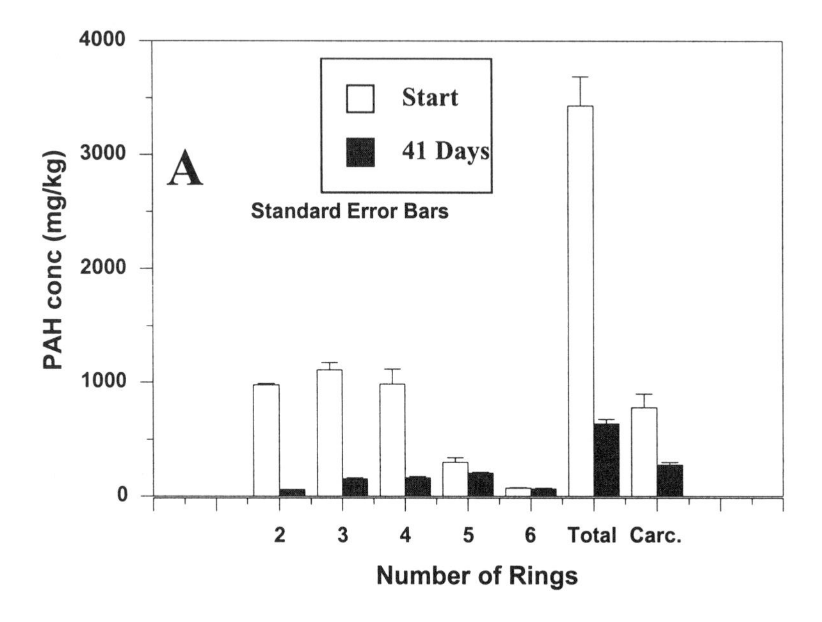

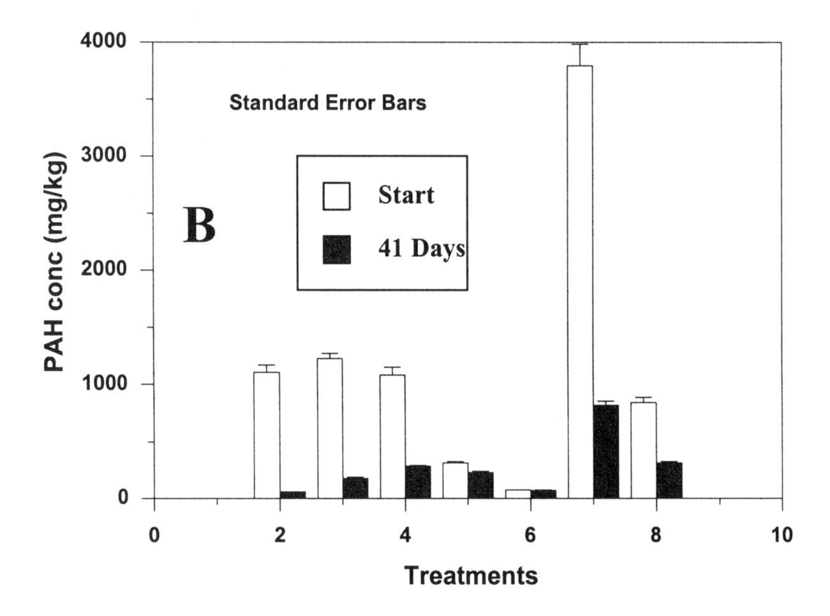

Figure 1. Most Effect Treatments in Experimental Design

degradation of PAHs and viable population regrowth indicated that only more drastic disinfection techniques offer the expected conditions of a killed control.

Supplemental information derived from phospholipid analysis of the biomass for these two reactors showed them to be significantly different than the other members of the experimental design (White et al., 1979). The comparison of short and long term stress markers in the phospholipid analysis lead to a greater understanding of bioavailability limitations and the "hockey stick" phenomena associated with many forms of bioremediation (Brouwer et al, 1994). The interpretation of these results offer additional understanding of the limitations of microbial processes to treat soil contamination. These ideas may offer opportunities to improve bioremediation processes to overcome the bioavailability conditions that have been suggested as controlling the extent of treatment possible.

REFERENCES

Bouwer E, Durant N, Wilson L, Zhang W, Cunningham A. 1994. "Degradation of xenobiotic compounds in situ: capabilities and limits", *FEMS Microbiol. Rev.* *15*(2-3):307-17

Cutright, T. J. and S. Lee 1994. "Bioremediation: A competitive alternative for the cleanup of contaminated MGP sites." *Energy Sources 16*(2): 269-277.

Ginn, J. S., R. C. Sims, et al. 1995. "Evaluation of biological treatability of soil contaminated with manufactured gas plant waste." *Haz. Waste. Haz. Mater.* *12*(3): 221-232.

Luthy, R.G., D.A. Dzombak, C.A. Peters, S.B. Roy, A. Ramaswami, D.V. Nakles, B.R. Nott. 1994. "Remediating Tar-Contaminated Soils at Manufactured Gas Plant Sites", *Environ. Sci. Technol. 28*:266A-276A.

USEPA 1998. Calhoun Park Superfund Site, Fact Sheet, Region IV, Atlanta, GA.

Pradhan, S. P., J. R. Paterek, et al. 1997. "Pilot-Scale Bioremediation of PAH-Contaminated Soils." *Appl. Biochem.Biotechnol. 63*:759-773.

Srivastava, V. J., R. L. Kelley, et al. 1994. "A field-scale demonstration of a novel bioremediation process for MGP sites." *Appl. Biochem.Biotechnol. 45-46*: 741-756.

Tabak, H.H. 1998. Personal communication.

White, D. C., W. M. Davis, J. S. Mickels, J. D. King, and R. J. Bobbie, 1979. Determination of the sedimentary microbial biomass by extractable lipid phosphate, Oecologia 40:51-62.

DEMONSTRATION OF THE NATURAL ATTENUATION OF PHENOLS AT A COKING FACILITY

Mary L. Presecan (Parsons Engineering Science, Denver, Colorado)
Kent A. Friesen (Parsons Engineering Science, Denver, Colorado)

ABSTRACT: A recent study conducted at a coking facility in the Western U.S. demonstrated the natural attenuation of site-related phenolic compounds in a shallow groundwater plume. Contamination of the soil and shallow groundwater at the coking facility by phenolic compounds is the result of historical disposal of tar and tarry water from the plant's coking process. Evidence of biodegradation of phenolics documented during the field investigation included changes in contaminant concentrations over time; biodegradation-signature distributions of geochemical indicators such as reduction/oxidation potential, dissolved oxygen, nitrate, ferrous iron, sulfate, methane, and carbon dioxide; and characterization of an *in-situ* microbial mass through a phospholipid-ester-linked fatty acid (PFLA) analysis.

INTRODUCTION

When plant operations at a coking facility in the Western U.S. began in 1960, an unlined 22 acres [89,030 square meters (m^2)] evaporation pond was constructed for disposal of tar, decanter tank sludge, and tarry water associated with the coking process. Discharge of tarry water to the pond was discontinued in 1985, and the tarry water pond was closed in accordance with the Resource Conservation and Recovery Act (RCRA). A RCRA facility investigation (RFI) was conducted in 1998 to assess the nature and extent of contamination.

Site Description. The plant is located in a geographic area characterized by gently sloping ridges and draws. The local stratigraphy is characterized by a highly weathered to unweathered shale ranging in thickness from 3,000 to 5,000 feet (ft) [914 to 1524 meters (m)], which acts as a confining unit to shallow, unconfined to semi-confined groundwater. Groundwater at the plant flows from north north-west to south-south east in a single groundwater system. The depth to groundwater at the plant ranges from 5 to 50 ft below ground surface (bgs) (1.5 to 15.2 m bgs).

BACKGROUND

Phenolic compounds of concern encountered in the groundwater at the coking facility include phenol, o-cresol, m-cresol, p-cresol, and 2,4-dimethylphenol. Biodegradation of phenol and phenolic compounds, which are similar to fuel hydrocarbons in chemical structure, is expected to proceed in the environment resulting in methane, carbon dioxide (CO^2), water (H_20), and other simple inorganic compounds such as ammonia (NH^3) and phosphate (PO_4^{3-})

(Ehrlich *et al.*, 1982). The areal and temporal distribution of contaminants and byproducts of microbially mediated biodegradation reactions can be used to indicate the occurrence and pathways of biodegradation of phenols.

RESULTS

Phenolic compounds were detected in 25 of the 75 groundwater samples, defining a broad plume emanating from the former tarry water pond. Maximum detections for individual phenols were: 9,300 μg/L 2,4-dimethylphenol, 77,000 μg/L co-eluted m- and p-cresols, 18,000 μg/L o-cresol, 64,000 μg/L phenol, and 170,000 μg/L total phenolics. Neither polynuclear aromatic hydrocarbons (PAHs) nor xylenes were detected in the groundwater.

Temporal Trends. The areal distribution of total phenols in groundwater in 1985 and 1998 are presented in Figure 1. Examination of the areal distribution and concentrations of total phenols in groundwater indicate that the concentrations and mass of total phenols in groundwater within and downgradient of the source area decreased between 1985 and 1998.

a)

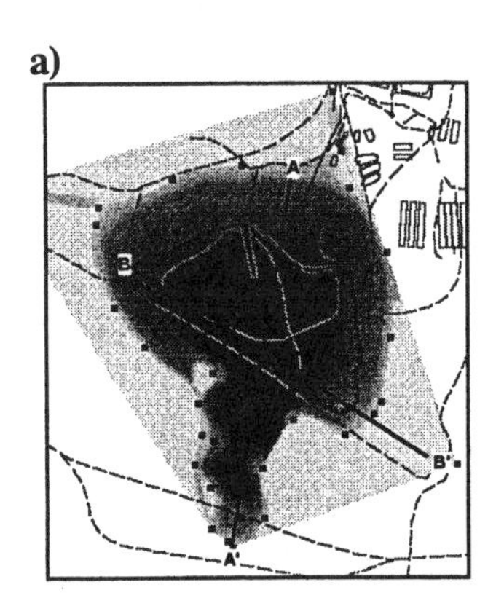

b)

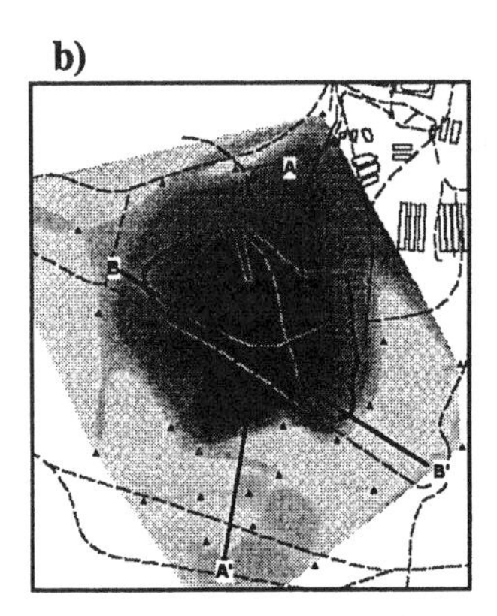

FIGURE 1. Areal distribution of total phenols in groundwater in a) 1985 and b) 1998, with cross-section A-A' and B-B' locations.

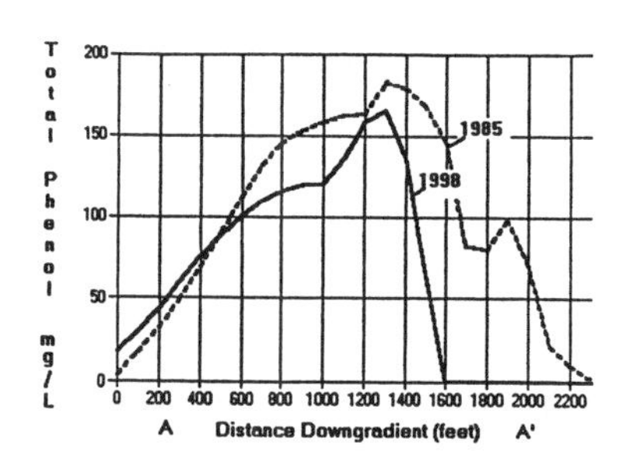

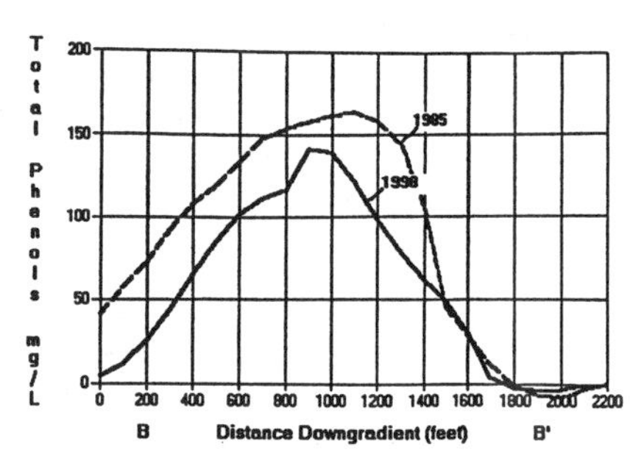

FIGURE 2. Vertical profile of total phenols in groundwater in 1985 and 1998 along cross-sections A-A' and B-B'.

Comparison of the concentrations of phenolic constituents detected during the two sampling events along the axis of southerly migration (cross-sections A-A' and B-B'; Figure 2), confirms that contaminant mass decreased between 1985 and 1998, and that total phenols did not migrate further to the south (cross-section A-A') or southeast (cross-section B-B') during that period.

Geochemical Evidence. Groundwater monitoring data demonstrate the occurrence of *in-situ* biodegradation of phenolic compounds based on spatial trends of dissolved oxygen (DO), nitrate, ferric iron, sulfate, CO^2, methane, alkalinity, pH, temperature, and oxidation/reduction potential (ORP).

DO concentrations collected during the latest round of groundwater sampling at the coking facility ranged from 0.0 to 5.1 mg/L, with the highest concentrations of DO generally measured in wells outside the area affected by phenolic compounds. Figure 3(a) illustrates the distribution of total phenolics in groundwater and Figure 3(b) illustrates the distribution of DO in groundwater at the coking facility. Comparison of the distribution of total phenolics with the areal distribution of DO in groundwater indicates decreased DO concentrations throughout most of the phenolics plume, providing strong evidence that DO is being utilized as an electron acceptor in the biodegradation of phenolic compounds. The higher DO concentrations in groundwater downgradient from the source area suggest that phenolic compounds migrating to the southern part of the coking facility property are degrading aerobically.

After DO has been consumed, anaerobic conditions prevail. Figure 3(c) shows the distribution of nitrate/nitrite (as N) in shallow groundwater at the coking facility. Comparison of the distributions of total phenolics and nitrate reveals that the area of depleted nitrate/nitrate (as N) corresponds to the area of elevated concentrations of phenolic compounds, expected as a result of denitrification. Similarly, elevated ferrous iron concentrations [Figure 3(c)] correspond to the area of elevated concentrations of total phenolics, suggesting that iron (III) is being reduced to ferrous iron by biodegradation processes. Depleted sulfate concentrations [Figure 3(d)] also correspond with the area of elevated concentrations of phenolic compounds, suggesting that sulfate is being reduced to sulfide via biodegradation.

Methanogenesis generally occurs after oxygen, nitrate, iron (III), and sulfate have been depleted in the subsurface. Under the strongly reducing conditions of methanogenesis, carbon dioxide is utilized as an electron acceptor, producing methane. As shown on Figure 3(e), carbon dioxide concentrations are elevated above background concentrations throughout the area of the phenolics plume. The elevated carbon dioxide concentrations are the result of aerobic and anaerobic phenolics biodegradation processes. In addition, elevated methane concentrations are present in groundwater samples containing elevated concentrations of phenolic constituents, suggesting that methanogenesis is occurring within the source area. The occurrence of methanogenesis is further supported by the presence of ethane and ethene in groundwater samples corresponding to elevated concentrations of methane. The presence of ethane and ethene in groundwater suggest that, at least in some areas, the biodegradation of phenolic compounds is proceeding to completion.

Negative ORPs at the coking facility occur directly downgradient from the source area [Figure 3(f)]. The positive ORP values measured in the former tarry water pond may be a result of the toxic effects of elevated concentrations of phenolic compounds on the microbial population. Downgradient from the former tarry water pond, within the area containing more dilute concentrations of phenolic constituents, ORPs are negative; DO, nitrate, and sulfate are depleted; and concentrations of ferrous iron, carbon dioxide, and methane are elevated, indicating that phenolic compounds in shallow groundwater are being biodegraded anaerobically.

Groundwater temperature at the coking facility ranged from 6.7 to 10.7°C, which is suitable for subsurface microorganisms. The groundwater pH measured at the site ranged from 5.67 to 7.85 standard units, which is within the optimal range for most microbial populations that degrade organic matter.

Microcosm Evaluation. Microcosm studies were performed on groundwater samples at the coking facility to evaluate whether a viable microbial community capable of degrading phenolic compounds was locally present, and to identify potential stresses that may be affecting the microbial community. Groundwater samples were collected from locations upgradient, downgradient, and within the plume, and were analyzed using molecular techniques to identify and characterize the phospholipid fatty acid (PLFA) characteristics of the microbial community, thereby providing additional evidence of natural attenuation of phenolic compounds by microorganisms (Microbial Insights, 1998).

Results from the PFLA microbial evaluation indicated that downgradient of the source are, where substrate is readily available, there exists a diverse biomass characterized by a high turnover rate and an active growth phase. This downgradient sample was the only groundwater sample collected at the coking facility to contain significant quantities of sulfate- or iron-reducing bacteria. Conversely, within the source area, the microbial community appears to be inhibited by the toxic effects of elevated contaminant concentrations. While biodegradation is occurring within the source area, the process is not as efficient as observed at downgradient locations, where decreased concentrations of phenolic compounds are presumably less toxic to the microbial communities. As would be expected, the microbial community upgradient exhibits a slow growth/turnover indicative of a stationary growth phase where the microbial community is not experiencing drastic change as a result of the introduction of foreign material.

Estimate of Decay Rates. Chemical concentration information was used to estimate first-order rate constants for phenolic compounds at the coking facility. The Buscheck and Alcantar (1995) method for determining first-order rate constants for BTEX compounds was used because of the molecular similarities between BTEX and phenolic compounds. This method involves coupling the regression of contaminant concentration versus distance with an analytical advection-dispersion equation for one-dimensional, steady-state contaminant transport (Bear 1979).

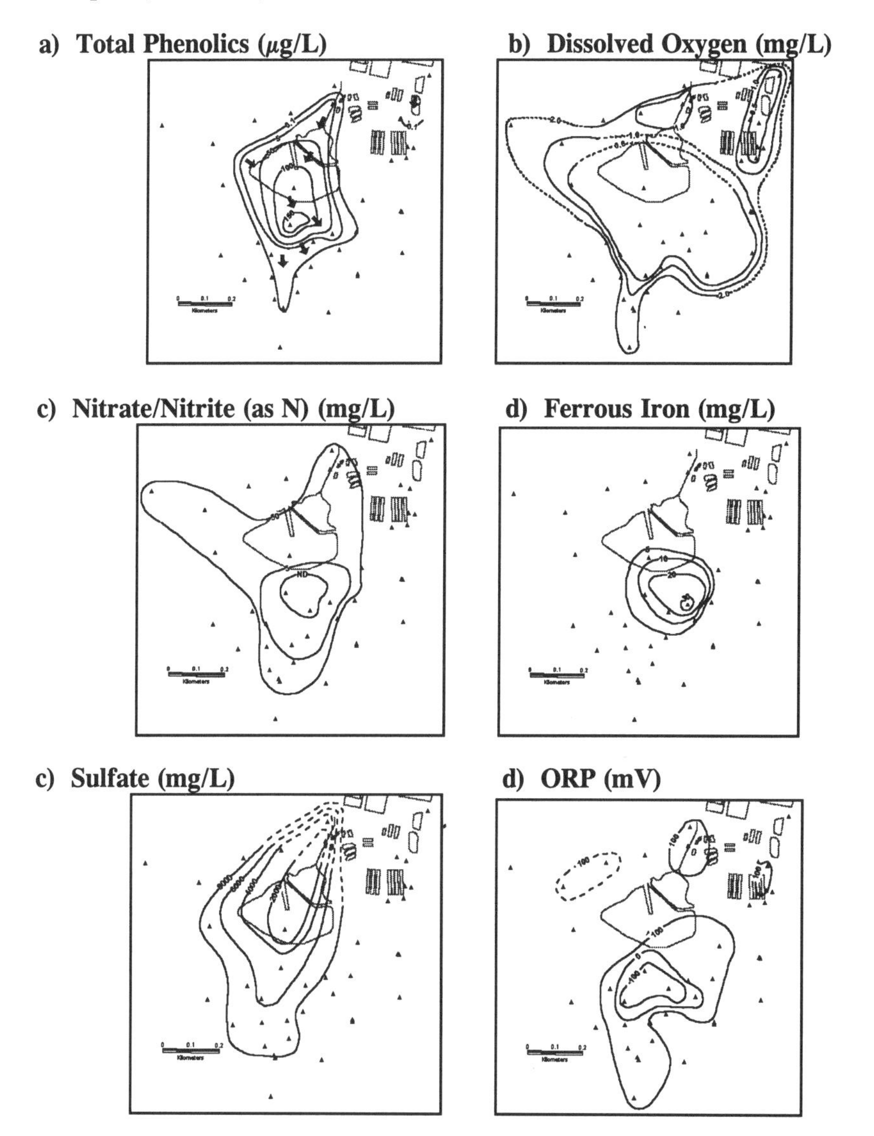

FIGURE 3. Areal distribution of geochemical indicators of biodegradation including a) total phenolics, b) DO c) nitrate d) ferrous iron e) sulfate and f) ORP.

Table 1 summarizes the first-order, steady-state degradation rates and half-lives for site related contaminants. Decay rates calculated using the results of analyses of groundwater samples are about two orders of magnitude lower than rates observed in the laboratory. (Howard *et al.*, 1990) This may be a consequence of local geologic and hydrogeologic conditions, limited groundwater recharge, poor background water quality, community structure and metabolic status of the local microbial population, or conditions locally toxic to the microbial community as a result of elevated concentrations of phenolic constituents.

TABLE 1 Calculated biodegradation rates of selected phenolic compounds.

Compound	Decay Rate ($days^{-1}$)	Half-Life (days)	Half-Life (years)
Total Phenols	$1.45x10^{-4}$	5525	15.2
Phenol	$7.72x10^{-4}$	897	2.5
o-Cresol	$3.03x10^{-4}$	2290	6.3
2,4-Dimethylphenol	$4.35x10^{-4}$	2380	6.5

CONCLUSIONS

Several lines of evidence indicate that phenolic compounds are undergoing biodegradation within and downgradient from the source area at the coking facility. The results of historic groundwater monitoring demonstrate that the concentrations of total phenolic constituents within and downgradient of the source area have decreased through time, suggesting that degradation mechanisms are attenuating phenolic constituents in groundwater. The areal distribution of phenolic constituents, electron acceptors, and end products of biodegradation indicates that aerobic and anaerobic degradation of phenolic compounds is occurring in the subsurface at the site. Microcosm evaluations of the site indicate that microbial communities capable of degrading phenolic compounds are present, and that those communities are thriving in downgradient locations.

REFERENCES

Bear, J., 1979, *Hydraulics of Groundwater*. McGraw-Hill, Inc., New York, New York, 569p.

Buscheck, T. E., and Alcantar, C. M., 1995, "Regression Techniques and Analytical Solutions to Demonstrate Intrinsic Bioremediation." In: Proceedings of the 1995 Battelle International Symposium on In Situ and On-Site Bioreclamation, April 1995.

Ehrlich, G. G., D. F. Goerlitz, E. M. Godsy, and M. F. Hult. 1982. "Degradation of Phenolic Contaminants in Groundwater by Anaerobic Bacteria: St. Louis Park, Minnesota," *Ground Water,* 20(6):703-710.

Howard, Phillip H. 1990. *Handbook of Environmental Fate and Exposure Data for Organic Chemicals - Volume I. Large Production of Priority Pollutants.* Lewis Publishers Inc.

Microbial Insights, Inc. *Applying Analytical Chemistry for Solutions in Microbiology.* 1998.

INTRINSIC BIOREMEDIATION OF PAH COMPOUNDS AT A FUEL-CONTAMINATED SITE

J. Steven Brauner, Mark A. Widdowson, John T. Novak, and Nancy G. Love
(Virginia Polytechnic Institute and State University, Blacksburg, VA 24061 USA)

ABSTRACT: The bioremediation potential of selected polycyclic aromatic hydrocarbon (PAH) compounds at a fuel oil-contaminated site was evaluated by conducting long-term (200+ day), unamended laboratory microcosm experiments. Biotic losses were differentiated from abiotic losses by comparing the measured rate of PAH loss in the live microcosms with the rate for autoclaved control microcosms. After an extended acclimation period of approximately two and a half months, phenanthrene, fluorene, and pyrene were biodegraded by the collected microbial consortia at estimated first-order biodegradation rates of 2.41 $year^{-1}$, 3.28 $year^{-1}$, and 2.98 $year^{-1}$, respectively. This same microbial population did not degrade benzo(b)fluoranthene or acenaphthene. Measured concentrations of dissolved oxygen and sulfate in the live microcosms were depressed relative to concentrations measured in autoclaved controls. Stoichiometric mass balance calculations, incorporating the observed mass loss of PAHs and electron acceptors, indicate that aerobic and sulfate-based biodegradation account for 70 percent and 10 percent of the observed PAH loss, respectively. Results demonstrate that feasibility studies designed to investigate intrinsic bioremediation for relatively recalcitrant compounds should account for extended lag periods prior to the onset of biodegradation.

INTRODUCTION

Human exposure to polycyclic aromatic hydrocarbons (PAHs) is a significant health risk due to the carcinogenic and mutagenic properties of these compounds. Although natural biodegradation rates of PAHs are often slow, in situ biodegradation under natural conditions may offer an effective remediation strategy for PAH removal when 1) risks associated with excavation of contaminated soils are greater than the risk to the general public if the contaminants are left in place and 2) appropriate quantities of electron acceptor are available for contaminant oxidation over time.

Soil for the present study was collected in a residential area from the Atlantic Coastal Plain of Northern Virginia where fuel oil had leaked from a large number of underground storage tanks. Previous site investigation showed that hydrocarbon contamination was widely distributed throughout the site, and that acenaphthene, phenanthrene, fluorene, pyrene, and benzo(b)fluoranthene were present in measurable quantities. The wide hydrocarbon contaminant distribution at the site made excavation and ex situ treatment of all contaminated soil technically infeasible and prohibitively expensive. The close proximity of the contaminant to private residences meant that contaminant excavation would also significantly increase exposure risk to local residents. To evaluate the effectiveness of natural in situ bioremedation as a primary or secondary remediation strategy for this site, the current study examines the ability of indigenous microorganisms to degrade five different PAHs under simulated natural conditions.

MATERIALS AND METHODS

Soil Collection and Characterization. A soil sample was collected approximately 2 ft (0.6m) below land surface near a recently excavated, leaky UST. The sample was placed in a sterilized jar and transported on ice back to the laboratory where it was stored at 20°C until used in the microcosm experiment. Analysis using an HP 5890 Gas Chromatograph with Mass Spectrometer indicated that the collected soil was contaminated with a series of alkane hydrocarbons, but no detectable PAHs.

Chromatographs of the soil sample were consistent with those typically seen in fuel contaminated soils, implying that the indigenous biota have been exposed to some xenobiotic petroleum hydrocarbon compounds.

Microcosms Construction. Microcosms were prepared in sterilized, 10mL threaded test tubes with solid, Teflon-lined caps. The soil sample was split into two components, with approximately 40 percent of the soil sample receiving PAH contamination via the addition of acenaphthene, phenanthrene, fluorene, pyrene, and benzo(b)fluoranthene dissolved in hexane. Hexane was allowed to evaporate and the PAH-spiked soil was combined with the remaining 60% of soil, which had been stored in the original jar at 20°C to maintain the indigenous microbial population. The soil mixture was ground with a mortar and pestle, and 5g of soil mixture along with 5mL of autoclaved, distilled water was added to each microcosm. Control microcosms were created by repeatedly autoclaving soils prior to addition of the PAH/Hexane mixture described above. Microcosms were kept in the dark at a constant temperature of 12°C until sacrificed.

Analytical Methods. PAH compounds were extracted from the microcosm soils by combining the microcosm contents with 15mL CH_2Cl_2 in 40mL amber vials. The 40mL vials were sealed with screw-on caps containing Teflon coated septa and rotated to promote thorough contact between the microcosm contents and CH_2Cl_2. After rotation, the 40mL vials were stored in the dark at 4°C until analysis. Samples were analyzed for PAHs using an HP-5890 Gas Chromatograph with a Flame Ionization Detector (GC-FID). The GC-FID was calibrated using external standards of known PAH concentrations dissolved in CH_2Cl_2.

Aqueous phase dissolved oxygen measurements were performed using a Diamond Electric Microsensor II connected to a PO_2 Needle Electrode. A two point calibration procedure was performed by successively bubbling nitrogen gas (0% Oxygen) and breathing air (21% Oxygen) through a calibration cell. Electrode calibration was monitored and adjusted (when needed) before and after each set of measurements to account for the effects of drift inherent to the electrode.

Anion concentrations were measured via ion chromatography (IC). Prior to CH_2Cl_2 extraction, 2mL of water were pipetted from the microcosm for IC analysis. Each sample was filtered prior to injection through a 45μm filter to remove suspended solids from the sample. External standards of known anion concentrations were used to calibrate the IC.

Mass Balance using Stoichiometric Calculations. The mass of hydrocarbon contaminant which can be mineralized to CO_2 by a microbial population using a given electron acceptor (EA) can be estimated using equation 1.

$$\begin{matrix}\text{Potential Mass} \\ \text{of Degraded PAH}\end{matrix} = \frac{\text{Mass of EA}}{\text{EA Molecular Weight}} \times \frac{\text{Moles of PAH}}{\text{Moles of EA}} \times \text{PAH Molecular Weight} \qquad (1)$$

This formulation assumes that 1) hydrocarbon oxidation to CO_2 is complete and 2) microorganisms that mediate the redox reaction do not require energy for growth or cell maintenance. If hydrocarbon oxidation is not complete and stops at an intermediate compound, this method may underpredict the mass of degraded PAH. Additionally, microorganisms are not perfectly efficient and will need more EA than predicted by the above expression for cell functions other than the redox reaction. Note that biodegradation can not occur without an appropriate microbial population and environmental conditions. Thus, developing the stoichiometric equations for contaminant oxidation does not imply that the reaction will take place, but simply provides an estimate of how much hydrocarbon mass can be biologically degraded to CO_2 given a known mass of EA. The molar ratio of PAH to EA is stoichiometrically determined by combining electron donor and EA half reactions to produce complete

TABLE 1. Stoichiometric reactions for selected PAHs and under aerobic and sulfate-reducing conditions (after McFarland and Sims, 1991; Stumm and Morgan, 1981).

Electron Donor	Redox Reaction
Fluorene	$C_{13}H_{10} + 15.5\ O_2 \rightarrow 13\ CO_2 + 5\ H_2O$
	$C_{13}H_{10} + 7.75\ SO_4^{2-} + 11.625\ H^+ \rightarrow 13\ CO_2 + 3.875\ HS_2 + 5\ H_2O$
Phenanthrene	$C_{14}H_{10} + 16.5\ O_2 \rightarrow 14\ CO_2 + 5\ H_2O$
	$C_{14}H_{10} + 8.25\ SO_4^{2-} + 12.375\ H^+ \rightarrow 14\ CO_2 + 4.125\ HS_2 + 5\ H_2O$
Pyrene	$C_{16}H_{10} + 18.5\ O_2 \rightarrow 16\ CO_2 + 5\ H_2O$
	$C_{16}H_{10} + 9.25\ SO_4^{2-} + 13.875\ H^+ \rightarrow 16\ CO_2 + 4.625\ HS_2 + 5\ H_2O$

redox reactions for each hydrocarbon contaminant. The complete redox reactions for selected PAHs under aerobic and sulfate-reducing conditions are given in Table 1. Using phenanthrene as an example, the stoichiometric ratios shown in Table 1 indicate that 16.5 mol of Oxygen are required to oxidize 1 mol of phenanthrene. Using this ratio and the formulation developed above, the mass of phenanthrene that can be completely mineralized to CO_2 by 1 mg of oxygen can be calculated using equation 2.

$$\text{Mass of Phenanthrene} = 1\,\text{mg}\,O_2 \times \frac{1\,\text{mol}\,O_2}{32\,\text{grams}} \times \frac{1\,\text{mol Phenanthrene}}{16.5\,\text{mol}\,O_2} \times \frac{178.2\,\text{grams}}{1\,\text{mol Phenanthrene}} = \begin{matrix}0.33\,\text{mg of}\\ \text{Phenanthrene}\end{matrix} \qquad (2)$$

The mass of contaminant which can be oxidized to CO_2 by a given mass of other EAs can be calculated in a similar manner, with the total degradable hydrocarbon mass estimated by summing the mass of hydrocarbon degraded by each EA process.

RESULTS

PAH Degradation. After an initial period of inactivity, a significant loss of fluorene, phenanthrene, and pyrene was observed between Day 75 and Day 133 in the live microcosms, which was not observed in the autoclaved controls. Pyrene degradation (Figure 1) was typical for these three PAHs. No significant difference in PAH loss was observed for either acenaphthene (Figure 2) or benzo(b)fluoranthene (data not shown) when comparing losses in the live and control microcosms. Although the loss of fluorene, phenanthrene, and pyrene in the live microcosms visually appeared to be greater than losses in the controls, statistical testing was used to clearly demonstrate that losses in the live microcosms were greater than losses in the controls.

To statistically show whether the rate of fluorene, phenanthrene, and pyrene loss in the live microcosms was greater than that of the controls, first-order degradation rates and 95 percent confidence intervals were calculated for both the control and live microcosms using a protocol similar to Wilson et al. (1996). By plotting concentration versus time on a semi-log scale, processes that follow first-order kinetics plot as linear trends. Since the 95 percent confidence interval of the slope for acenaphthene and benzo(b)fluoranthene in the live microcosms includes the slope estimate from the control microcosms, these slopes are not considered statistically different and it can be concluded that losses in the live microcosms are not greater than those in the controls. Conversely, the 95 percent confidence interval of the slope for fluorene, phenanthrene, and pyrene from the live microcosms does not include the slope of the control microcosms, and it is

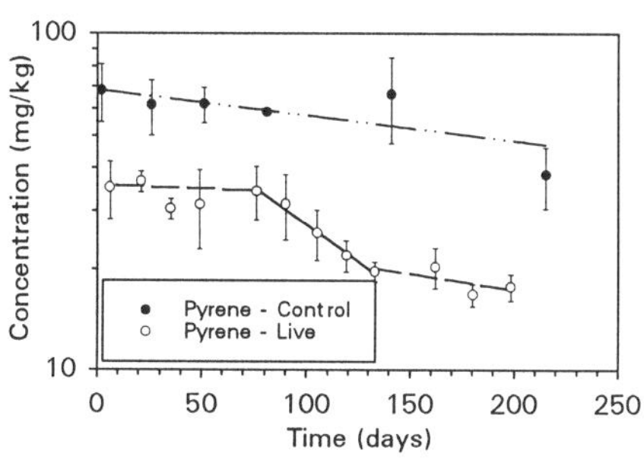

FIGURE 1. Concentration versus time for pyrene. The dash dotted (— · ·) line represents the best fit regression line through the control microcosms. The solid (—) line represents the best fit line through the live microcosms during the 'active' phase, while the dashed (— —) line represents the best fit line during periods when the live microcosms were 'inactive'.

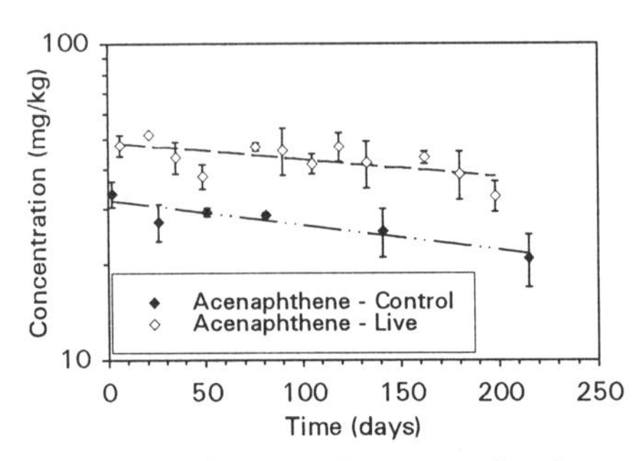

FIGURE 2. Concentration versus time for acenaphthalene. The dash dotted (— · ·) line represents the best fit regression line through the control microcosms, while the dashed (— —) line represents the best fit line through the live microcosms.

therefore concluded that the degradation rate in the live microcosms is statistically greater than that of the controls. By subtracting the first-order degradation rate of the control microcosms from the rate calculated for the live samples, the rate of biodegradation of fluorene, phenanthrene, and pyrene is estimated as shown in the last column of Table 2.

The presence of slope in the control microcosms (as shown in Table 2, Column 3) indicates that some PAH loss in the experiment was due to abiotic processes. Several studies have shown that long contact times between PAHs and soil may result in decreasing extractable hydrocarbon mass with time (e.g. Karickhoff, 1980; Hatzinger and Alexander, 1995). These researchers attribute this phenomenon to diffusion of hydrophobic compounds into soil micropores when contact times are on the order of several days to several months. The solvent extraction procedure used in this study measured the combined mass of aqueous and 'reversibly' sorbed PAH per dry soil mass, as the portion of PAH mass located in the soil micropores (i.e. irreveribly sorbed PAH) is not likely to come into contact with CH_2Cl_2. Thus, decreasing extraction efficiency due to micropore diffusion may have caused the abiotic losses observed in the control microcosms.

Electron Acceptor Disappearance. Triplicate dissolved oxygen measurements collected at Day 105 yielded a mean DO concentration in the live microcosms of 1.2 mg/L (σ_{SD} = 0.4mg/L), while the average DO concentration in the autoclaved controls was 5.5 mg/L (σ_{SD} = 1.8mg/L). Triplicate DO measurements recorded on Day 119 indicated that aqueous phase dissolved oxygen levels in the live microcosms had decreased to 0.3 mg/L (σ_{SD} = 0.2 mg/L). Decreased DO levels in the live microcosms suggests that oxygen was biologically consumed in the non-autoclaved microcosms, while the measurement of higher DO levels in the autoclaved microcosms suggests that autoclaving of the soil samples prior to contamination with PAHs was an effective method for inhibiting

TABLE 2. First-order rate constants (year^{-1}) for disappearence of PAHs. Degradation rates for the live microcosms were calculating using five data points between Day 75 and Day 133. Degradation rates in the autoclave controls were developed using all available data points.

	Live (Non-autoclaved) Microcosms	Control (Autoclaved) Microcosms	Net Rate of Removal (1/year)
	First-Order Rate (1/year)		
Acenaphthene	-0.46	-0.67	n/a
95% Confidence Interval	±0.38	±0.38	
R Squared	0.41	0.86	
Fluorene	-2.89	-0.48	-2.41
95% Confidence Interval	±2.38	±0.41	
R Squared	0.74	0.72	
Phenanthrene	-3.47	-0.19	-3.28
95% Confidence Interval	±2.18	±0.30	
R Squared	0.83	0.44	
Pyrene	-3.74	-0.76	-2.98
95% Confidence Interval	±0.73	±0.84	
R Squared	0.99	0.61	
Benzo(b)fluoranthene	-1.37	-0.83	n/a
95% Confidence Interval	±0.85	±1.24	
R Squared	0.59	0.46	

biological activity. Considering the low DO concentration in the live microcosms at Day 119, conditions in the microcosm vials were considered oxygen-limited after Day 119. Further hydrocarbon degradation would therefore require a different EA, such as nitrate or sulfate.

Concentrations for two other potential EAs, nitrate and sulfate, were measured at the end of the experiment using ion chromatography. IC analysis showed a significant decrease in sulfate concentration (>30 mg/L) when comparing water samples collected from the live microcosms with the autoclaved controls. Additionally, low concentrations of nitrate (0.71 mg/L) were measured in the autoclaved controls, while nitrate levels in the live microcosms were below detection. The depressed sulfate concentration measured in the live microcosm samples suggests that sulfate-reducing microorganisms were active during at least some portion of the study. Conversely, the high sulfate concentration in the control samples suggests that sulfate-reducing microorganisms were inhibited by autoclaving the soil. The absence of nitrate in the live samples suggests that nitrate removal was due to biological processes which were inhibited by repeated autoclaving of the soil. Elevated chloride concentrations were also measured in the live microcosms, although the reason an increase in chloride concentration is not known.

Estimation of Available Electron Acceptor Mass. The mass of oxygen at the start of the experiment was estimated using the volume of water and air present in each microcosm vial. After adding 5 grams of soil and 5 mL of water to each vial, a headspace of approximately 2.5 mL remained. The mass of oxygen in each microcosm vial was then calculated assuming an initial dissolved oxygen concentration of 5.5 mg/L in the aqueous phase and 21% O_2 in the air-filled headspace. Using these volumes and concentrations, the total mass of oxygen in each microcosm is estimated as 0.74 mg.

The total mass of sulfate consumed during the experiment was also calculated by assuming that the difference in sulfate concentration between the control and live microcosms was due to biological processes. The measured concentration difference of approximately 32 mg/L implies that up to 0.16 mg of sulfate was consumed in each live microcosm vial.

PAH Mass Balance. To determine the relative importance of each degradation process during the microcosm experiment, mass balance calculations were performed assuming that abiotic, aerobic, and anaerobic (i.e. sulfate-reducing) processes contributed to PAH loss. Using the observed changes in PAH concentration between Day 75 and Day 133, the mass of fluorene, phenanthrene, and pyrene lost in the live microcosm experiment was estimated as 0.12 mg, 0.16 mg and 0.075 mg, respectively, corresponding to a total PAH mass loss of 0.355 mg per microcosm. Total PAH losses due to abiotic processes were estimated at 0.035 mg PAH (10% of the total observed PAH loss) over this 58 day period by using the PAH degradation rates in the control microcosms. Based on the stoichiometric equations developed in Table 1 and that an estimated 0.74 mg O_2 was consumed in each live microcosm, aerobic biodegradation processes have the potential to mineralize approximately 0.25 mg of PAH to carbon dioxide, corresponding to 70% of the observed PAH loss. Using a similar procedure for sulfate, the potential mass of PAH which could be mineralized via sulfate-reduction is estimated at 0.035 mg or 10% of the total observed PAH loss. This mass balance procedure leaves approximately 10% of the PAH lost during the microcosm experiment unaccounted for, which is attributed to the conservative assumption that the PAHs were complete mineralization to CO_2.

CONCLUSIONS

Results from this study indicate that a native microbial consortia collected from a petroleum contaminated site can biodegrade fluorene, phenanthrene, and pyrene under intrinsic (i.e. aerobic, non-amended) conditions after a lag period of approximately two and a half months. Degradation rates in the live microcosms returned to levels similar to those observed during the initial lag period within two weeks of recording dissolved oxygen levels in the live samples as below detection. Evidence that loss of these three compounds was biologically mediated, rather than abiotic, is supported by conservation of these three PAHs in a separate set of autoclaved control microcosms. Oxygen and sulfate concentrations were significantly lower in the live microcosms when compared to the autoclaved controls, lending further support to a biologically-mediated explanation for PAH loss. Biodegradation of acenaphthene and benzo(b)fluoranthene was not observed during this experiment.

Stoichiometric calculations, based on EA and electron donor redox reactions, indicate that aerobic biodegradation can account for approximately 70 percent of the observed PAH loss and is the dominant biodegradation process. Similar stoichiometric calculations estimate that the mass of consumed sulfate has the potential to oxidize approximately 10 percent of the total mass PAH lost in the live microcosms. Evidence that sulfate-reduction may have been coupled with PAH oxidation is therefore indirectly supported by an improved PAH mass balance when comparing the observed PAH mass loss with stoichiometric calculations. The stoichiometric mass balance also suggests that PAH biodegradation in the microcosms stopped due to EA limited conditions.

This study illustrates the importance of identifying the degradation potential for individual compounds, rather than assuming uniform biodegradation rates in contaminated soil. The extended lag period and relatively slow degradation rates observed in this study support the recommendation made by Wilson et al. (1996) and others that the duration of microcosm studies for slowly degrading compounds in the natural environment should be on the order of months or, in some cases, years. Continued investigation into possible PAH oxidation under sulfate-reducing conditions by the microbial consortia collected at this site is also merited, as Coates et al. (1996, 1997) have shown that a microbial population collected from a marine sediment biodegraded naphthalene, fluorene, phenanthrene, and fluoranthene under sulfate-reducing conditions.

REFERENCES

Coates, J.D., J. Woodward, J. Allen, P. Philip, and D.R. Lovley. 1997. "Anaerobic degradation of polycyclic aromatic hydrocarbons and alkanes in petroleum-contaminated marine harbor sediments." *Appl. Environ. Microbiol. 63*(9):3589-3593.

Coates, J.D., R.T. Anderson, and D.R. Lovley. 1996. "Oxidation of polycyclic aromatic hydrocarbons under sulfate-reducing conditions." *Appl. Environ. Microbiol. 62*(3):1099-1101.

Hatzinger, P.B. and M. Alexander. 1995. "Effect of aging of chemicals in soil on their biodegradability and extractability." *Environ. Sci. Tech. 29*(2):537-545.

Karickhoff, S.W. 1980. "Sorption kinetics of hydrophobic pollutants in natural sediments." In R.A. Baker (Ed.), *Contaminants and Sediments. Volume 2. Fate and Transport Studies, Modeling, Toxicity,* pp. 193-205. Ann Arbor Science, Ann Arbor, MI.

McFarland, M.J. and R.C. Sims. 1991. Thermodynamic framework for evaluating PAH degradation in the subsurface. *Ground Water. 29*(6):885-896.

Stumm,W., and J.J. Morgan. 1995. *Aquatic Chemistry.* 3rd ed. John H. Wiley & Sons, New York.

Wilson, B.H., J.T. Wilson, and D. Luce. 1996. "Data and interpretation of microcosm studies for chlorinated compounds." *Proceedings of the Symposium on Natural Attenuation of Chlorinated Organics in Ground Water.* Dallas, TX, September 11-13.

BIOREMEDIATION POTENTIAL OF PAHs IN COMPOST

Brian J. Reid; Kevin C. Jones;
Kirk T. Semple (Lancaster University, Lancaster, UK)
Terry R. Fermor (HRI, Wellesbourne, UK)

ABSTRACT: This study investigated the potential for Phase 2 mushroom compost to degrade phenanthrene (a representative PAH) in both slurry and soil systems. Initially, induction of catabolic ability within the compost was assessed after increasing compost-phenanthrene contact times. This was achieved by monitoring the evolution of $^{14}CO_2$ from the mineralization of freshly added ^{14}C-9-phenanthrene to aqueous compost slurries in respirometers. A subsequent experiment to assess the potential of the compost to bioremediate aged phenanthrene (300 d) from soil was then conducted in aerated microcosms. The extent of mineralization using induced compost was compared to that where uninduced and no compost was used. The results suggested that with increasing compost-phenanthrene contact time there was an increase in catabolic capability. Initial, (0 d compost-phenanthrene contact time) mineralization was less than 2%. However after 7 weeks compost-phenanthrene contact time greater that 65% mineralization was observed. When remediation of aged phenanthrene from soil was attempted the induced compost produced the greatest extent of mineralization initially. However, after 21 d remediation time the cumulative extents of mineralization were similar in all microcosms.

INTRODUCTION

Composting (i.e. the processes by which compost is produced from raw materials) has frequently been employed to ameliorate contaminated land. The use of compost has however been given less attention. Previous studies have examined the degradation of organic pollutants in other composts. For example, it has been shown that windrow composts are capable of mineralizing pentachlorophenol (Valo and Salkinoja-Salonen, 1986). The importance of the solid organic matrix of composts was highlighted by Semple and Fermor (1997), where compost isolates did not degrade PCP in liquid culture to the extent that the composts did. Catabolic enrichment of compost has been shown to facilitate mineralization of chlorophenols, particularly pentachlorophenol (Semple and Fermor., 1985). There were two aims to this investigation. Firstly, to characterise the induction of any catabolic ability within phenanthrene-amended compost. Secondly, to assess the suitability of SMC either induced or uninduced to remediate aged phenanthrene-contaminated soil.

MATERIALS AND METHODS

Compost Production: Phase 2 compost was obtained from Horticulture Research International (Wellesbourne, UK). Its production is described elsewhere (Fermor et al. 1985).

Compost Spiking: Phase 2 mushroom compost was spiked with phenanthrene to a concentration of 400 mg kg^{-1} (wet weight). Spiking was crudely achieved by incrementally sprinkling 2 g of phenanthrene crystals onto 5 kg of compost and mixing. This procedure was repeated until 25 kg of spiked compost had been produced. The spiked compost was placed in an incubation bin, consisting of a polythene lined steel basket (0.5 m x 0.5 m x 0.5m) fitted with an air sparger at the base. The bin was housed within a polystyrene lined unit (1m x 1m x 1m) with a similarly lined lid. The temperature of the compost was regulated by a feed back probe to provide a constant temperature of 30°C. The compost was sparged for 30 minutes in every hour with humidified air at a flow rate of 10 L min^{-1} throughout the incubation period. A spiked compost was thus produced, along with a control compost to which no phenanthrene was added.

Respirometric assessment of catabolic potential: Compost samples (150 g) were removed from the incubation bins initially on a weekly basis (up to 4 weeks), then after 7 and 9 weeks, and finally after a total of 27 weeks phenanthrene-compost contact time for assessment of catabolism potential. The mineralization of ^{14}C-9-phenanthrene was assessed using respirometers (modified Erlenmeyer flasks) (Semple and Fermor, 1997). To the respirometers compost (10 g) was added along with 30 ml of a minimal basal salts solution, containing between 1100 Bq and 1600 Bq of ^{14}C-9-phenanthrene. No additional non-^{14}C-radiolabelled phenanthrene was added to the respirometers. $^{14}CO_2$ liberated by the mineralization of the phenanthrene was trapped using 1M KOH (1 ml) present in a glass scintillation vial (solvent washed) suspended from the top of the respirometer by a stainless steel clip. The vials were removed and replaced at regular intervals and analysed by liquid scintillation counting (Canberra Packard Tri-carb CA2250CA) after mixing with Ultima Gold scintillation fluid. Respirometers were run in triplicate.

Spiking of soil with ^{14}C-9-phenanthrene: Subsurface soil (5-15 cm) was collected from a rural hillside environment (Lancaster University, Hazelrigg, Field Station, UK - O.S. sheet 97, [493578]). The soil was passed through a 10 mm gauge sieve, air dried for 14 d and subsequently passed through a 1.7 mm gauge sieve to remove roots. The soil was artificially contaminated with both ^{14}C-radiolabelled and non-radiolabelled phenanthrene using a single step spiking/rehydration procedure described elsewhere (Reid et al., 1998). A phenanthrene concentration of 10 mg kg^{-1} (wet weight) was achieved, providing an associated radioactivity of approximately 50 Bq g^{-1} (wet weight). Soil was stored in amber glass jars (500 ml) and allowed to age for 300 d. Residual activity after the ageing period was determined by sample oxidation of 15 (1 g) samples using a Packard 307 sample oxidiser. The resultant CO_2 generated was trapped in carbosorb and eluted with permafluor-E scintillation cocktail. The resultant samples were counted as before.

Bioremediation of ^{14}C-phenanthrene contaminated soil: Aged soil (75 g) was mixed with induced compost (75 g) and the mixtures placed in Kilner jars (1 L). The jars were adapted to incorporate air inlet and outlet lines. The inlet line was inserted to the base of the soil compost mixture to allow sparging with humidified air through a bed of washed gravel (6 L air d^{-1}). The outgas was bubbled through two 1 M KOH traps. The first trap was a Dreshcel bottle with scintered sparger (grade 1) containing 70 mL of KOH. The second was a quickfit screw thread vented receiver (VT24/8) with a pasture pipette bubbler containing 20 mL of KOH. In this way $^{14}CO_2$ was trapped from the mineralization of ^{14}C-9-phenanthrene. The KOH traps were changed after 5 d and 21 d and the KOH analysed for radioactivity following the addition of Ultima Gold scintillation fluid as described previously. Similar microcosms were produced using uninduced compost with spiked soil and with spiked soil only. A blank microcosm containing unspiked soil was also produced.

RESULTS AND DISCUSSION

Induction of Phase 2 mushroom compost (Figure 1): The extents of mineralization shown are cumulative totals after 96 h assay time. Initially (0 d phenanthrene-compost contact time) it was observed that minimal biodegradation of freshly added ^{14}C-9-phenanthrene was achieved after 96 h assay time in either the induced or uninduced composts (Figure 1a and b). After 2 weeks compost-phenanthrene contact time, there was a 45 % increase in the extent of mineralization (Figure 1a). The extent of mineralization continued to increase with increasing compost-phenanthrene contact time until it peaked after 7 weeks (Figure 1a). The maximum extent of mineralization was approximately 65% under the experimental conditions described above. After 9 weeks compost-phenanthrene contact time, the extent of mineralization dropped to approximately half the maximum value achieved after 7 weeks (Figure 1a). After 27 weeks compost-phenanthrene contact time, the extent of mineralization had dropped slightly further to approximately 30% (Figure 1a). This decrease in catabolic ability may reflect a change in compost conditions. It was noted that as the induction time increased the water content of the compost decreased. It is quite possible that the lack of water within the compost inhibited induced microorganisms after the optimal 7 week period.

Mineralization by the uninduced compost indicates an increase in catabolic ability with increased aeration time from 0 to 4 weeks (Figure 1b). Mineralization was at its maximum after 4 weeks; approximately 6 % (Figure 1b). This increase may reflect a change in physical conditions within the induction bins due to aeration. After this time the extent of mineralization decreased to approximately

2% (Figure 1b). Again this may reflect a change in physical conditions within the compost from optimal to sub-optimal. It should be stressed that the increase in catabolic ability in the uninduced compost was an order of magnitude less than the induced compost.

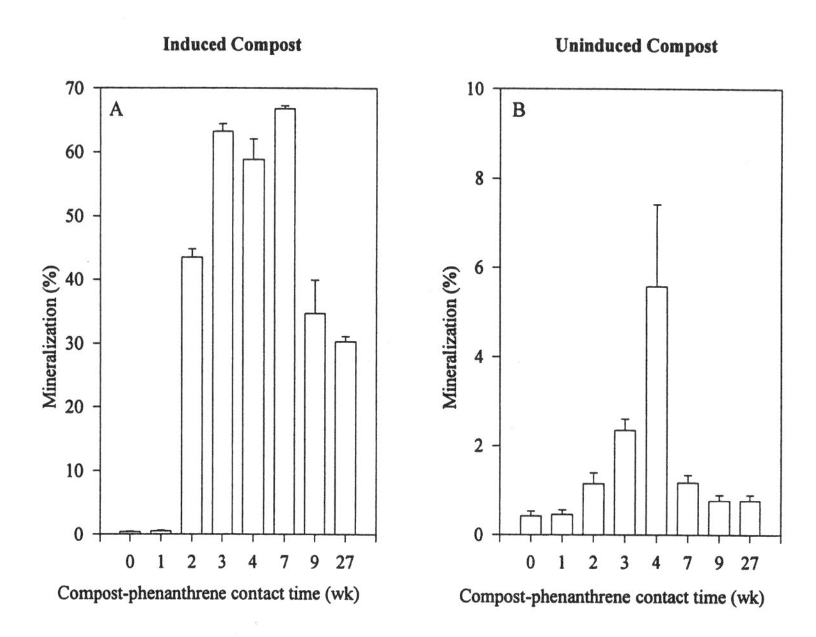

FIGURE 1. Assessment of catabolic induction within SMC for (A) phenanthrene spiked compost and (B) unspiked compost.

Bioremediation of phenanthrene contaminated soil: After 5 d remediation time, minimal degradation had occurred in all of the microcosms (Figure 2). The extents of mineralization were approximately 3.5%, 0.6% and 0.3% for the induced compost, uninduced compost and soil only systems, respectively. These results suggest that the induction of the compost made a significant (although minor) impact on the extent of mineralization achieved. After 21 d remediation time, the cumulative extents of mineralization in all of the microcosms was very similar; 10% to 12% mineralization in all cases (Figure 2). These results suggest that it may not be the introduction of catabolically active microorganisms which governs the extent of mineralization, but rather the stimulation of indigenous microflora. It is quite possible that this stimulation may have been achieved by the aeration applied to the microcosm for sparging. Origgi et al. (1997) reported a siginficant remediation of naphtha, toluene and xylenes as a result of a 2 month bioventing treatment.

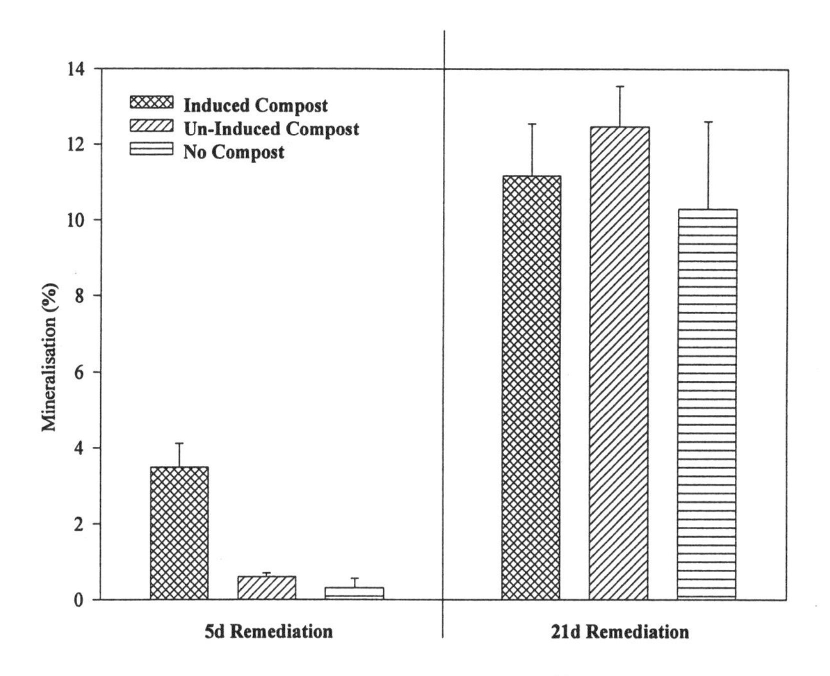

FIGURE 2. Mineralization of soil associated ^{14}C-9phenanthrene in the presence of induced and uninduced composts, and with no compost present.

CONCLUSIONS

This investigation indicates that Phase 2 mushroom compost can be induced to degrade phenanthrene. Furthermore the extent of mineralization using induced compost (cf. 65%) is an order of magnitude greater than for uninduced compost (cf. 6%). However, the application of mushroom compost - whether induced or not - does not appear to influence the extent of aged phenanthrene removal from soil achieved in the 21 d trial period. It appears from the results that degradation by the indigenous soil microflora plays the most significant role under these air sparged conditions. This experiment is ongoing and will continue to be monitored until mineralization has terminated.

ACKNOWLEGEMENTS

The authors would like to acknowledge the support for this project from the Natural Environment Research Council (NERC) UK.

REFERENCES

Fermor, T. R.; P. E. Randle. and J.F. Smith. 1985. "Compost as a substrate and its preparation" in *The Biology and Technology of the Cultivated Mushroom*. (P. B. Flegg, D. M. Spencer and D. A. Wood) (pp 81-109). John Wiley and Sons,Chichester, UK.

Origgi G.; M. Colombo.; F. DePalma.; P. Rossi. and V. Andreoni. 1997. "Bioventing of hydrocarbon-contaminated soil and biofiltration of the off-gas: Results of a field scale investigation". *J. Envron. Sci. Health. 32*: 2289-2310.

Reid, B.J.; G.L. Northcott.; K.C. Jones. and K.T. Semple. 1998. "Evaluation of spiking procedures for the introduction of poorly water soluble contaminants (PAHs) into soil". *Environ. Science and Technol. 32:* 3224-3227.

Semple, K.T. and T.R. Fermor. 1997 "Enhanced mineralization of [U-^{14}C]PCP in mushroom composts". *Res. Microbiol. 148*: 795-798.

Valo, R. and M. Salkinoja-Salonen. 1986 "Bioreclamation of chlorophenol-contaminated land". *Appl. Microbiol. Biotechnol. 25*: 68-75.

REMEDIATION OF CONTAMINATED EAST RIVER SEDIMENT BY COMPOSTING TECHNOLOGY

Carl L. Potter, John A. Glaser, Ronald Herrmann (U.S. EPA, National Risk Management Research Laboratory)
Majid A. Dosani (IT Corporation).

ABSTRACT: Contaminated sediments sampled from the New York Bay area near the East River between New York, NY and Jersey City, NJ were selected for a series of treatability studies by compost treatment. The sediment solids content averaged 41.8% with total organic carbon content of 42.7 g/kg and density of 1.33 kg/L. Due to high water content, it was necessary to adjust the moisture content for compost treatment. Perlite (expanded volcanic material) was examined as one amendment to the sediment to provide structural stability of the compost mixture. Initial concentrations of total polycyclic aromatic hydrocarbons (PAHs) ranged from 800-1400 mg/kg in the compost mixtures in six instrumented composters. The treatment conditions giving the greatest PAH removal were found to be sediment, rabbit chow, perlite, and crushed limestone where the concentration of total PAHs was reduced by 61% over a 12-week treatment period. The 2 & 3 and 4 ring PAHs were reduced by 73% and 49% respectively. Large (5 & 6 ring) PAHs were not removed by any of the compost treatments. The composting treatment period was characterized by increased temperatures, reaching 50°C to 60°C in reactors containing rabbit chow, but temperature remained at ambient levels in the reactors containing only sediment and perlite. During early temperature increases, biomass declined and then increased rapidly as determined by lipid-associated phosphate measurements. Removal of PAHs did not follow biomass changes during composting, an *ex situ* process.

INTRODUCTION

Polycyclic aromatic hydrocarbon (PAH) contamination is a concern in soil and sediment at many environmental sites. Past research by EPA's Office of Research and Development has focused on assessment rather than the remediation of sediments. Evidence of ecological risk associated with contaminated sediments has stimulated interest in remediation of contaminated sediments.

Contaminated sediment poses a significant problem to many metropolitan areas since there has been a major restructuring of disposal options. Land disposal can be a reasonable option but the landfill option is viewed with greater skepticism when conducted without any treatment strategy. The contaminant composition of sediments differs widely depending on the location and conditions leading to deposition of the sediment. The highly variable nature of sediments has offered perplexing problems for candidate treatment technologies.

A suitable treatment option may be composting since it has been used to degrade a wide range of solid waste materials such as yard waste, sewage sludge, and food wastes for many years. More recently, composting has been evaluated as a remediation technology for munition waste (U.S. Army Corps of Engineers, 1991; Ziegenfuss *et al.*, 1991; Williams *et al.*, 1992).

Composting differs from other *ex-situ* sediment treatment systems in that bulking agents are added to the compost mixture to increase porosity and serve as sources of easily assimilated carbon for biomass growth. Aerobic metabolism generates heat resulting in significant temperature increases that bring about changes in the microbial ecology in the compost mixture. The conventional aerobic compost process involves four distinct microbiological phases defined by temperature changes: mesophilic (30° - 45°c); thermophilic (45° - 75°C); cooling; and maturation. We have controlled conditions to emphasize the mesophilic and low thermophilic stages since the greatest microbial diversity is encountered in this temperature range and microbial PAH degraders were found to disappear when temperature exceeded 60°C.

Due to the relatively poor water solubility of the PAH class of contamination, the sediment at many environmental sites is highly contaminated with these pollutants. Evidence of ecological risk associated with contaminated sediments has stimulated interest in remediation of contaminated sediments. Sediment treatment processes may involve removal and *ex situ* cleanup treatment. Treating contaminated sediments *in situ* may be less expensive than sediment removal, and, in special cases, *in situ* sediment treatment may be possible. We have undertaken an investigation of compost treatment of PAH contaminated sediments from the New York Harbor.

OBJECTIVE

The objective of this research was to evaluate the potential of composting to reduce PAH contamination in marine sediment with different bulking agents. We have developed bench-scale compost reactors to evaluate factors controlling compost treatment under controlled conditions. These small-scale treatability studies may provide insight into compost processes that can be applied to field-scale compost treatment of sediments.

MATERIALS AND METHODS

Sediment contaminated with PAHs was obtained from the East River between New York, NY and Jersey City, NJ for use in this study. Sediment was excavated and placed in 55-gal drums for shipment to the U.S. EPA Test and Evaluation (T&E) Facility in Cincinnati, OH for treatability testing. Test sediment was analyzed for 19 PAHs during the study. Origin of the PAH contaminated was unknown, but probably includes numerous sources. The sediment solids content averaged 41.8% with total organic carbon content of 42.7 g/kg and density of 1.33 kg/L. Sediment was dewatered and screened to one-half inch particle size prior to mixing for compost treatment.

Eight drums of excavated contaminated sediment were separated into pairs and placed pairwise into a paddle wheel mixer at the T&E Facility and mixed for 30 min. After 4 pairs of drums were mixed, the process was repeated using non-paired drums. This process was repeated once more to obtain homogeneous sediment for treatment. The study design included 3 replicated treatment conditions involving different bulking agents and nutrient sources. Each reactor contained 100 lb sediment and bulking agent in a 1/2 (sediment/bulking agent, v/v) ratio. Bulking agents consisted of perlite, corn cobs, rabbit chow or a combination thereof. When rabbit chow was added, it was mixed with corn cobs or perlite 50/50 (v/v). The following conditions were established for the treatability study.

1. Sediment and perlite, inorganic nutrients, and 1% (w/w) agricultural limestone. Sediment/perlite (1/2, v/v) provided inadequate bulking volume for composting and was changed to sediment/perlite (1/3, v/v) at the beginning of the treatability study.
2. Sediment, perlite and rabbit chow , inorganic nutrients, and 1% (w/w) agricultural limestone.
3. Sediment, corn cobs and rabbit chow , inorganic nutrients, and 1% (w/w) agricultural limestone.

Perlite is an inert bulking agent (expanded volcanic material) used to promote oxygen and water movement in soil. Agricultural limestone was added to the compost mixture at the beginning of the study to provide buffering capacity. C:N:P (Carbon:nitrogen:phosphate) was adjusted to a ratio of 100:5:1 and moisture was maintained near 40%. Air flow rates were 10 L/min for each reactor. The treatability study was conducted for 12 weeks.

PAH results are presented as mg PAH / kg ash to correct for dilution addition of bulking agents. Samples were collected at week 0,1,4,8, and 12. PAH analysis was performed by gas chromatogrphy (GC-FID) according to EPA Method 8100.

Reactor Design. Twelve 55-gal, insulated, stainless steel compost reactors have been fabricated to provide closely monitored and controlled conditions required for treatability studies (Glaser and Potter, 1996). Six of the 12 compost reactors (1&8, 2,&7, and 3&6) were selected at random for this treatability study. These fully enclosed, computer monitored, bench-scale reactors are of 1/4 cubic yard total internal volume. The bottom of each reactor contains a conical collection system for periodic sampling of any leachate leaving the reaction mixture. The space above the leachate collection system holds 2 inches of gravel.

Insulation between the reactor core and outer shell reduces heat loss from the reactor during aerobic activity. Each composter houses five thermocouples connected to a central computer for online temperature measurements. Thermocouples reside at four equally spaced locations within the compost mixture and a fifth thermocouple tracks ambient temperature outside the reactor.

Preliminary studies in this system indicated disappearance of PAH degrading microorganisms when compost temperature exceeded 60°C. Therefore, these reactors were equipped to progress through a cooling cycle if mean temperature of the middle two reactor thermocouples exceeded 60°C. When a compost reactor exceeded 60°C, the computer switched that unit to high air flow (50 L/min) until the reaction mixture cooled to 55°C. The computer then switched the unit back to standard air flow (10 L/min) to reduce further heat loss from the reaction mixture.

The reactor units stood upright with air flowing vertically up through the compost mixture for 23.5 h/day. Cylindrical reactor design permitted mixing of reactor contents by rolling each unit on a drum roller for 30 min/day. Mixing was used to break up anaerobic pockets and to avoid packing of the compost mixture.

Biomass. Biomass was estimated as total phospholipid-associated phosphate in the samples. Total lipid analysis was performed on 2.5 g of compost, collected from each reactor, by a modified technique of Findlay and Dobbs (1993). Total biomass in a sample was considered proportional to total lipid phosphate.

RESULTS AND DISCUSSION

Composting of East River sediment with perlite without an additional carbon source resulted in an average 31% total PAH removal in the two reactors over 12 weeks of treatment (Table 1). The only PAHs removed were 2 & 3 ring compounds (52%). Little, if any, removal of 4, 5, or 6 ring PAHs occurred during compost treatment.

Addition of rabbit chow to the sediment-perlite mixture resulted in removal of 52% of total PAHs after 12 weeks of compost treatment. Compost processes containing both perlite and rabbit chow removed 68% of 2 & 3 ring PAHs, 41% of 4 ring PAHs, and 30% of 5 & 6 ring PAHs.

TABLE 1: Start and finish sediment PAH concentrations in compost mixtures. Values represent mg PAH / kg ASH from the average of duplicate reactors

Experimental Conditions	Week	Number of PAH Rings			
		2 & 3	4	5&6	Total
Perlite Only	0	410 +/- 9	265 +/- 9	136 +/- 7	810 +/- 25
	12	198 +/- 1	226 +/- 5	135 +/- 0.3	559 +/- 7
% Reduction		52%	15%	0.7%	31%
Perlite + Rabbit Chow	0	584 +/- 65	374 +/- 38	244 +/- 5	1202 +/- 108
	12	188 +/- 28	222 +/- 31	171 +/- 56	581 +/- 116
% Reduction		68%	41%	30%	52%
Corn Cobs + Rabbit Chow	0	629 +/- 71	357 +/- 45	261 +/- 32	1246 +/- 149
	12	226 +/- 8	396 +/- 18	348 +/- 8	970 +/- 34
% Reduction		64%	-11%	-33%	22%

Combination of rabbit chow and corn cobs in the sediment compost mixture resulted in removal of only 22 % of total PAHs. However, 64% of 2 & 3 ring PAHs were removed. The low total value (22%) resulted from analytical detection of higher concentrations of 4, 5, and 6 ring PAHs after 12 weeks of compost treatment than at the beginning of treatment. It's not clear why 4 ring PAHs were not removed the rabbit chow-corn cob mixture. Previous compost studies in this system using corn cobs as bulking agent, combined with cow manure or activated municipal sludge, showed significant reduction in concentrations of 4 ring PAHs after 12 weeks of compost treatment (Potter *et al.*, 1997).

All reactors that contained rabbit chow underwent temperature increase to between 50°C and 60°C during the first 2 days of compost treatment and remained there for 6 weeks. This indicated that these reactors experienced significant aerobic microbial metabolic activity. Reactors containing no rabbit chow (perlite only) remained at ambient temperature of 18°C to 22°C throughout the entire 12 weeks of compost treatment. By week 7, all compost reactors were operating at ambient temperature.

During the first week of composting, biomass concentrations decreased in all reactors that contained rabbit chow (Figure 1). This corresponding temporally to rising temperatures. Biomass then increased to above starting concentrations, and generally remained elevated during weeks 4 through 8. Biomass in reactors that contained only perlite started less than 10% of that in reactors containing rabbit chow, and remained low throughout the study.

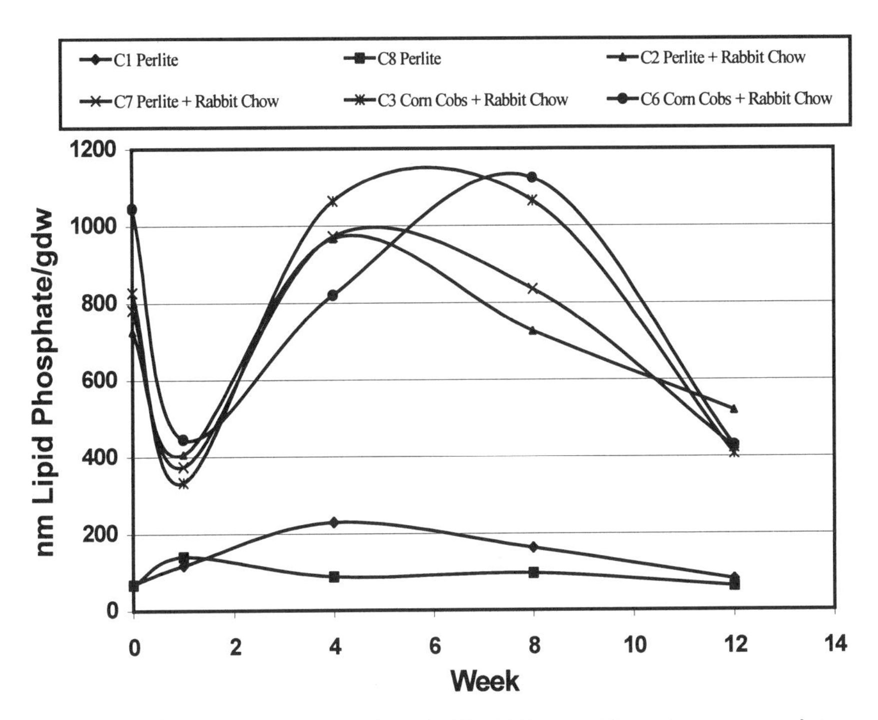

FIGURE 1. Biomass concentrations in East River sediment compost mixtures. Biomass was estimated as a proportion of total lipid associated phosphate.

In this study, PAH removal did not appear to follow changes in biomass concentration. Enumeration of phenanthrene degraders revealed population changes that were similar in all treatment conditions, including those with perlite only, even though PAH removal rates differed between treatment conditions.. Phenanthrene degrader counts decreased during the first week of composting and then became slightly elevated by week 4. This pattern of population change in phenanthrene degraders appeared to follow total biomass changes in reactors containing rabbit chow shown in Figure 1. Relying on phenanthrene degrader counts for information about degradation activity presents problems since only about 1% of a microbial population can be isolated from a complex environment like composting.

Phospholipid data collected from this study will be analyzed to determine FAME (fatty acid methyl ester) patterns that may reveal types and physiological status of microorganisms present in each compost mixture. FAME analysis might provide insights into population differences that account for differences in PAH degradation. It has become clear that simply increasing biomass in the compost mixture is insufficient to promote greater PAH removal, and enumeration of specific degraders can be misleading. FAME analysis can provide descriptions of microbial ecosystems and their physiological status that might be linked to more extensive PAH degradation. This could then be used to track microbial conditions while optimized treatment parameters.

REFERENCES

Findlay, R.H. and F.C Dobbs.. 1993. "Quantitative description of microbial communities using lipid analysis." In: Handbook of Methods in Aquatic Microbial Ecology. Kemp, P.F., Sherr, B.F., Sherr, E.B., and Cole, J.J. Lewis Publishers, Boca Raton, FL. Chapter 32, pp. 271-284.

Glaser, J.A. and C.L. Potter. 1996. "Sediment bioremediation research at EPA." Biocycle January 1996: 50-53.

Potter, C.L., J.A. Glaser, M.A. Dosani, S. Krishnan, and E.R. Krishnan. 1997. "Design and testing of a bench-scale composting system for treatment of hazardous waste." Proceedings of the Fourth International *In-situ* and On-site Bioremediation Symposium, New Orleans, April 28 – May 1, 1997. *4*(2):85-90.

U.S. Army Corps of Engineers. *Toxic and Hazardous Materials Agency. Optimization of composting for explosives contaminated soil, Final Report.* Report no. CETHA-TS-CR-91053, November 1991.

Williams, R.T., P.S. Zeigenfuss, and W.E. Sisk. 1992. "Composting of Explosives and Propellant Contaminated Soils Under Thermophilic and Mesophilic Conditions." J. Indust. Microbiol. 9:137-144.

Ziegenfuss, P.S., R.T. Williams, and C.A. Myler. 1991. "Hazardous materials composting." *J. Haz. Mater. 28*:91-99.

A LANDFARMING FIELD STUDY OF CREOSOTE-CONTAMINATED SOIL

Jason Winningham (EnSafe, Memphis, Tennessee)
Ronald Britto, Madhu Patel, Frank McInturff (EnSafe, Memphis, TN)

ABSTRACT: A field study was conducted at a wood-treating plant in Arkansas to evaluate the effectiveness of ex-situ landfarming creosote-contaminated soil. Polynuclear aromatic hydrocarbons (PAHs) had been detected in surface soil and required remedial action. Two test cells were constructed to test landfarming on soil from the area of concern. Initial soil concentrations in the cells averaged approximately 1,300 milligrams per kilogram (mg/kg). Soil was augmented with chicken manure and exogenous microorganisms. The study showed that up to 80.8% of PAH concentrations were removed by this method, and site remediation levels could be attained compared to tilling and irrigation alone. These results have been used to remediate contaminated surface soil at the site.

INTRODUCTION

A field bioremediation study was conducted to evaluate the effectiveness of landfarming creosote-contaminated soil at a wood-treating plant in Arkansas. The field study followed bench-scale treatability tests that concluded that site soil could be biologically treated for PAHs. The field study was performed to provide technical insight and engineering guidance for full-scale landfarming of site soil exceeding site remediation levels (SRLs). The SRL was established at combined total of 450 mg/kg for a select list of priority PAH compounds. As much as 11,000 cubic yards (8,410 cubic meters) of soil in the old dragout area required remedial action.

Approach. The field study was conducted in the area of concern, which was a dragout area of the former wood-treating plant used to dry and store creosote treated poles. Details of the field system setup, construction and mobilization activities, system design conditions (such as nutrient and soil moisture requirements), operation and monitoring details (such as tilling and irrigation frequency), and water management were predetermined in a work plan. A sampling plan was predefined that outlined the strategy used to collect samples from the landfarming cells, the soil parameters that were measured, and the analytical methods used to evaluate the samples.

Two test cells, Cell A and Cell B, contained soil that required remedial action. They were regularly irrigated and tilled (municipal water was made available for irrigation). Additionally, Cell B was bioaugmented with PAH microbial degraders and chicken manure to promote PAH degradation. Chicken manure was added for several reasons: it is known to improve soil structure, it contains slow-releasing inorganic nutrients essential for microbial activity, and it is a possible source of microbial strains that can withstand organic compounds that are difficult to break down (EPA/600/R-93/164). Exogenous microbial PAH degraders were added (bioaugmentation) to compare degradation rates and organic compound breakdown in Cell B with those in Cell A in which indigenous microorganisms were allowed to develop.

Each cell was 15 feet long by 30 feet wide (4.6 by 9.1 meters), contained approximately 100 cubic yards (76 cubic meters), and was enclosed by a 2-foot (0.61 meters) berm with a 1:1 slope. Each cell bed was graded to a slope of 3% in the south-north direction along its long axis, with a drainage trench running its entire length. The drainage system included collection pipes, drainage rock, and filter fabric so that water flowed from cells into a 150-gallon underground sump (568 liters). Water was pumped from the sump to aboveground tanks with a total capacity of 3,000 gallons (11,355 liters). The drainage water in the aboveground tanks was then recirculated to Cell B but not A, because exogenous microorganisms could have been present in the recirculated water.

SUMMARY OF FIELD ACTIVITIES

Landfarming operations and field activities began when Day 0 sampling was completed in September and concluded in January on the date of the last sampling event. Operations included:

- Addition of chicken manure, exogenous microorganisms, and nutrients
- Tilling and irrigation
- Field measurements of temperature, pH, and moisture content
- Sampling events

Bioaugmentation. Cell A was periodically irrigated and tilled only, with no biostimulating agents added. Cell B was supplemented with chicken manure and exogenous microorganisms at the beginning of the study. Chicken manure addition, at a concentration of 1.5% by weight of soil, was a one-time event at the beginning of the study. Exogenous microorganisms (PAH degraders) were supplied by a commercial dealer which has been developing specialized microbial cultures for industrial, municipal, and sanitary applications for more than 40 years. Based on preliminary information on PAH concentrations, it was decided to add 6 pounds (2.7 kg) of dry microbial culture containing a blend of PAH-degrading microorganisms to Cell B. Microorganisms were added to Cell B using a spreader to ensure uniform distribution.

Tilling, Irrigation, and Field Measurements. Soon after organic and inorganic amendments were added to the cells, they were tilled with a manual rototiller to thoroughly mix the soil and the supplemented into each cell. Each cell was tilled twice a week thereafter, using three up-and-down passes with the rototiller, to aerate and mix the soil.

Moisture added to the cells came from rainfall an irrigation system which was built for this study. Soil porosity, unit weight, and moisture content were used to determine how much irrigation water was needed. The mean soil porosity was 0.42 and the mean unit weight was 117 lbs per cubic foot (1,874 kg/m^3), as measured onsite at the start of the study. The desired moisture content was set at approximately 40% of the moisture content of the saturated soil.

Temperature, pH, and moisture content were measured twice a week in the field on the day before tilling and irrigation were scheduled. Soil pH was measured with a standard pH meter, and moisture and pH were measured, following standard guidelines (SW-846, Vol. I), in three independent samples collected from each of the cells.

Sampling Methods. After the soil was placed in the cells, it was sampled to determine baseline concentrations. A two-dimensional grid was established for each test cell to create discrete sampling locations for performance monitoring; intervals between grid nodes were equal in length and width (3 feet or 0.91 meters). Internal grid nodes for each cell were individually numbered, and nine nodal points were randomly selected and prioritized for sampling. Of these nine samples collected from each cell, six were analyzed by the laboratory, and the remaining three were preserved for possible future use based on statistical analysis. Each sample was homogenized in a stainless-steel bowl and containerized appropriately. Quality assurance/quality control samples were collected in accordance with the site-specific Quality Assurance Project Plan. The samples were shipped to a certified laboratory for analysis.

DATA ANALYSIS

Five rounds of soil samples were collected to monitor performance of the test cells. PAH concentrations found in these samples are illustrated in Figure 1. Soil samples from rounds 2, 3, 4, and 5 were analyzed for PAHs, ammonia-nitrogen, nitrate-nitrogen, and inorganic phosphorus.

Summary of Laboratory Analytical Results. Figure 1 shows that mean PAH concentrations in Cell B fell below SRLs (450 mg/kg) within 71 days after the study began. On Day 114, all six samples from Cell B were below SRLs for PAHs. The removal efficiency was 80.8% for Cell B, but only 63.4% for Cell A. Results showed a rapid initial decrease in the proportion of two- and three-ringed PAH compounds followed by a more steady decrease. PAH compounds greater than three rings degraded less rapidly as expected, than did two- and three-ringed compounds.

Statistical analysis for the final sampling event (Day 114) determined that the number of samples analyzed was adequate to characterize Cell B soil at the recommended 80% confidence interval. Statistical analysis also indicated that the number of samples necessary to demonstrate SRL attainment in Cell A was inadequate (39 samples required compared with six analyzed). This is due to inconsistent contaminant reduction in Cell A, with some locations remaining much higher than the SRL.

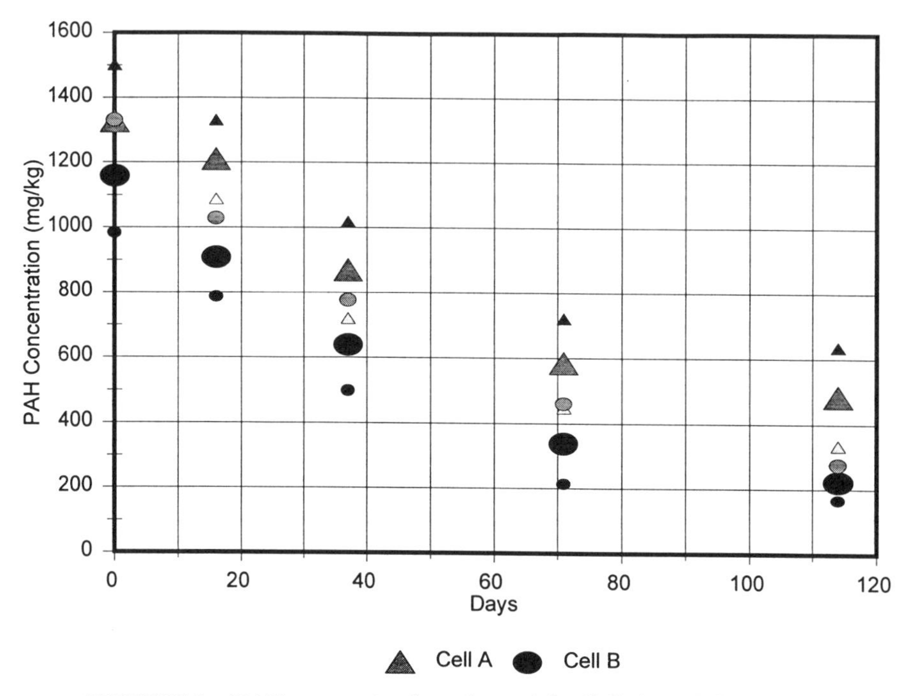

FIGURE 1. PAH concentrations (mean) in Cell A and Cell B. Smaller markers indicate one standard deviation.

Laboratory Results of Inorganic Compounds. A gradual decrease in ammonia-nitrogen and inorganic phosphorus concentrations was observed as the study progressed, possibly due to nutrient incorporation into the microbial cell mass. The ammonia-nitrogen concentration at the end of the study in Cell B was 93 mg/kg and the inorganic phosphorus concentration was 34 mg/kg. These concentrations are above the limiting concentrations for microbial growth and survival, indicating that the quantity of chicken manure added was sufficient.

Microbial numbers (total bacterial plate counts) indicate the robustness of microbial activity. These numbers indicate a healthy heterotrophic microbial population in Cell B throughout the study ($>1x10^6$), which increased during the study period.

Bioremediation Kinetic Analysis. Degradation rates for PAH compounds vary with the smaller-ringed compounds biodegrading more easily than the larger-ringed compounds. However, when all other environmental conditions (such as nutrient availability and pH) are optimized, a first-order (exponential decay) expression can be used to model the overall degradation pattern. When analyzing degradation rates and fitting analytical data to a prescribed model (such as the first-order model), a level of confidence must be assigned to the analysis. Because the medium is heterogeneous – particularly clayey soil and complex organic compounds (PAHs) – nonparametric methods were used, since they are not based on any statistical assumptions, such as normality, which could bias the study's true outcome.

Nonparametric tests typically involve the median rather than the mean as the descriptive mean for a set of samples. For a particular sample size, the confidence interval for the available data is obtained at or close to the desired level of significance (Helsel and Hirsch, 1992). Six soil samples were analyzed for PAHs in each of the study's data sets. Using nonparametric statistics, the confidence interval for the population median lies between the second and the fifth value (for each set of six samples) at a confidence level of approximately 80%. Nonlinear regression PAH analysis was performed on the data from each cell. Data which covered the confidence limit were used to fit a first-order regression model.

Table 1 indicates that the PAH degradation rate achieved in Cell B was 0.015 per day^{-1}, and Cell A was 0.010 per day^{-1}. The coefficient of determination (R^2), which indicates how accurate the data are as compared with the model, was high (>0.8) for both cells, but the model for Cell B showed a coefficient of determination greater than 0.90. The theoretical time necessary to attain SRLs for PAHs was lower in Cell B, the equivalent of 61 days. Cell A showed the theoretical time to attain SRLs was 106 days, but this was not the case.

Table 1. Summary of biodegradation rate and regression analysis

Parameter	Cell A	Cell B
Regression Model $C_t = C_o e^{-kt}$	$C_t = 1300e^{-0.010t}$	$C_t = 1119e^{-0.015t}$
First-Order Degradation Rate (k, day^{-1})	0.010	0.015
R^2 Coefficient of Determination	0.85	0.93
Time to Attain SRLs (days)	106	61

Notes:
C_o = Starting concentration, mg/kg
C_t = Concentration at time t, mg/kg

RESULTS, CONCLUSIONS, AND APPLICATIONS

The field study demonstrated that site soil is amenable to landfarming to reduce PAH concentrations below SRLs in the time frame of the study (114 days) under specific operating conditions. Cell B (which contained chicken manure and exogenous microorganisms) achieved a significantly higher percentage of PAH removal than Cell A. All six samples analyzed in Cell B were less than the SRL in the final sampling round. The nonlinear regression model for PAH degradation in Cell B had a high coefficient of determination (R^2>0.90 at an 80% data confidence interval determined by nonparametric methods), showing 61 days to attain SRLs for PAHs, and the first-order removal rate, k, to be 0.015 day^{-1}. Not surprisingly, Cell B sustained a robust microbial population throughout the study.

The analyses and results of the field test have provided specific and broad-based conclusions which can be used in full-scale landfarming. First-order removal rates(k), calculated in this study, compare favorably with rates from the bench-scale study. Two- and three-ringed PAHs degraded at a faster rate and to a greater extent than PAH compounds with three or more rings. At no time in the study was there any indication that pH was fluctuating out of the range necessary to sustain viable microbial growth, which indicated that site soil is buffered and suitable for full-scale application. PAH compounds were not detected in drainage water from the cells. From mass balances performed, the amount of PAHs lost in drainage water was negligible compared to the amount of PAHs in the soil at the start of the study. Biological degradation was therefore unaffected by leaching of contaminants from the cells.

Based on the results and conclusions, the following applications were made for full-scale landfarming. The recommended quantity of chicken manure to be added to site soil is 1 to 2% by weight, and the recommended quantity of exogenous microorganisms to be added is 0.25 lb (113 grams) per cubic yard of soil. Because the field study has demonstrated that landfarming is a feasible alternative, the form of land treatment (in-place or constructed landfarming units) can be altered to suit existing site conditions. Similarly, other parameters — such as depth of soil lifts, frequency of field monitoring, and analytical measurements - were all adjusted to meet full-scale objectives and for cost savings. This included initiation of full scale *in-situ* landfarming in the dragout area in 3-foot lifts, which has resulted in approximately 6,000 cubic yards (4,587 cubic meters) of soil remediated to below SRLs. No further action is required at the remediated areas, and the remainder of the soil is expected to reach the SRLs this year.

REFERENCES

USEPA. *Bioremediation Using the Land Treatment Concept*, EPA/600/R-93/164. August 1993.

USEPA. *Test Methods for Evaluating Solid Waste*, SW-846, Volumes I and II, November 1986.

D.R. Helsel and R.M. Hirsch. 1992. *Statistical Methods in Water Resources*,, Elsevier Publications, The Netherlands.

FULL SCALE BIOREMEDIATION OF PAH

Mark Connolly (Connolly Environmental, Melbourne, Australia)
Frances Howe (Connolly Environmental, Melbourne, Australia)
Margaret Mazur (Connolly Environmental, Melbourne, Australia)

ABSTRACT: Full scale bioremediation of polyaromatic hydrocarbon (PAH) contaminated fill was attempted using a combination of active landfarming and intrinsic bioremediation, with an emphasis on inexpensive methods. Treatments comprised regular mechanical mixing, addition of fertiliser and wood chips and irrigation during summer. Sampling over 18 months showed total PAH concentrations decreased in four of eight sampling areas with confidence >95% and in two further areas with confidence >90%. The average total mass of PAH was estimated to have decreased by greater than one third over the same period. However the clean up criteria were not met. Mechanical turning appeared to have dispersed PAH aggregates throughout the soil, causing variation in results. Laboratory analysis of different soil particle size classes showed bioremediation may be limited by low concentrations of the smaller PAH molecules, the presence of PAH aggregates, suboptimal moisture and nutrient concentrations and elevated soil pH. We concluded that more active treatment involving more frequent soil turning and moisture and nutrient control were necessary to achieve the cleanup criteria.

INTRODUCTION

Site contamination by the import of fill containing PAH is a widespread problem. Much work has studied the potential for bioremediation of PAH. However few studies have involved full scale bioremediation using methods which are economically viable.

Approximately 1,600 m^3 of PAH contaminated soil was detected at a former government depot at Geelong, Victoria. Initial sampling and analysis during December 1996 detected the full range of PAH compounds, with total PAH concentrations in the range 0.8 to 174 mg/kg and benzo(a)pyrene (BaP) concentrations up to 20 mg/kg. The site was free from significant concentrations of other potential contaminants. The PAH was associated with fill imported to the site. The fill comprised gravely sand and clay with charcoal fragments. Underlying the fill was uncontaminated natural clay.

Remediation of the PAH was required to enable residential development of the 9,000 m^2 site. The initial concentration ranges and clean up criteria for average concentrations are shown in table 1.

Three cleanup options were considered. Costs estimates were as follows:

(a) Excavation, disposal and backfilling with clean soil: $180,000.
(b) Burial 1 m deep on site: $100,000.
(c) Bioremediation: $38,000.

TABLE 1. Initial concentrations and clean up criteria for PAH and BaP

Contaminant	Initial concentration (mg/kg)	Clean up criteria (mg/kg)
BaP	<0.1 to 20	<1.5
Total PAH	0.8 to 174	<20

Bioremediation was the preferred option because of cost and the availability of time. Bioremediation was attempted using landfarming methods similar to those commonly used to treat petroleum contamination. The method combined active landfarming with passive intrinsic bioremediation. The emphasis of this bioremediation project was to use simple, cost effective and practical methods.

MATERIALS AND METHODS

The landfarm was established in February 1997 as follows. The contaminated fill was mixed and aerated thoroughly using an excavator and road grader and spread out across the site in a layer approximately 0.25 m thick. Initially one tonne of NPK fertiliser and 400 m^3 of wood chips were added and mixed thoroughly into the soil to increase permeability and provide additional carbon and nutrients. The landfarm was then turned using earthmoving equipment every three months approximately to increase the supply of oxygen. Turning methods aimed to minimise movement of soil around the site. During the dry summer months the landfarm was irrigated. In March 1998 one tonne of fertiliser was again added.

Soil Sampling and Analysis. Six rounds of sampling at 24 locations across the landfarm were conducted between February 1997 and August 1998. Each sampling round was conducted immediately following turning of the landfarm. Due to the expected large spatial variability in concentrations the same sampling locations were sampled in each round to enable comparison of results for successive rounds. Additional samples were collected in April 1998 in areas where high variation in QA/QC results was observed.

At each location approximately 0.5 kg of soil was collected from 0-0.2 m depth using a sampling trowel. The samples were well homogenised then placed into clean glass jars. As field observations and analytical results showed the contamination to be highly heterogeneous, in August 1998 a different sampling method was used in an attempt to collect more representative samples. Approximately 15 kg of soil was collected at each sample location using a shovel following thorough mixing of soil over an area of 1 x 1 m and 0.5 m deep. The sample was well homogenised. Half the sample was discarded and the remainder further homogenised. Two 250 ml sample jars were subseqently collected from this material.

All samples from each round of sampling were analysed for PAH concentration. Samples were composited up to four into one within each sample area. To assess the conditions for PAH degrading organisms in the landfarm several rounds of analysis were conducted for pH, moisture, total organic carbon, nitrogen, phosphorus and potassium. To estimate the population of heterotrophic microorganisms in the soil the laboratory conducted two plate counts from soil collected in August 1998. To investigate any relationship between the soil particle size and the PAH concentration two samples were sifted before analysis. The grain size classes presented for analysis were <0.06 mm, 0.6-2.0 mm and >2 mm.

RESULTS AND DISCUSSION

Average Individual PAH Concentrations. The results for concentrations of individual PAH compounds are shown in figure 1. No PAH degradation was observed for compounds with initially low concentrations (below 2 mg/kg), such as naphthalene, acenaphthylene, acenaphthene, fluorene, antracene and dibenz(a,h)anthracene. This suggested that low concentrations of PAH were not bioavailable, possibly due to binding to the soil (Weissenfells *et al.* 1992, quoted in Prince *et al.* 1997).

The remaining PAH compounds showed an overall decrease in concentrations. The largest decreases in concentration observed were for fluoranthene, pyrene, benzo(p)fluoranthene, benzo(a)pyrene and indeno(1,2,3-cd)pyrene. An initial increase in PAH concentrations was observed between February and March 1997. This may have resulted from material heterogeneity or contaminant redistribution due to turning of the soil.

Total PAH Concentrations by Sample Area. Each sample area was defined by the composite sample relating to that area. In areas where individual samples were analysed the total PAH concentration for the area was calculated as the average of individual samples within the area. Results are shown in figure 2.

A non parametric Mann-Kendall test for trend with time (Gilbert 1987) was performed for each sample area to determine if the observed downward trend was statistically significant.

Six of the eight areas showed an overall decrease in total PAH concentrations during the remediation period. Four of these areas (areas 2, 4, 5 and 8) showed a downward trend in PAH concentrations with greater than 95% confidence. The confidence for significant decrease in areas 6 and 7 was 90%. Area 6 showed low PAH concentrations, generally below 10 mg/kg, with very small changes observed during the remediation period. This indicated that the bioavailablity of PAH in these concentrations could be low.

There was no significant decrease in total PAH concentrations in areas 3 or 5. It was considered likely that the slight variations in PAH concentrations in area 3 were due to heterogeneity in the sampled material. In area 5 the low PAH concentrations reported in the first three sampling rounds increased markedly for the second three rounds. It was considered likely that this increase was due to increasingly effective dispersal of PAH aggregates throughout the soil and/or inadvertant intermixing of soil from adjacent areas.

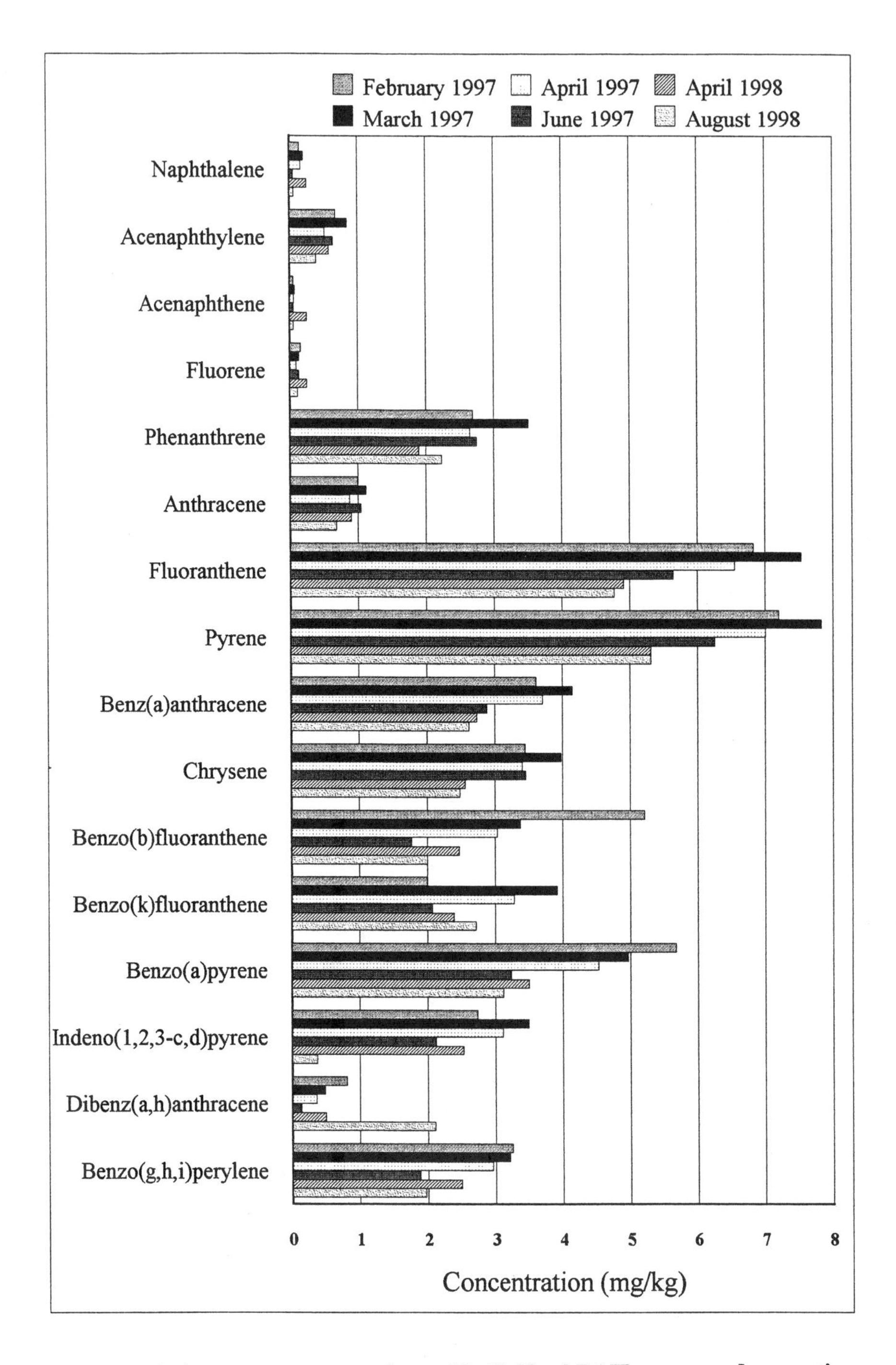

FIGURE 1. Average concentrations of individual PAH compounds over time

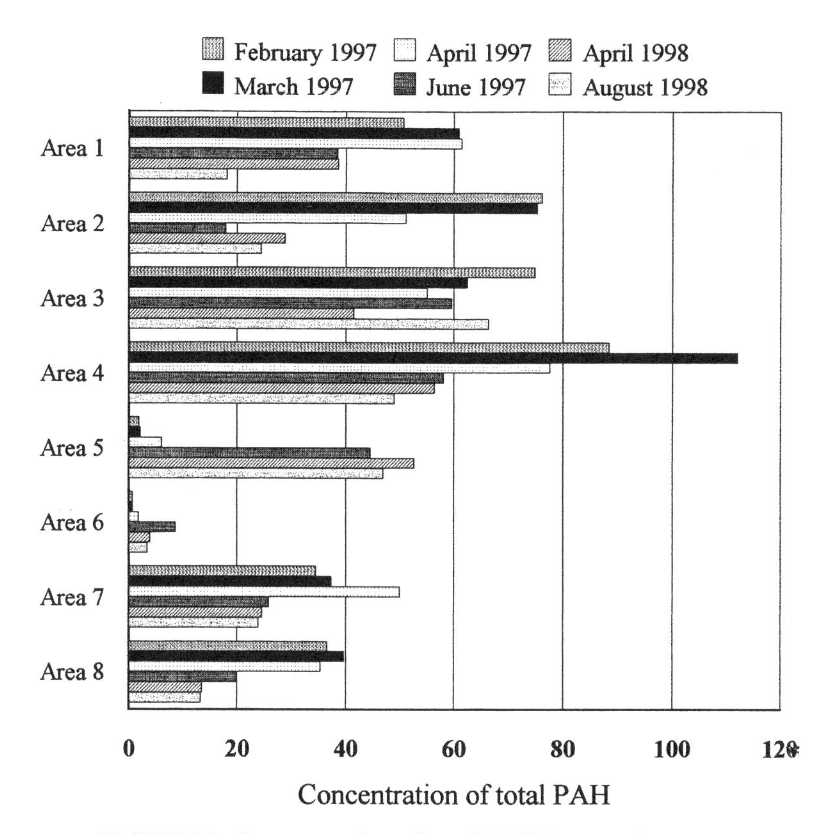

FIGURE 2. Concentration of total PAH by sample area

Changes in Average Total PAH Mass. The ratio of total PAH mass to the unit thickness of the landfarm was estimated using the weighted average of PAH concentrations. Concentrations were weighted by an estimate of the proportion of the landfarm area represented by a given composite sample. The variation in total PAH mass with time is shown in figure 3.

The average mass of PAH was calculated to have decreased by over one third during the remediation period. The largest decrease occurred between March 1997 and June 1997. The lower subsequent rate of decrease may have resulted from reduced bioremediation and/or detection of higher PAH concentrations due to contaminant redistribution through turning of the soil.

PAH Distribution in the Soil Mass. The >2 mm soil grain size fraction comprised 50% of the total landfarm mass. Much of the PAH in this fraction comprised PAH aggregates (eg. charcoal and tar fragments). The bioavailablility of PAH in this soil fraction was likely to be limited by:

(a) The low dissolution rate of hydrophobic PAH molecules in soil aggregates.

(b) The large particle size (and resulting small surface area relative to volume) of PAH aggregates decreasing the opportunity for microbial attack.

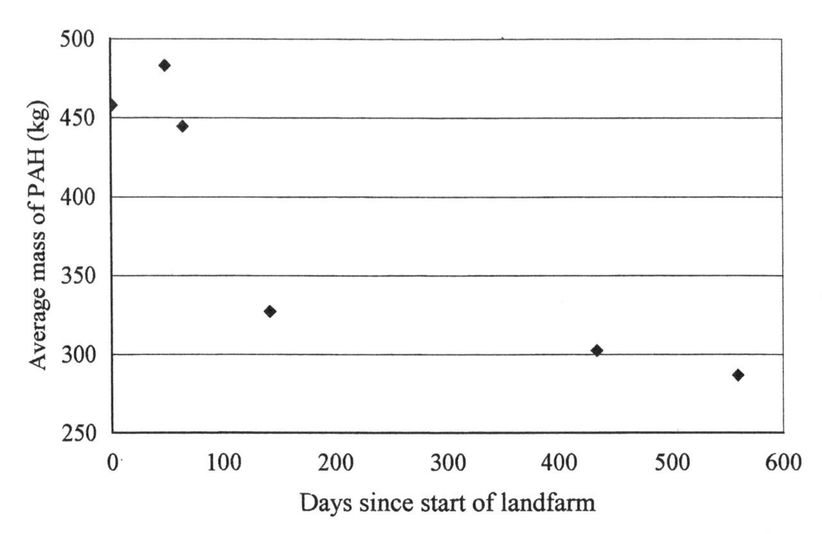

FIGURE 3. Change over time of average mass of PAH across the site

About 40% of the total PAH mass was fine-grained material (particle size <0.06 mm diameter). Although this fine PAH material was likely to be strongly adsorbed to the soil, its small particle size would enhance its bioavailablity.

The regular mixing of soil reduced the size of soil peds and particles, potentially increasing the bio-availability of contaminants previously not available for uptake by soil microorganisms and increasing landfarm homogeneity.

Heterotrophic Microorganisms. Two plate counts showed 15 million and 40 million heterotrophic microorganisms per gram of soil following 48 hours at 37^0C. The results indicated that microbial population was not likely to be a limiting factor for aerobic biodegradation.

Nutrients, Moisture and pH. The carbon:nitrogen:phosphorus (C:N:P) ratio was measured twice with results 100:1.3:0.6 and 100:1:0.7. Peramaki (1997) recommended a nutrient ratio of 100:10:1. Soil pH ranged from 8.4 to 8.9, above the recommended optimum range of 6-8 (Peramaki, 1997). Average moisture content varied from 5% in April 1997 (the end of a hot dry summer) to 15.3% in August 1998, and was generally below the optimum range of 15-20%.

Comparison with Clean Up Criteria. To compare the August 1998 sampling results with the clean up criteria the site was divided into grid blocks with corners defined by the sampling locations. For each grid block the average total PAH and BaP concentrations were calculated and compared to the criteria. One of fourteen grid blocks (G12) met the clean up criteria for total PAH and BaP and a further two grid blocks (G1 and G14) met the criteria for total PAH.

CONCLUSIONS

There was an overall decrease in PAH concentrations during the remediation period. Total PAH concentrations decreased in four of eight sampling areas with confidence of greater than 95% and in two further areas with confidence greater than 90%. Average total PAH mass decreased by over one third during the remediation period.

However despite the observed decreases in PAH concentrations and total PAH mass most of the site did not meet the clean up criteria. We concluded that bioremediation had been limited by:

(a) Low bioavailablility of PAH compounds due to the relatively low solubility and PAH distribution within the soil mass.

(b) Limited presence of PAH compounds with less than four carbon rings which could be used as growth substrates to stimulate the metabolism of the higher molecular weight PAH compounds.

(c) Difficulty in maintaining the optimum soil moisture during summer.

(d) Elevated soil pH.

(e) Difficulties in maintaining even distribution of nutrients.

Mechanical soil turning aerated the soil and aided dispersal and distribution of the contaminants throughout the soil volume, thereby increasing PAH bioavailability. We concluded that PAH compounds will continue to biodegrade at a relatively slow rate and that the landfarm would have to operate for at least several more years to achieve cleanup criteria. We also concluded that more active treatment involving more frequent turning and better moisture and nutrient control would hasten bioremediation.

REFERENCES

Gilbert, R. O. 1987. *Statistical Methods for Environmental Pollution Monitoring.* Van Nostrand Reinhold, New York, NY.

Peramaki, M. P. and K. R. Blomker. 1997. "Practical Design Considerations for Composting Contaminated Soil." *Fourth International In Situ and On-site Bioremediation Symposium (April 28-May 1, 1997. New Orleans). 4(2): 103-112.* Battelle Press, Columbus, OH.

Prince, R. C., E. N. Drake, P. C. Madden, and G. S. Douglas. 1997. "Biodegradation of Polycyclic Aromatic Hydrocarbons in a Historically Contaminated Soil." *Fourth International In Situ and On-site Bioremediation Symposium (April 28-May 1, 1997. New Orleans). 4(2): 205-210.* Battelle Press, Columbus, OH.

BIODEGRADATION OF POLYCYCLIC AROMATIC HYDROCARBONS IN SOILS CONTAINING WEATHERED CRUDE OIL

James L. Brown, Roy F. Weston/REAC, Edison, NJ
John Syslo, Roy F. Weston/REAC, Edison, NJ
Royal Nadeau, U.S. EPA, Environmental Response Team Center, Edison, NJ

ABSTRACT: A laboratory treatability study was conducted to measure the biodegradation of polycyclic aromatic hydrocarbons (PAHs) in soil containing weathered crude oil residues from three abandoned sites. Total petroleum hydrocarbon (TPH) levels before treatment ranged from 3 to 16 %; after treatment, residual TPHs of 1.2 to 2.5 % remained. All PAH data were normalized to hopanes as an internal standard. Losses of dibenzothiophenes, phenenthrenes, pyrenes and chrysenes did not conform to first order kinetics. Loss of three-ring dibenzothiophenes exceeded 95% for all three soils. Losses of four-ring pyrenes ranged from 71 to 87%, while loss of chrysenes ranged from 63 to 88%. Dimethyl homologues of pyrene and chrysene were the most persistent. Losses of these compounds ranged from 54 to 62%, with residual levels of 1 to 5 mg/kg remaining after treatment. A risk assessment at one of the sites showed an excess cancer risk to humans of less than 10^{-6} from potentially carciogenic PAHs. In spite of this, all three sites will require require remediation because their high soil TPH levels represent a threat to surface water. Active land treatment bioremediation followed by low maintenance, plant-mediated bioremediation has been implemented at one of the sites, and is proposed for the others.

INTRODUCTION

Weathered crude oil wastes are moderately recalcitrant, and residual TPHs of 1 to 3 percent (10,000 to 30,000 mg/kg) may remain after bioremediation treatment of contaminated soil. Polycyclic aromatic hydrocarbon (PAHs) in these petroleum residues could present an unacceptable health risk, as many of these compounds are both carcinogenic and persistent. Persistence of PAHs in soils may be due to structural recalcitrance, reduced bioavailability from sorption, or some other factor. In this study, crude oil-contaminated soils were collected from three abandoned sites in Wyoming, Kansas and Kentucky. The soils are fine-textured and contain crude oil residues that have weathered in place for between 10 and 40 years. As a result, they have a high potential for reduced bioavailability due to sorption by soil colloids. The goal of this treatability study was to evaluate the rate and extent of biodegradation of moderately recalcitrant dibenzothiophenes, pyrenes and chrysenes over a 6 to 12 month period. The more recalcitrant five- and six-ring PAHs were not studied because their concentrations were too low to monitor quantitatively.

MATERIALS & METHODS

Soil for this investigation was obtained from two inactive refinery sites in Kansas and Wyoming and an abandoned oil well field in Kentucky. Levels of TPH in soil ranged from 3 to 16%. Soil characteristics and bench-scale test conditions for the petroleum biodegradation study are presented in Table 1. Soils from all three sites were fine-textured and required addition of bulking agents to improve their physical condition. Soils from the Lovell Refinery site in Wyoming were amended with wood chips, composted cattle manure and sawdust as bulking agents prior to treatment. Soil from the Marco Refinery site in Kansas were amended with either chopped grain straw or sunflower hulls. Soils from the Taffy site in Kentucky were amended with sand and chopped grain straw. Differences in petroleum biodegradation from bulking agent treatments were very slight. For this reason, some of the treatments were considered as replicates for purposes of data evaluation. Soil samples were analyzed using either gas chromatography/mass spectrometry (GC/MS) in the selection ion monitoring (SIM) mode, or gas chromatography/flame ion detection (GC/FID) methodology. The GC/MS SIM method is a compilation of U.S. Environmental Protection Agency (U.S. EPA) Method 8015B of SW846, D3328, Marine Safety Laboratories Notice 5200.9, and several other methods used by the U.S. Coast Guard, Environment Canada and other laboratories currently developing methods for SIM oil analysis and oil characterization techniques. Samples were extracted using the Soxhlet method outlined in U.S. EPA SW-846. All PAH data were normalized to hopanes. Two compounds, C29 hopane (17α(H),21β(H)-30-norhopane) and C30 hopane (17α(H),21β(H)-hopane) were selected for this purpose because they were most abundant. Hopanes are resistant to biodegradation and represent a stable internal standard. Use of hopane-normalized data has been shown to eliminate most of the data variability caused by uneven spatial distribution of contaminants (Butler, et al., 1991).

RESULTS

Losses of dibenzothiophenes, pyrenes and chrysenes from Lovell soil are presented in Figures 1a, 1b and 1c, respectively. Data points in the figures represent averages from at least three replicates. Normalization of data in Figure 1 to hopanes successfully reduced data variability. Loss patterns of PAHs from Marco and Taffy soils are not shown graphically, but were similar to those for Lovell soils. There was no evidence of hopane biodegradation during the during the 6 and 12 month test periods. Soil and bench scale study characteristics are shown in Table 1.

DISCUSSION

Loss of PAHs appeared to be a two-stage process, with a rapid initial phase followed by a much slower decline. Lack of linearity of the natural logarithmic plots precludes use of first order kinetics to evaluate PAH losses, and use of these data to predict treatment endpoints would be inappropriate. Bulman et al., (1985) found reaction orders that ranged between 0 and 2, depending on the PAH compound. They concluded that use of a first order model to predict PAH disappearance from soil is clearly not adequate for some PAHs, and if used will severely underestimate their persistence in soil. An additional concern about the use of laboratory kinetic data was expressed by Blackburn (1998). He noted that laboratory data greatly overestimate the time required to achieve a treatment endpoint in the field.

Total PAH losses during the treatability study were 98, 95 and 98% for Lovell, Marco and Taffy soils, respectively. Figure 1 clearly shows the effect of methyl groups on PAH biodegradation. Within each PAH group, parent compounds degraded more rapidly than their structurally complex methylated homologues. These results are in agreement with the findings of both Alexander (1994) and Huesemann (1997). Loss of three-ring dibenzothiophenes exceeded 95% for all three soils. Losses of four-ring pyrenes ranged from 71 to 87%, while chrysene losses ranged from 63 to 88%. Dimethyl homologues of pyrene and chrysene were more persistent, with losses ranging from 54 to 62%. There was no indication that pyrenes were less susceptible to biodegradation than chrysenes based on their more condensed structure.

Table 1. Soil Characteristics and Bench Scale Conditions for Petroleum Biodegradation Study

	Lovell	Marco	Taffy
Soil pH	7.0 – 7.6	6.8 – 7.4	5.9 – 7.2
Soil Texture	Silty Clay	Silt Loam	Silty Clay
Cation Exchange Capacity (meq/100 g) [a]	58	21	21
Total Kjeldahl N (mg/kg) [b]	560	3,250	1,720
Available Phosphorus [c]	Medium to Very High	High	Medium to High
Available Potassium [c]	High to Very High	High	Medium
Sulfate (mg/kg) [d]	2,100	1,060	100
Soil Temperature	17–22 °C	17–22 °C	17–22 °C
Soil Moisture (percent dry wt.)	20–25	20–25	20–25
TPH (mg/kg)	30,000	160,000	100,000
Total PAHs (mg/kg)	390	1,240	330
Carbon:Nitrogen [e]	30:1	25:1	40:1

[a] Cation exchange capacity analyzed using USEPA Method SW-846-9081.
[b] Total Kjeldahl nitrogen analyzed using USEPA Method 351.
[c] Available phosphorus and potassium extractable in a Mehlich 2 solution.
[d] Sulfate analyzed using USEPA Method SW-846-9038.
[e] Based on 75 percent estimated carbon in petroleum hydrocarbons; supplemental diammonium phosphate fertilizer added to all soils.

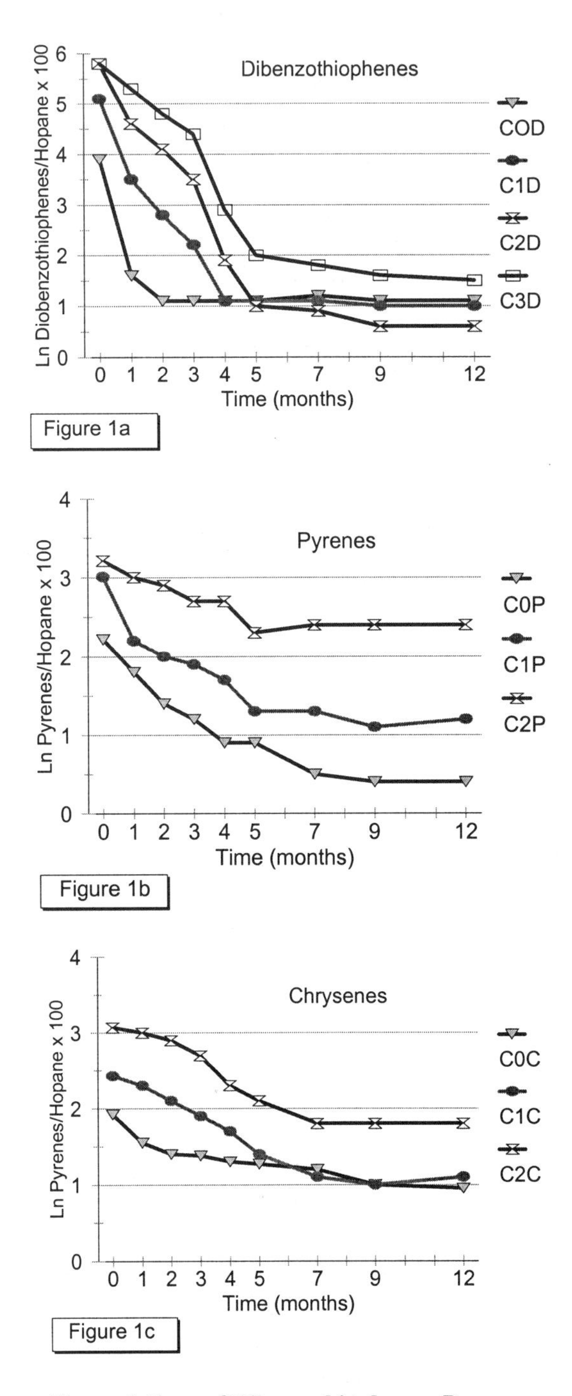

Figure 1. Loss of Dibenzothiophenes, Pyrenes, and Chrysenes from Lovell Soil.

Loss of PAHs (Figure 1) occurred in two relatively distinct phases. The first phase was rapid, and extended for the first four to five months. This was followed by a slower second phase, during which concentrations of most PAHs declined to analytical method detection limit (MDL) levels. The dimethyl homologues of pyrene and chrysene declined slowly, and then attained stable concentrations ranging from 1 to 5 mg/kg. The stability of these compounds may be due to either structural recalcitrance or limited bioavailability, or both. Alexander (1996) reported both sorption and desorption of some soil contaminants follow a two-step process, a rapid phase followed by a much slower phase. In this study, a slow PAH desorption phase may have limited biodegradation. Crude oil residues in fine textured soils from the three test sites had aged for periods of 10 to 40 years, during which PAH sorption should have been complete. Similar PAH loss patterns occurred for Marco soils, except that final concentrations were slightly higher. This could be due to much higher initial PAH concentrations in Marco than in Lovell soils (Table 1). Treatment effectiveness for Taffy soils was very high for all PAHs. Only the dimethyl homologues of pyrene and chrysene were above MDLs after six months of treatment.

Huesemann (1997) evaluated both contaminant sequestration and inherent recalcitrance models as limits to PAH biodegradation under optimum conditions. He concluded inherent recalcitrance best described the cause of incomplete PAH biodegradation. This could explain why the structurally complex, di- and trimethyl PAH homologues are the most persistent in our study. Unfortunately, we cannot distinguish between limited bioavailability due to sorption and inherent recalcitrance as mechanisms limiting PAH biodegradation. Biodegradation of structurally complex PAHs could also be influenced by co-metabolism.

Loehr and Webster (1996) reported that aging reduced contaminant bioavailability in soil. In our study, the aging effect was minimal even though soils had been in contact with oil residues for prolonged periods. The presence of oily coatings on soil colloid surfaces could have impeded PAH sorption and contributed to bioavailability. This effect may be most pronounced for soils with very high TPH levels.

SUMMARY AND RECOMMENDATIONS

Loss of selected PAHs was a two-stage process which did not follow first order kinetics. The slow second stage, which occurred from 5 to 7 months, may have been due to limited bioavailability or inherent recalcitrance, or both. Parent compounds degraded more rapidly than highly methylated homologues. Loss of three-ring dibenzothiophenes exceeded 95% for all three soils. Losses of 4-ring pyrenes ranged from 71 to 87%, while losses of chrysenes ranged from 63 to 88%. Dimethyl homologues of pyrene and chrysene were the most persistent. Losses of these compounds ranged from 54 to 62%, with residual levels of 1 to 5 mg/kg remaining after treatment. Residual TPH levels in soil after 6 to 12 months of treatment ranged from 1.2 to 2.5%, which greatly exceed numerical cleanup standards. For petroleum contaminated sites, it is important to explore other options for negotiating a successful cleanup. Berg et al., (1998) recently introduced the concept of an environmentally acceptable endpoint (EAE) for aged petroleum residues in soils, where a large portion of petroleum residue is resistant to biodegradation. An EAE may be considered when chemicals remaining in soil do not have an adverse impact on human health and the environment. At one of the three sites studied in this investigation, a risk assessment conducted after one year of full scale land treatment bioremediation showed the excess cancer risk was well below the commonly accepted value of 10^{-6}, and the threat to surface water was mitigated. We recommend conducting site-specific risk assessments at crude oil contaminated sites. Carcinogenic PAHs are typically low in crude oil wastes, and residual soil TPH levels in excess of 1% may be acceptable. If further treatment is desired, a phased approach can be implemented, in which active land treatment bioremediation is followed by low-maintenance plant-mediated bioremediation.

REFERENCES

Alexander, M. 1994. *Biodegradation and Bioremediation.* Academic Press, New York, NY.

Alexander, M. 1996. "Sequestration and Bioavailability of Organic Compounds in Soil." In *Environmentally Acceptable Endpoints in Soil: Risk-Based Approach to Contaminated Site Management Based on Availability of Chemicals in Soil.* Gas Research Institute, Chicago, IL.

Berg, M. S., R. C. Loehr, and M. T. Webster. 1998. "Release of Petroleum Hydrocarbons from Bioremediated Soils." *Journal of Soil Contamination.* 7(6): 675–695.

Blackburn, J. W. 1998. "Bioremediation Scaleup Effectiveness: a Review." *Bioremediation Journal* 1(3): 265–282.

Bulman, T. L., S. Lesage, P. J. A. Fowlie and M. D. Webber. 1985. "The Persistence of Polynuclear Aromatic Hydrocarbons in Soil." *PACE Report No. 85-2.* Ottawa, Ontario.

Butler, E. L., Douglas, G. S., Steinhauer, W. G., Prince, R. G., Aczel, T., Hsu, C. S., Bronson, M. T., Clark, J. R., and Lindstrom, J. E. 1991. "Hopane, a Chemical Tool for Measuring Oil Biodegradation." In R. E. Hinchee and R. F. Olfenbuttel (Eds.), *In Situ Bioreclamation: Applications and Investigations for Hydrocarbon and Contaminated Site Remediation.* 539 p. Butterworth-Heinemann, Stoneham, MA.

Huesemann, M. H. 1997. "Incomplete Hydrocarbon Biodegradation in Contaminated Soils: Limitations in Bioavailability or Inherent Recalcitrance?" *Bioremediation Journal.* 1(1): 27–39.

Loehr, R. C. and M. T. Webster. 1996. "Behavior of Fresh vs. Aged Chemicals in Soil." *Journal of Soil Contamination.* 5(4): 361–383.

MEASURED BIOAVAILABILITY AS A TOOL FOR MANAGING CLEAN-UP AND RISKS ON LANDFARMS

Joop Harmsen (SC-DLO, Wageningen, The Netherlands)
Marijke Ferdinandy (RIZA, Lelystad, The Netherlands)

ABSTRACT: Four different sediments, polluted with PAH and mineral oil, were studied, which were already treated with intensive and intrinsic landfarming at demonstration scale during 4 to 8 years. The bioavailability was assessed using chemical and physical tests. The ecological risks were examined using bioassays and bioaccumulation tests. The results of these tests were compared with the chemical composition and experimental data from the demonstration fields.
The partial extraction tests turned out to distinguish three fractions of pollutants: fast, slow and very slow desorbing. These fractions could be related to the results of the bioassays, the bioaccumulation tests and the bioremediation on the experimental fields. The measured bioavailability is therefor an excellent tool to manage clean-up results and risks on a landfarm.

INTRODUCTION

Landfarming is a well-known and relatively simple and inexpensive biological treatment technique for soils and sediments contaminated with organic pollutants. The biodegradation of these compounds requires oxygen. Since sediments have an anaerobic origin, due to the water filled pores, landfarming of sediments starts with dewatering and ripening. As soon as air-filled pores exist, the degradation can start.

Bioremediation of soils and sediments often follows a characteristic pattern: starting with high degradation rates of the bioavailable part (phase 1) followed by slow rates for the strong adsorbed part (phase 2). This part can be degraded, after (slow) desorption from the solid phase to the water-phase.

This two-phased character of bioremediation affects the way in which a landfarm can be managed most effectively. In the Netherlands we do distinguish three different landfarming techniques (see table 1).

TABLE 1. Different landfarming techniques

Steps in landfarm	Goal of process control	Technique
1. dewatering and ripening	providing aerobic conditions (ploughing)	intensive landfarming
2. fast biodegradation (phase 1)	limited by process conditions	intensive landfarming or greenhouse farming
3. slow biodegradation (phase 2)	limited by availability	intrinsic landfarming

In step 1 and 2 of the landfarm, bioremediation is limited by biological activity; which can be increased by cultivation (intensive landfarming or greenhouse farming). In step 3 the sediment is fully ripened and aerobic, but biodegradation is limited by bioavailability. A long time research was aimed at increasing the bioavailability, which did not lead to a breakthrough. It was shown that the strongly absorbed fraction could be degraded, by enlarging the residence time till 5-30 years (Harmsen et al., 1997).

To manage an intrinsic landfarm it is necessary to have quantitative knowledge on the bioavailability. This study investigated the applicability of different techniques for the quantification of bioavailability (mobility) of organic pollutants and their usefulness in guiding the management and controlling the environmental risks of landfarming.

MATERERIALS AND METHODS

It is known that the distribution of available/non-available pollutants is depending on the type of sediment as well as the phase in bioremediation. Therefore, the experiments were performed with four different sediments (polluted with PAH and mineral oil) that were in different phases of landfarming in a demonstration field (300 - 1000 m^3 per sediment) (Harmsen et al. 1997). Table 2 gives an overview. While the landfarming continued, the PAH and mineral oil contents were measured to determine the degradation rates. The samples for the tests described in this paper were taken in September 1997.

TABLE 2. Origin and phase of treatment of the investigated sediments.

Sediment	*Type of Treatment*	*Time of Treatment*	*Oxygen Situation*	*Phase in Bioremediation* (1)	*PAH (mg/kg d.m.)* (2)	*Mineral oil (mg/kg d.m.)*
PET	(start)	0 year	Anaerobic	Start	550	12,000
	Cultivation	3 years	Aerobic	End of phase 1	35	3,000
	Vegetation	3 years	Aerobic	End of phase 1	45	4,000
	Vegetation	3 years	Partly aerobic	Phase 1	120	6,000
WEM	-	0 years	Anaerobic	Start	45	1,750
	Cultivation	3 years	Aerobic	End of phase 1	30	700
	Vegetation	3 years	Aerobic	End of phase 1	30	700
	Vegetation	3 years	Partly aerobic	Phase 1	30	1.100
GH	Cultivation vegetation	8 years	Aerobic	Far in phase 2	2	300
ZZ	Cultivation vegetation	8 years	Aerobic	Far in phase 2	15	300

(1) phase 1: fast bioremedation, phase 2: slow bioremedation. (2) expressed as 10 PAH of the Dutch National List. PET = Petroleumharbor, WEM = Wemeldinge harbor, GH = Geulharbor, ZZ = Zierikzee

The bioavailability was assessed using chemical and physical tests:

- *Leaching test*, PAH were measured in the drainagewater of the landfarm, in pore water taken in the soil just below the sediment and in water from a leaching test (CEN-test)
- *Partial extraction*, the sediments were extracted with 70% acetic acid. These results can be correlated with the biodegradable fraction (Doddema et al., 1998).
- *Solid phase extraction,* desorption through the waterphase was measured using Tenax as adsorption medium. This gives the bioavailable fraction but also information on the kinetics of desorption (Cornelissen et al., 1997).

The ecological risks were examined using bioassays and bioaccumulation tests. It was necessary to distinguish between fresh and saline sediments (table 3).

TABEL 3. Organisms used with the bioassays and bioaccumulation tests

Saline sediments, Petroleumharbor Wemeldinge harbor	Fresh sediments, Geulhaven harbor, Zierikzee harbor
Vibrio fischeri	*Vibrio fischeri*
Crassostrea gigas	*Daphnia magna*
Corophium volulato	*Chironomus riparius*
Oligochaetes	*Oligochaetes*

RESULTS AND DISCUSSION

Biodegradation. Most of the PAH degradation already had taken place, prior to this study, in the first phase of bioremediation. For the present study, the degradation rate was calculated for the period September 1996 till September 1998. In this way the present rates could be related to the tests for bioavailability and risks (see below). Degradation rates were only (relatively) high in the partly aerobic layers of PET and WEM sediments (see Table 4). These were the lower layers of the not cultivated fields covered with vegetation. In all the other sediments and layers, bioremediation already reached the second phase of bioremediation.

Because the low molecular PAH are easier to degrade, the composition of the individual PAH changes during the degradation. In figure 1 this is illustrated for PET and WEM sediments, treated in the cultivated landfarms. After three years the contribution of the four rings PAH becomes more important.

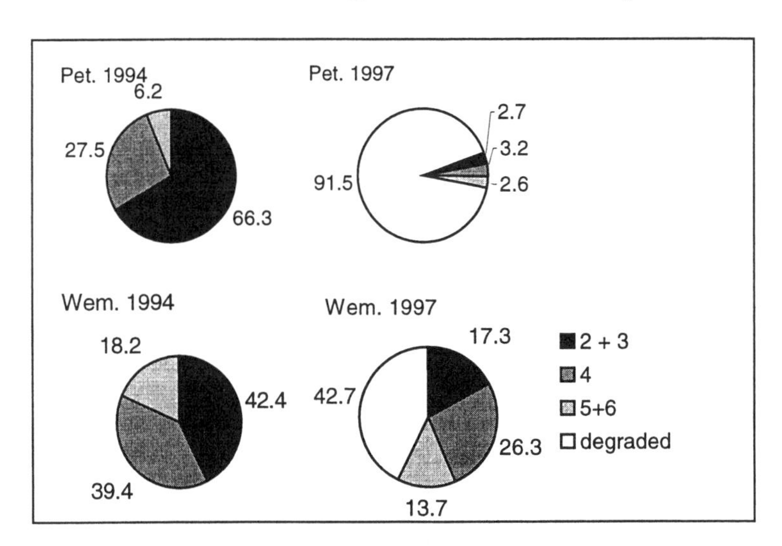

FIGURE 1. Composition of PAH in original (1994) and treated sediments on the cultivated landfarm (1997).

Bioavailability. Bioavailability was measured with physical tests (Leaching tests, pore water, drainage water) and in chemical tests (mild extraction with acetic acid, solid phase extraction with Tenax).

TABLE 4. Degradation of PAH on the landfarm, measured bioavailability, results bioassays and bioaccumulation

Sediment	treatment	Degradation rate		HAc test	Tenax test			bioassys					bioaccumulation
					Fast fraction	Slow fraction	Very slow fraction	*Vibrio Fisheri*	*Daphnia Magna*	*Corophium volulator*	*Crassostrea gigas*	*Chironomus riparius*	
		mg/kg/year	%	%	%	%	%	EC_{20}	LC_{50}	Mortality (%)	PNR (%)	Mortality (%)	mg/kg d.m.
Blank								>45	>100	4.5	-5,4	7	1.3
PET	Original sediment	400	82	87	80	16	4	8.4		100	99		
	Cultivation	5	15	34	9	13	78	>45		17	58		30
	Vegetation, aerobic	10	22		34	41	25	>45		76	98		27
	Vegetation, partly aerobic	60	50		45	26	39	30		98	96		79
WEM	Original sediment	10	22	41	21	13	66	>45		77	39		
	Cultivation	5	16					>45		6	13		6.8
	Vegetation, aerobic	5	16		12	39	49	>45		10	14		6.8
	Vegetation, partly aerobic	5	16		10	5	85	>45		12	24		48
GH	Cultivation followed by vegetation	0.4	20	29	6	10	84	>45	>100			10	2.0
ZZ	Cultivation followed by vegetation	0.4	2.5	12	3	5	92	>45	>100			14	5.6

EC20= volume % of the original sample where the bioluminescence decreases with 20%; LC_{50} = volume % where 50 % of the organisms shows an effect. PNR= Percent Netto Response, which is the amount death, misshapen and retarded animals calculated with the formula of Abbott (0 = no effect, 100 = maximum effect) and in % mortality

The physical tests gave little information related to the biodegradation rates. Leaching tests turned out not to be useful in determining the bioavailability. Concentrations of PAH and mineral oil were below detection level, with exception of the most polluted PET sediments where mineral oil was detected. In the drainage water from the PET and WEM landfarms, very little PAH could be detected (<0.1 ug/L, 10 National List). The concentrations in the pore water just below the sediment were slightly higher (ca. 0.15 μg/L). This was expected, since the pore water is more in equilibrium with the sediment than is the drainage water. The partition coefficients between pore water and organic matter for the individual PAH expressed as K_{oc} were 7-7.5. This means that the availability of PAH was lower than expected (K_{oc} values of 4.5-6.5). The presence of mineral oil could not be established.

The chemical tests turned out to have good correlation's with the PAH degradation (see table 4 and figure 2). The results of both tests were comparable and made it possible to distinguish between individual PAH. The solid phase extraction furthermore results in differences in desorption rates: in a fast, slow and very slow fraction (table 4). The rate constants (first order kinetics) were between 0.12 and 0.65 hr^{-1} for the fast fraction and $0.42 \cdot 10^{-3}$ and $3.95 \cdot 10^{-3}$ hr^{-1} for the slow fraction.

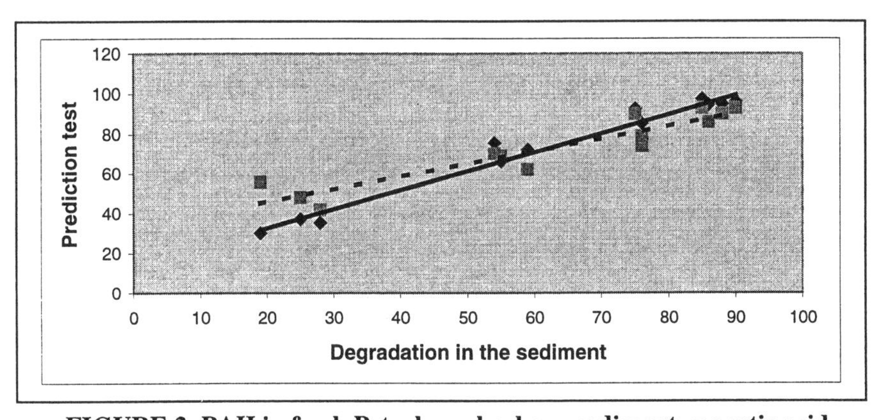

FIGURE 2. PAH in fresh Petroleum harbour sediment. □ acetic acid ◆ tenax. At the right low molecular PAH, at the left high molecular ones

Ecotoxicological risks. The toxicological risks were measured using bioassays and bioaccumulation tests. *Bioassays* were carried out with organisms that live in the water phase (*Vibrio Fisheri and Daphnia Magna)* and the solid phase (*Crassostrea gigas, Corophium volulator and Chironomus riparius*).

The water assays only showed toxicity in the most polluted PET sediment, in the untreated sample and in the partly aerobic sample (table 4), where the bioavailability was measured to be high. The other samples were non toxic. Therefore it can be concluded that these assays are not very sensitive for sediments. Bioassays for the solid phase of the sediment showed more variation in toxicity and had better relations with the degradation rate and the solid phase extraction test (size of fast desorbing fraction) (see table 4).

Bioaccumulation of PAH was highest in PET sediments and lowest in GH and ZZ sediments, which is in accordance with the high fast bioavailable fraction. In two of the three WEM sediments bioaccumulation is related to the fast bioavailable PAH. In the partly aerobic layer the bioaccumulation is increased. This is not agreement with the measured available part, but can be explained by the limited possibility for degradation caused by partly anaerobic circumstances. Representative sampling in this non-homogeneous layer is very difficult. The four rings PAH, fluoranthene and pyrene, have become the most important accumulating PAH.

CONCLUSIONS

It is possible to measure the available part of PAH with chemical extraction tests. The results can be related to both the biodegradation rates as well as to the ecotoxicological risks for in sediment living organisms. Therefore: measuring the bioavailability holds a key in the process control of landfarming, for optimization of the biodegradation and for minimization of the risks.

When bioavailability is high, the potential risks and the potential of biodegradation are high. By cultivation (intensive landfarming) bioremediation is stimulated: available pollutants are quickly degraded and risks are minimised. When the availability is low, stimulation is ineffective. The degradation is limited by sorption and the risks are low. Cheaper intrinsic landarming is most effective.

The tests shows that risks for water living organisms and leaching of PAH from the landfarm are low even at high contaminant concentrations. This means that the influence of the landfarm on groundwater and ditches will be small.

With regard to the environmental risks, a very important point of discussion however remains. What will be our 'reference levels'? At which level of bioavailability and eco-toxicity for different organisms, the risks are nil or acceptable? These items will need further discussions and studies.

REFERENCES

Cornelissen, G., P.C.M. van Noort en A.J. Govers, 1997. "Desorption Kinetics of Chlorobenzenes, Polycyclic Aromatic Hydrocarbons, and Polychlorinated Biphenyls: Sediment Extraction with TENAX and Effects on Contact Time and Solute Hydophobicity." *Environmental Toxicology an Chemistry, 16(*7): 1351-1357.

H. J. Doddema, M.P. Cuypers, G.B. Derksen, J.T.C. Grotenhuis, M.P. Harkes, J. Harmsen, W.H. Rulkens en A.J. Zweers, 1998. *Characterization of PAH Polluted Sediments for Bioremediation (in Dutch).* STOWA report, 98.32, Utrecht, The Netherlands.

Harmsen, J., H.J.J. Wieggers, J.J.H. van den Akker, O.M. van Dijk-Hooyer, A. van den Toorn and A.J. Zweers. 1997. "Intensive and Extensive Treatment of Dredged Sediments on Landfarms." In: *In Situ and On-Site Bioremediation: Volume 2*: pp. 153-158. Battelle Press, Columbus.

RECYCLING OF CONCENTRATED REFINERY TAR RESIDUES USING BIOREMEDIATION

Barry Ellis Paul Harold and John F Rees
CELTIC Technologies Ltd., Cambrian Building, Cardiff CF1 6DL, UK

ABSTRACT: Concentrated tar waste has been treated by bioremediation and converted to soil at a disused refinery in Scotland, UK. Starting concentrations of contaminants in the raw sludge were approximately 300,000 mg/kg oil hydrocarbons in the range C_{10} – C_{40} and 600 mg/kg polynuclear aromatic hydrocarbons (PAHs). Risk-based clean-up criteria were agreed for PAHs including 6 mg/kg, dibenz (a,h) anthracene; 17 mg/kg, benzo (a) pyrene and 5,000 mg/kg hydrocarbons (diesel range organics DROs). Following a short laboratory and field feasibility programme, full-scale remediation was initiated in August 1997 and completed in three phases during 1998. A total of 7000 m^3 (approximately 10,000 tonnes) of tar was successfully treated and met validation requirements. Key elements of the project comprised: **(i)** Rapid on site treatment with no off site disposal; **(ii)** Conversion of tar into soil cover for golf course construction; **(iii)** No measurable environmental impact. During treatment the biopiles and the underlying groundwater aquifer were monitored and sampled and results of treatment over a period of 8-10 weeks indicated typical final mean concentrations were as follows: 2,385 mg/kg alkanes; 42.2 mg/kg total PAHs; 4.7 mg/kg benz (a) pyrene; 2.2 mg/kg dibenz (a,h) anthracene.

INTRODUCTION

Concentrated tar waste has been treated by bioremediation and converted to soil at a disused refinery in Scotland, UK. The waste materials for bioremediation comprised tar sludge, produced as a by-product from oil shale refining activities and historically deposited at various sites constructed around the area of the site. Tars comprised wet black viscous sludge, with percentage concentrations of diesel range oil hydrocarbons (DRO) and elevated PAHs at several hundred mg/kg. Tars additionally contained slightly elevated concentrations of volatile organic compounds (BTEX) and phenols, as well as certain heavy metals. The tar contained a range of contaminants including polynuclear aromatic hydrocarbons (PAHs) and oil hydrocarbons in the range C_{10} – C_{40}. Starting concentrations in the raw sludge were approximately 300,000 mg/kg oil hydrocarbons and 600 mg/kg PAHs. Risk-based clean-up criteria were agreed for PAHs including 6 mg/kg, dibenz (a,h) anthracene; 17 mg/kg, benzo (a) pyrene and 5,000 mg/kg DRO. Following a short laboratory and field feasibility programme, full-scale remediation was undertaken from August 1997 to October 1998 in three phases of 10-12 weeks treatment.

OBJECTIVES OF THE TREATMENT

The remediation goals for the site were set by the client in a commercial contract in agreement with the Scottish Environment Protection Agency (SEPA), and were calculated using assumptions based upon treated tars being used as a soil cover within the final site development as a golf course. Specifically, treatment objectives were as follows:

- Production of a soil like material with physical and chemical characteristics suitable for re-use within the final restoration program.
- Achievement of specific target concentrations (as below) set using computerised risk assessment techniques to validate the treatment:
    1. Reduction of Diesel Range Hydrocarbons to below 5000 mg/kg.
    2. Reduction of Benz (a) pyrene to below 17 mg/kg.
    3. Reduction of Dibenz (ah) anthracene to below 6 mg/kg.
- Achievement of results within a time-scale that conformed to the overall program of integrated site restoration, and within a commercially viable budget, in relation to site restoration costs and alternative remedial options.

SUMMARY OF TREATMENT OPERATIONS

Treatment works comprised a program of physical waste processing, aerobic biological treatment, together with treatment monitoring and management. Waste processing involved combining tar sludge with specific admixtures, in order to produce a suitable matrix to enable aerobic biodegradation of organic contamination within soil bed (biopile) systems employing specialist and conventional agricultural equipment.

An area of approximately 2 hectares was available for treatment activities at the site, and this was subjected to initial preparation, involving leveling and installation of a basal drainage layer. In order to accommodate the volume of sludge to be treated, bioremediation works had to be undertaken in 3 phases, with each phase requiring construction of 14 to 16 trapezoid biopiles, each measuring roughly 50 x 15 x 0.8 m. Specifically, the proposed treatment involved several stages comprising:

- Importation and preparation of admixtures, including shale, soil and organic materials,
- Construction of treatment beds involving controlled application of tars and admixtures in specifically sequenced layers:
    1 Blaes (spent shale)
    2 "Sub soil" (fill material from the site)
    3 Organic additive (manure, yeast extract)
    4 Tar
- Biological treatment comprising: mechanical mixing and aeration, amendment with further organic and inorganic fertilisers to provide necessary macro and micro nutrients.
- Treatment bed and ambient monitoring, to validate bioremedial processes, maintain optimal treatment matrix environments and determine environmental and occupational impacts. Specific monitoring included:

1. Routine analysis of treatment parameters: moisture, pH, temperature, oxygen and soil gases,
2. A comprehensive sampling and analysis program for soil contaminants including DRO (as normal alkanes), PAHs, phenols and metals. Together with evaluation of degradation products such as acids and alcohols.
3. Routine ambient monitoring including airborne dusts and VOCs across the site, within treatment areas and in respect of exposure for site workers, together with groundwater quality at defined downstream sampling locations.

Each treatment phase was scheduled to be completed within a 16 week program, comprising 6 weeks for preparation and construction works and 10 weeks intensive bioremediation. After each phase, soil beds were stockpiled, where materials were subjected to a further period of secondary passive biological treatment and weathering.

RESULTS

Results of chemical analysis found that all 3 bioremediation phases had achieved specified remediation targets within the program timescales. Final concentrations of contaminants ranged between 2277 and 2728 mg/kg n-alkanes, 4.4 to 4.7 mg/kg benz(a)pyrene, 1.2 and 2.2 mg/kg dibenz(ah)anthracene and 39 to 48.7 mg/kg total PAHs (mean results are indicated in Figs 1 & 2). Final samples in each phase of work (n = 96) suggested comprehensive degradation throughout the entire material and the absence of any hot-spots, with standard error measurements found to lie within 20% of mean values. In addition none of the individual final validation samples exceeded the remediation goal targets.

Fig 1: Bioremediation: mean n-alkane concentrations

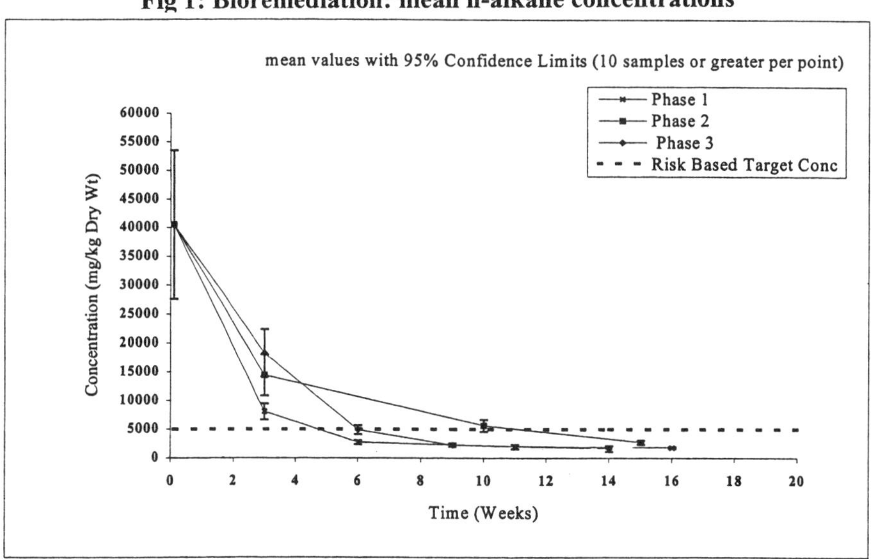

Fig 2: Bioremediation Phase III, PAH Degradation

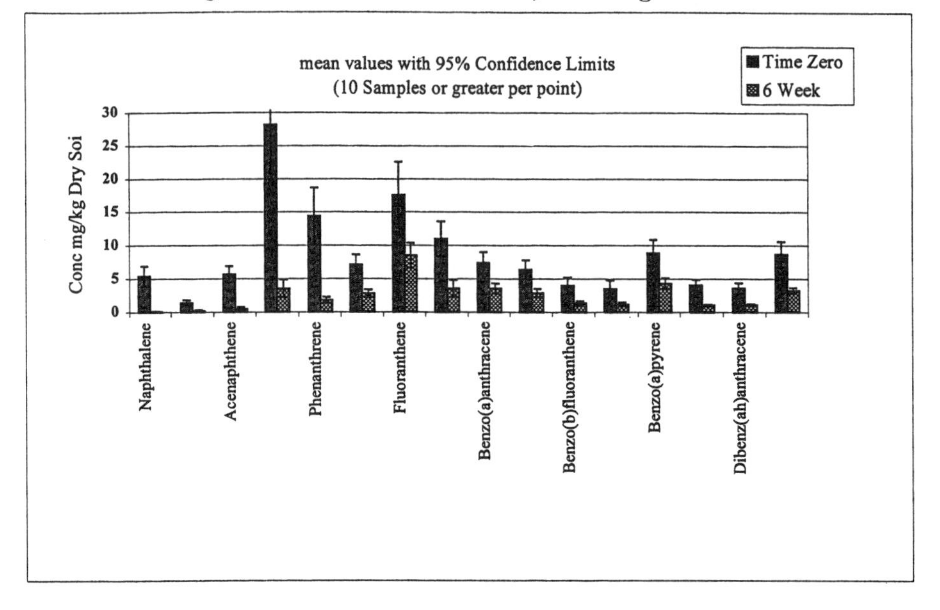

Mean reductions in excess of 80% and 70% of starting concentrations were recorded for n-alkanes and for PAHs respectively. Data indicated that the most significant reductions were due to biodegradation and not dilution effects. Mass balance calculations for metal analytes were used to verify that mass dilution to the original tar approximated to 1:3 and results expressed in this paper are reported following that physical dilution (mixing) excersize. Aerobic enzyme activity assays (calylase) undertaken on the biopile materials reflected contaminant degradation rates. GCMS profiles indicated mineralisation (or possible humification) of contaminants with no obvious build up of intermediates.

Routine inspection and monitoring of soil beds throughout the treatment process indicated production of a well-mixed soil-like material, forming an excellent environment for aerobic microbial activity. This was corroborated by rapid establishment of elevated temperatures, ranging up to around 40°C (see Figure 3) and soil gas measurements indicating utilisation of oxygen and carbon dioxide production (Fig 4). Throughout the treatment process soil bed pH, moisture, nutrient and oxygen levels were maintained at optimal conditions for biological activity by careful management of treatment activities.

Ambient site monitoring involving measurement of VOCs, total and respirable dusts and groundwater, were found to indicate little environmental impact. VOCs continually remained below HSE exposure limits for benzene (5 ppm), as did respirable dusts (5 mg/m^3). Slightly elevated concentrations of total dusts were apparent intermittently during the works but these did not appear to be directly attributable to treatment activities and were not problematic. Groundwater monitoring did not indicate any obvious impact from treatment activities, with concentrations remaining within ranges routinely determined at the site prior to the onset of treatment works. Odours, although not problematic during the works, were found to reduce dramatically during treatment.

Chemical analysis results suggested additional reduction of contaminants within treated materials stock-piled during the winter. The low mobility and volatility of residual contaminants determined from leaching tests and headspace analysis, suggest that losses may be due to further biological activity, with evidence suggesting the possibility of iron being utilised as an alternative electron acceptor in this process.

Laboratory and field seeding tests undertaken during the program have indicated that the treated material can support a variety of grasses and other vegetation, with little difference compared to garden topsoil. Additionally, initial earthworm studies have indicated reduced toxicity within treated materials compared to pure tars, while further testing of this is ongoing.

DISCUSSION AND CONCLUSIONS

The treatment operations undertaken by CELTIC Technologies Ltd during in 1997 and 1998, successfully remediated approximately 10,000 tonnes of waste tars to agreed targets. Furthermore, the treatment process resulted in the production of approximately 30,000 m^3 of soil-like material, which tests have shown to be appropriate for its desired end-use, being relatively inert in respect of leachate and VOC production and suitable to support a range of vegetation. Additionally, from results of initial earthworm studies, the treated material appears have a reduced biotoxicity, compared to the original waste tars.

Key elements of the project comprised: **(i)** Rapid on site treatment with no off site disposal; **(ii)** Conversion of tar into soil cover for golf course construction; **(iii)** Utilisation of local resources as admixtures; **(iv)** Validation of extraction and analysis methods; **(v)** Provision of data with statistical validity; **(vi)** Agreement with the Scottish Environment Protection Agency (SEPA) for methods and associated monitoring; **(vii)** No measurable environmental impact.

The remediation comprised: **(a)** An appropriate Health and Safety plan to protect human health and the environment; **(b)** Biopile Construction and Primary Treatment were undertaken in an area approximately 120 m x 150 m. 14 treatment biopiles (16 biopiles during 1998) were constructed, of dimensions 15 m x 50 m x approximately 0.8 m height. Each biopile was constructed in 17 sequential layers of admixtures and tar. They were intensely intermixed using soil engineering equipment. Biopile construction and primary treatment was undertaken in a four-week period; **(c)** Secondary treatment of the biopiles comprised physical aeration, supplementation with nutrients (organic and inorganic), moisture (to approximately 65% moisture-holding capacity) and oxygen (>10% v/v, via physical aeration). Biopile temperatures were typically in the region of 40°C.

During treatment the biopiles and the underlying groundwater aquifer were monitored and sampled. Each biopile was samples sufficiently to allow adequate statistical interpretation (8-10 samples per biopile per sampling point). Analyses included moisture content. pH, PAHs (GC-MS), SVOCs and VOCs (also by GC-MS), Oil Hydrocarbons (GC-FID), metals, phosphorous, nitrogen and other inorganic components. Soil leaching tests were also undertaken.

Results of treatment showed tars containing over 600 mg/kg PAHs and over 300,000 mg/kg oil hydrocarbons was completed in a treatment period of 8-12 weeks. Typical final mean concentrations were as follows: 2,385 mg/kg alkanes; 42.2 mg/kg total PAHs; 4.7 mg/kg benz (a) pyrene; 2.2 mg/kg dibenz (a,h) anthracene. . A total of 7000 m^3 (approximately 10,000 tonnes) of Tar was successfully treated and met validation requirements.

Fig 3. Bioremediation, Phase III: mean biopile temperatures

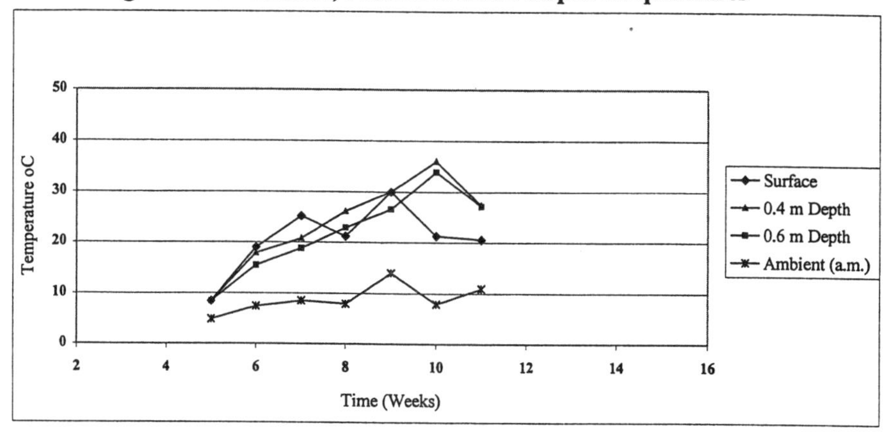

Fig. 4, Bioremediation, Phase III: monitoring of *in-situ* respiration within the Biopile

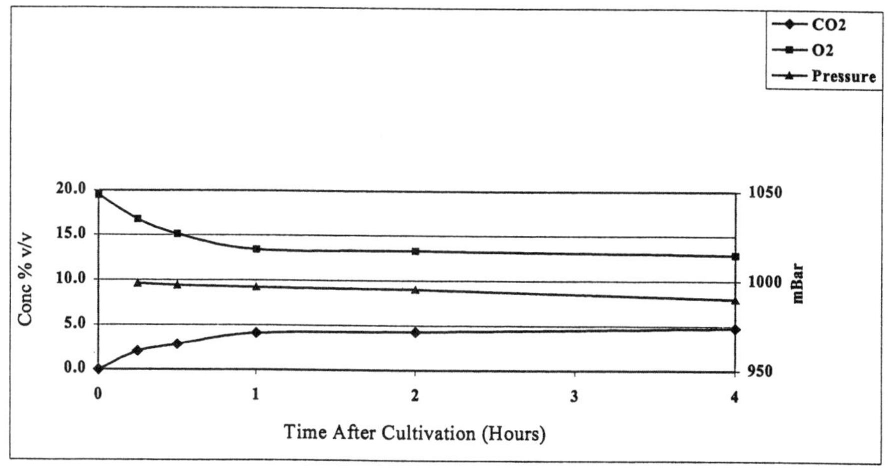

DEVELOPMENT OF WHITE ROT FUNGAL TECHNOLOGY FOR PAH DEGRADATION

Michiel Kotterman, Johan van Lieshout, Tim Grotenhuis, and Jim Field
Wageningen Agricultural University, Wageningen, The Netherlands.

ABSTRACT: In this project, we have optimized the parameters for PAH degradation by the white rot fungi *Bjerkandera* sp. strain BOS55 and *Pleurotus ostreatus*. *Bjerkandera* sp. strain BOS55 was previously shown to be an outstanding PAH degrading isolate, which could degrade PAHs at extremely high rates in liquid cultures.

In artificially contaminated soils, benzo[*a*]pyrene was also rapidly degraded by *Bjerkandera* sp. strain BOS55, with only low residual concentrations. In a size-fractionated soil, the degradation rate of benzo[*a*]pyrene increased with the smaller fractions. The rate and extent of benzo[*a*]pyrene degradation could be enhanced by the addition of surfactants (1-7.5 g L^{-1}). The performance of both white rot fungi was then tested in bench-scale experiments with aged PAH-contaminated soils. At low, economical amounts of 15% wood chips inoculum, *Bjerkandera* sp. strain BOS55 survived for only a few weeks. In contrast, *Pleurotus ostreatus* grew very well with as low as 10% inoculum. In spite of high fungal activity and high ligninolytic enzyme titers, the PAH elimination rate was only increased for a short period. In a soil with high endogenous PAH-degrading activity, the addition of the fungi even resulted in lower rates of PAH degradation. After a prolonged incubation period, no significant differences between the residual PAH concentrations in the fungal treated soils and in the biotic controls could be observed. Additions of surfactants (2.5-7.5 g L^{-1}) or acetone (10% v/v) did not increase the degradation of the residual PAHs. These results indicate that white rot fungal technology is not a solution for the poor bacterial bioremediation of some PAH-contaminated soils.

INTRODUCTION

The bioremediation of soils contaminated with polycyclic aromatic hydrocarbons (PAHs) is severely hampered by the low bioavailability of PAHs (Stucki and Alexander, 1987; Volkering et al., 1992; Mihelcic et al., 1993). The use of white rot fungi in bioremediation of PAH-contaminated soils is considered, as the extracellular enzymes of white rot fungi can rapidly degrade PAHs and this might circumvent the slow diffusion of PAH into bacterial cells.

The white rot fungus *Bjerkandera* sp. strain BOS55 is an outstanding PAH degrading strain. The most important PAH degradation rate limiting parameters in liquid cultures of this fungus were the endogenous hydrogen peroxide production rate and the bioavailability of PAH (Kotterman et al., 1996; Kotterman et al., 1998a). At optimized conditions, this fungus degraded PAHs as anthracene and benzo[*a*]pyrene at the extremely high rates of 1450 and 450 mg L^{-1} day^{-1}, respectively. Like other white rot fungi (Bezalel et al., 1996), *Bjerkandera* sp. strain BOS55 degraded ^{14}C-benzo[*a*]pyrene only partly to $^{14}CO_2$ (13%), and 61% accumulated as polar metabolites. Ames-tests showed that, unlike benzo[*a*]pyrene, these metabolites were neither mutagenic nor could they be activated to mutagens. Some of these accumulated metabolites were also easily mineralized by non-adapted microorganisms, as suggested by mineralization up to 34% after successive fungal-bacterial degradation (Kotterman et al., 1998b). This increased mineralization of PAH by successive fungal-bacterial treatment has also been observed for other white rot fungi (In der Wiesche et al., 1996).

Objective. The objective was to extend the PAH degradation by *Bjerkandera* sp. strain BOS55 from defined liquid cultures to aged PAH-contaminated soils. *Pleurotus ostreatus*, another well-known PAH-degrading white rot fungus, was also tested for its performance. Parameters that dominate the efficiency of PAH removal in soils by the white rot fungi were identified. The obtained results are discussed with respect to white rot fungal treatment of PAH-contaminated soils.

MATERIALS AND METHODS

Culture Conditions with PAH-contaminated Soils. The experiments with sterile, artificially PAH-contaminated soils were conducted as described by Field et al. (1995), the soil used was a sandy soil with 2.6% organic carbon. In these experiments, *Bjerkandera* sp. strain BOS55 was first pre-incubated on 0.7 g rice. After incubation, the rice was added to the bottles containing 4 g of benzo[*a*]pyrene spiked soils together with perlite (to promote aeration). To abiotic controls, autoclaved fungal cultures were added. In the soil-fractionation experiments, the soil was first sieved and then contaminated with benzo[*a*]pyrene.

For the bench-scale experiments with aged contaminated soils, the fungi were first pre-incubated for 4 to 6 weeks at 20 °C on an autoclaved woody substrate. Hereafter, the colonized substrate was mixed with PAH-contaminated soil in ratios varying from 10 to 20% (dry weight). Biotic controls consisted of either solely properly humidified soil or soil with autoclaved inoculum (10%). The abiotic controls consisted of soil (with 10% autoclaved inoculum) to which 5 g sodium azide kg^{-1} was added. Water was added to obtain a suitable water content, 17-35%, depending on the soil and amount of substrate used. Between 1 and 3 kg of soil was used, which was placed in a large bin (25L), aerated with humidified pressured air and incubated at room temperature in the dark. Soils from an old MGP-site (Schoonoord, NL), a creosote-contaminated site (Schijndel, NL), and a 3-years old ripened sediment from Petrol Harbor Amsterdam were used.

PAH-degradation. In experiments with artificially contaminated soils, triplicate bottles were sacrificed for PAH analysis. In the case of bench-scale experiments, 5 separate samples were drawn from every bin and thoroughly mixed.

PAHs were extracted by addition of 4 volumes of acetone, in soil-experiments with at least a liquid to soil ratio of 3:1 (v/w). The PAHs were analyzed by UV absorbance on HPLC equipped with a diode array detector.

RESULTS AND DISCUSSION

PAH Degradation in Artificially Contaminated Soils. In sterile, artificially contaminated soils, *Bjerkandera* sp. strain BOS555 caused rapid degradation of benzo[*a*]pyrene. The presence of organic matter did not affect the benzo[*a*]pyrene degradation, both in normal and ashed soils 90% of the benzo[*a*]pyrene (initial concentration 97 mg kg^{-1}) was degraded after 26 days. As shown in Table 1 for ashed soils, the size of the contaminated soil-particles was observed to be more important. The degradation rate of benzo[*a*]pyrene decreased with increasing soil particle size.

TABLE 1. The effect of soil particle size on the degradation of benzo[*a*]pyrene in artificially contaminated soils by *Bjerkandera* sp. strain BOS55. Values are shown as the average of triplicate samples with standard deviation, initial concentration benzo[*a*]pyrene 94 mg kg^{-1} soil.

	Residual Benzo[a]pyrene (%)			
Time	Soil particle size (μm)			
(days)	0-2000	45-90	125-500	500-2000
7	36.3 (4.3)	16.6 (3.2)	39.9 (1.4)	61.8 (4.4)
14	14.0 (2.1)	5.9 (3.8)	14.4 (6.4)	38.5 (3.7)
26	9.1 (0.2)	5.8 (0.8)	13.7 (2.5)	35.1 (1.4)

Similar effects on benzo[*a*]pyrene degradation rates were observed with contaminated glass-beads of different sizes (results not shown), indicating the slower degradation rate was not only due to slow diffusion of PAH from inside the large soil particles. During the process of contamination, benzo[*a*]pyrene precipitated from an acetone solution. In bottles containing small (soil)-particles, the benzo[*a*]pyrene could precipitate on a much larger surface area. A higher surface area of PAH particles can provide a higher PAH dissolution rate, which has been shown to limit PAH degradation rates (Volkering et al., 1992; Kotterman et al., 1998a). The effect of surfactants on the degradation of the residual benzo[*a*]pyrene in soil after 26 days was monitored in a separate experiment. The residual concentrations could be decreased from 7 mg kg^{-1} at day 26 down to 3 mg kg^{-1} at day 53 by additions of the surfactants Tween 80 and Polyoxyethylene (10) lauryl ether in concentrations from 2.5 up to 7.5 g L^{-1}. Clearly, the PAH oxidation rate in artificially contaminated soils was also limited by the low bioavailability of PAHs.

PAH Degradation in Aged PAH-Contaminated Soils. When added as a 15% chipped wood inoculum, *Bjerkandera* sp. strain BOS55 was not able to survive for more than 2 weeks in the bench-scale experiments as observed by the declining ligninolytic enzyme titers and the CO_2 production rates. Only in this initial two weeks period, the fungus increased the PAH oxidation rate (Figure 1.). In contrast, *Pleurotus ostreatus* grew well for 4 months when added at only 10% inoculum; high titers of laccase and MnP, and a high CO_2 production rate were observed during this period. However, again only in the first weeks a limited increase in the PAH degradation rate was observed. It was observed that *P. ostreatus* severely decreased the number of CFUs (to 0.1% after two months), possibly eliminating PAH-degrading microorganisms. After 2 months, the residual PAH-concentrations in both the fungal treated soils were not significantly lower than those in the biotic controls. Obviously, the extracellular ligninolytic enzymes could not degrade the non-bioavailable PAHs. Likewise, even the small oxidant persulfate could not oxidize more PAH in a soil than could be removed by a microbial treatment (Cuypers et al., 1999).

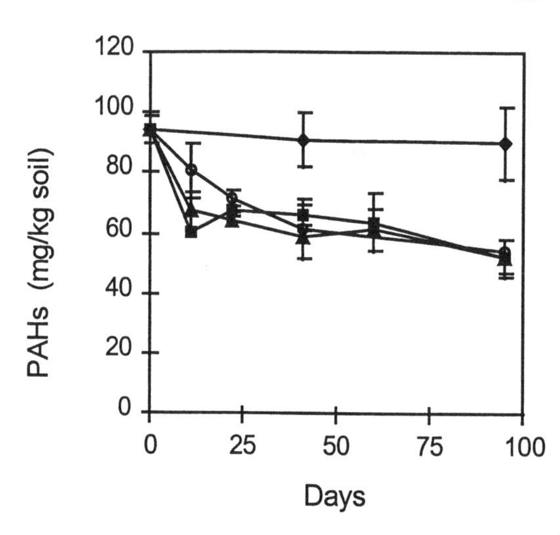

FIGURE 1. Degradation of PAHs in creosote-contaminated soil.

◆: Abiotic control

■: *P. ostreatus* (10% inoculum)

▲: *B.* sp. strain BOS55 (15% inoculum)

O: Biotic control (10% autoclaved inoculum)

The addition of white rot fungi to PAH contaminated soils could even decrease the rate of PAH degradation. As shown in Figure 2, the initial PAH degradation rate in a ripened sediment with high endogenous PAH degrading activity was decreased by fungal treatments. In this experiment with 20% inoculum, fungal activity of *Bjerkandera* sp. strain BOS55 was detected for 4 weeks while *P. ostreatus* was active during the whole incubation period.

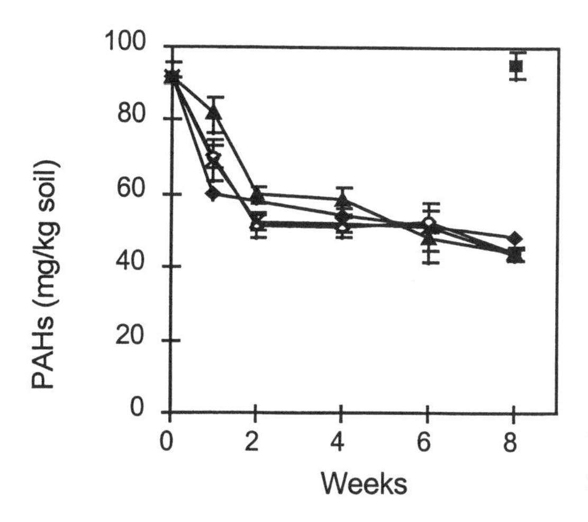

FIGURE 2: Effect of white rot fungi on the degradation of PAHs in ripened sediment.

■: Abiotic control (20% inoculum)
O: P. Ostreatus (20% inoculum)
▲: B. Sp. Strain BOS55 (20% inoculum)
★: Biotic control (20% autoclaved inoculum)
◆: Bacterial control (no additions)

The extent of PAH degradation also strongly depended on the soils used. In the sediment of Figure 2, the residual PAH concentration was 43 mg kg^{-1} soil, while in a soil from a MGP-site more than 80 mg PAH kg^{-1} soil remained after bioremediation (results not shown). As supported by results from Field et al. (1996), the PAH bioavailability and not the microbial activity limited the extent of PAH degradation in these experiments. Attempts to increase the degradation of the non-bioavailable PAHs by the addition of surfactants (2.5-7.5 g L^{-1}) or low amounts of acetone (10% v/v) were not successful in both creosote contaminated soils and soils from a MGP-site (results not shown).

CONCLUSION

The application of white rot fungi in soils can increase the degradation rate of the bioavailable fraction of PAHs compared to non-supplemented soils, when no high endogenous PAH-degrading activity is present. Unfortunately, the fungal extracellular ligninolytic enzymes can not degrade PAHs to a significantly lower level than the endogenous microflora. Therefore, it is unlikely that the limited benefits of white rot fungal treatment will compensate the increased costs of white rot fungal treatment.

ACKNOWLEDGMENTS

This project was funded by IOP project 91219 from Senter, an agency of the Dutch Ministry of Economics, by NOBIS project 96025 and by Ecotechniek (The Netherlands).

REFERENCES

Bezalel, L., Y. Hadar, and C. E. Cerniglia. 1996. Mineralization of Polycyclic Aromatic Hydrocarbons by the White Rot Fungus *Pleurotus ostreatus*. *Appl Environ. Microbiol. 62*(1): 292-295.

Cuypers, M. P., J. T. C. Grotenhuis, and W. H. Rulkens. 1999. "Prediction of PAH Bioavailability in Soils and Sediments by Persulfate Oxidation." In Situ and On-Site Bioremediation, April 19-22, San Diego.

Field, J. A., H. Feiken, A. Hage, and M. Kotterman. 1995. "Application of a white Rot Fungus to Biodegrade Benzo[a]pyrene in Soil." In R. E. Hinchee, J. Fredrickson and B. C. Alleman (Eds), Bioaugmentation for Site Remediation, Batelle Press, Columbus, Ohio, USA, pp. 165-171.

Field, J. A., J. T. A. Van Veldhoven, B. Weller, C. Soeter, R. Wasseveld, J. T. C. Grotenhuis, and W. Rulkens. 1996. "Degradation of PAH in Polluted Soils by the White Rot Fungus *Bjerkandera* sp. Strain BOS55." In R. A. Samsom, J. A. Stalpers, D. Van der Mei and A. H Stouthamer (Eds), *International Congress for Culture Collections,* pp. 336-343. The Netherlands.

In der Wiesche, C., R. Martens, and F. Zadrazil. 1996. Two-Step Degradation of Pyrene by White-Rot Fungi and Soil Microorganisms. *Appl. Microbiol. Biotechnol. 46*(5-6): 653-659.

Kotterman, M. J. J., R. A. Wasseveld, and J. A. Field. 1996. Hydrogen Peroxide as a Limiting Factor in Xenobiotic Compound Oxidation by Nitrogen-Sufficient Cultures of *Bjerkandera* sp. Strain BOS55 Overproducing Peroxidases. *Appl. Environ. Microbiol. 62*(3): 880-885.

Kotterman, M. J. J., H-J Rietberg, A. Hage and J. A. Field. 1998a. Polycyclic Aromatic Hydrocarbon Oxidation by the White Rot Fungus *Bjerkandera* sp. Strain BOS55 in the Presence of Nonionic Surfactants. *Biotechnol. Bioengin. 57*(2): 220-227.

Kotterman, M. J. J., E. H. Vis, and J. A. Field. 1998b. Successive Mineralization and Detoxification of Benzo[*a*]pyrene by the White Rot Fungus *Bjerkandera* sp. Strain BOS55 and Indigenous Microflora. *Appl. Environ. Microbiol. 64*(8): 2853-2858.

Mihelcic, J. R., Lueking, R. R., Mitzell, R. J., Stapleton, J. M. 1993. Bioavailability of Sorbed- and Separate-Phase Chemicals. *Biodegradation. 4*(3): 141-153.

Stucki, G., and M. Alexander. 1987. Role of Dissolution Rate and Solubility in Biodegradation of Aromatic Compounds. *Appl. Environ. Microbiol. 53*(2): 292-297.

Volkering, F., A. M. Breure, A. Sterkenburg, and J. G. van Andel. 1992. Microbial Degradation of Polycyclic Aromatic Hydrocarbons: Effect of Substrate Availability on Bacterial Growth Kinetics. *Appl. Microbiol. Biotechnol. 36*(4): 548-552.

PURE CULTURES OF A *PENICILLIUM* SPECIES METABOLIZE PYRENEQUINONES TO INEXTRACTABLE PRODUCTS.

L. A. Launen, P. Percival, S. Lam, L. Pinto and ***M. Moore***
Simon Fraser University, Burnaby, BC, Canada

ABSTRACT: Quinones are major metabolites resulting from the oxidation of polycyclic aromatic hydrocarbons (PAHs) by fungi. Previously we established that the soil fungus *Penicillium janthinellum* SFU403 (SFU403) oxidizes pyrene to 1,6- and 1,8-pyrenequinones (PQs). Since many PAH quinones, including PQs, are known mutagens, we investigated the fate of PQs in pure cultures of SFU403. SFU403 converts ≈100% of pyrene to PQs by three days of culture, after which PQs decline by 2/3 over one week, suggesting further metabolism. Using ^{14}C-pyrene radiotracer studies we observed that after prolonged culture (three weeks) in media optimized for PQ production, >95% of the ^{14}C formed inextractable cell-associated products (ICAP). Furthermore, significant reductions in the level of ICAP were found in SFU403 cultured in media known to repress pyrene oxidation, in metabolically-inactivated biomass of SFU403, and in a fungus that does not oxidize pyrene. Thus, the formation of ICAP is linked with the metabolism of pyrene to PQs.

We hypothesized that the formation of ICAP from PQs proceeds through the one-electron reduction of PQs to pyrenesemiquinones (PSQs) by intracellular reductants. PSQs could then polymerize and/or bind covalently to cellular macromolecules. In support of this hypothesis we utilized electron paramagnetic spectroscopy (EPR) to demonstrate that 1,6- and 1,8-pyrenequinones are reduced by NADPH to their corresponding pyrene semiquinone radical anions *in vitro*.

INTRODUCTION

Polycyclic aromatic hydrocarbons (PAHs) are highly persistent and toxic byproducts of the incomplete combustion of fossil fuels. Fungal degradation of PAHs by basidiomycete (wood-rot) fungi which degrade PAHs with extracellular ligninolytic enzymes (Lamar, 1992), or non-basidiomycetes which degrade PAHs via intracellular cytochrome P450 monooxygenases (van den Brink *et al.*, 1998), yields PAH quinones as major oxidation products. Since many PAH quinones are mutagenic (Flowers *et al.* 1997), it is important to understand the pathways by which these compounds can be further degraded. While the fate of PAH quinones in basidiomycetes has been documented, there is limited information regarding the fate of PAH quinones in non-basidiomycete fungi.

We previously isolated the deuteromycete fungus *Penicillium janthinellum* SFU403 (SFU403) from petroleum-contaminated, non-woody soils and demonstrated that this isolate possesses strong PAH metabolizing activity (Launen *et al.*, 1995). The metabolism of pyrene to 1,6- and 1,8-pyrenequinones (PQs) by SFU403 proceeds as shown in Fig. 3 (steps 1 – 3), similar to other reports of fungal pyrene metabolism (Cerniglia *et al.*, 1986). Once produced, PQs are neither conjugated nor mineralized by SFU403; however, recent studies in our laboratory suggested that PQs disappear from cultures over time and that added

^{14}C-pyrene eventually forms inextractable cell-associated products (ICAP). These data suggested that PQs are not terminal metabolites in SFU403 but are further metabolised to ICAP.

We wished to investigate the potential for one electron reduction of PQs by SFU403 to form highly reactive pyrene semiquinone (PSQ) radical anions, catalyzed by intracellular reductants. These PSQs could covalently bind with cell macromolecules and form ICAP. Thus, the purposes of this study were two-fold: 1) To determine whether metabolism of pyrene by SFU403 was linked to the formation of ICAP, and 2) to determine whether cellular reductants could reduce PQs to their corresponding PSQs via a one electron reduction.

MATERIALS AND METHODS

Fungal Strains. All experiments employed the soil deuteromycete filamentous fungus *P. janthinellum* SFU403 (ATCC # 201797) (Launen *et al.*, 1995). In addition a *Cladosporium* sp. and a *Fusarium* sp. isolated from a jet fuel-contaminated aquifer were used.

Fungal Growth Conditions. Culture conditions were as described in (Launen *et al.*, in press) including the media composition. Briefly, OPT medium results in the "optimal" level of pyrene oxidation (up to 100% pyrene conversion to PQs by 3 days) and good cell growth. MED medium results in ≅50% pyrene bioconversion to 1-pyrenol after 7 days and equivalent levels of cell growth relative to OPT (by wet weight). LOW medium suppressed pyrene metabolism to trace levels of 1-pyrenol formation, observed by 7 days and low levels of cell growth (20% that of OPT). Pyrene (Aldrich) and ^{14}C-pyrene (32.2 mCi/mmol; Chemsyn Laboratories, Lenexa, Kans.) were prepared and added as described in Launen et al. (1995). Control flasks with cells or pyrene (separately) in each culture medium were used. Where required, cultures of SFU403 were metabolically inactivated by autoclaving after 48 h of growth in OPT medium.

Extraction Protocol and HPLC Analysis. Cultures extractions (with ethyl acetate (EtOAc)) and HPLC analysis were done as described in (Launen *et al.*, in press).

Radio-tracer Methods. Complete mass balance analyses were conducted on triplicate cultures for all radiotracer experiments. ^{14}C-compounds in the cell and media EtOAc extracts, and media after extraction were quantified using liquid scintillation counting (LSC) on a Beckman LS 8000 liquid scintillation counter with quench correction. Where indicated, ^{14}C-pyrene and ^{14}C-pyrene metabolites (pooled 1-pyrenol and PQs) were quantified in the organic extracts using radio-thin layer chromatography (TLC) (Launen *et al.* 1995). The ^{14}C in the cell remainder after extraction was determined by combustion using a Harvey Biological Oxidizer (Ox 300) and quantified using LSC.

EPR Spectroscopy. EPR spectra were recorded on a Bruker ECS 106 spectrometer equipped with a T_m cavity and a quartz flat cell. Measurements were conducted in the X-band at a frequency of 976 GHz, at modulation amplitude 0.14 G, microwave power 2.53 mW, receiver gain 1 x 10^5, sweep width 20 G, resolution of field axis 2048 G, and conversion and time constants of 327 ms each. 1,6PQ (0.60 mM) and 1,8PQ (0.75 mM) were mixed with 20 mM NADPH in a 50:50 mixture of DMSO and Chelex-100 treated sodium phosphate buffer (5mM; pH 7.2) containing 100 mM NaCl (Flowers *et al.*, 1997) under anaerobic conditions at room temperature. The program WINEPR SimFonia (Version 1.25 (Bruker Analytische Messtechnik GmbH) was used to conduct the computer simulation modeling of both PSQs.

RESULTS AND DISCUSSION

Disappearance of PQs in Pure Cultures of SFU403. PQ concentrations peaked at 3 days (≅100% pyrene bioconversion), when cultures reached stationary phase, and then declined to 26% of the original pyrene by 7 days (Fig. 1). These data indicate 1) that pyrene metabolism is linked with culture growth phase and 2) that PQs are further metabolized in pure cultures of SFU403.

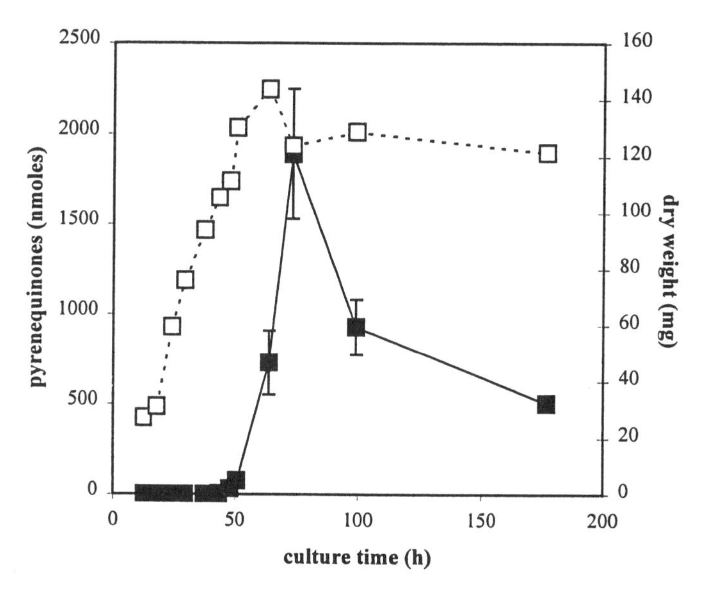

FIGURE 1. Pyrenequinones (1,6- and 1,8-pyrenequinone combined, closed black squares) and mycelial dry weight (separate parallel experiment, open squares) of *P. janthinellum* SFU403 incubated with pyrene (2000 nmoles added at t = 0) for one week in OPT medium. Values are the mean ± SD of three cultures.

Linking Pyrene Metabolism to PQs with the Formation of ICAP. To determine whether pyrene metabolism to PQs is linked to the formation of ICAP the level of ICAP was compared in several different sources of fungal biomass

TABLE 1. The formation if ICAP in fungal biomass according to the level of pyrene metabolism.

Biomass Source	Pyrene oxidizing ability	Culture Conditions	Pyrene[d] (%)	Metabolites[d] (%)	Cell-bound ^{14}C (%)	ICAP[e]	Recovery of ^{14}C (%)	Dry weight of cells (mg)
SFU403[a]	+/-	LOW 3d	79	0	40	3	85	27
		7d	52	2	41	7	80	20
		21d	17	20	47	30	95	19
SFU403[a]	+	MED 3d	83	1	63	1	96	66
		7d	21	17	93	52	105	186
		21d	<1	33	95	64	99	262
SFU403[a])	+++	OPT 3d	2	39	30	22	107	108
		7d	<1	27	78	53	115	114
		21d	<1	<1	80	95	96	76
SFU403[a]	+++	OPT 21 d	ND	ND	99	100	107	117
SFU403[b]	+++	OPT 21 d	ND	ND	108	100	121	109
SFU403[bc]	-	OPT 21 d	ND	ND	107	56	115	84
Cladospor. sp[a]	+	OPT 21 d	ND	ND	89	100	101	79
Fusarium sp.[a]	-	OPT 21 d	ND	ND	105	28	110	106

Note: All values are the mean of 3 cultures and % indicates the percent of initial ^{14}C pyrene. Cultures contained 0.025 mg/ml unlabelled and 0.20 μCi ^{14}C-pyrene. a = pyrene addition at t = 0 hours., b = pyrene addition at t = 48 hours, c = heat killed at 48 h. d = in culture EtOAc extracts. ND = not determined. Pyrene metabolism was ranked as (+++) = high levels of PQs only, (++) = 1-pyrenol + PQs, (+) = 1-pyrenol only, (+/-) = trace levels of 1-pyrenol and (-) = no metabolism. ICAP (e) calculated as

$$\text{ICAP (\%)} = \frac{\text{(dpm remaining in cells after extraction (determined by combustion)}}{\text{(total dpm associated with cells)}} \times 100$$

with differing abilities to oxidize pyrene. After 21 days of culture all of the ^{14}C partitioned into the biomass regardless of source (Table 1). Pyrene oxidation by SFU403 cultured in the three different media could be ranked as OPT>>MED>LOW. Similarly, the levels of ICAP formed by 21 days were 95%, 64% and 47% for live SFU403 cultured in OPT, MED and LOW respectively. By 21 days ICAP in the pyrene-oxidizing *Cladosporium sp.* cultured in OPT medium, was 100% (Table 1). In contrast, by 21 days the metabolically-inactivated biomass of SFU403, and the non-pyrene-oxidizing *Fusarium sp.* possessed only 56% and 28% ICAP respectively. Thus, the level of ICAP formed correlated with the level of pyrene metabolism occurring in the biomass. The ICAP in the *Fusarium sp.* and the non-active SFU403 demonstrates that a portion of ICAP (≅30 – 60%) is unreacted pyrene strongly sorbed to the fungal biomass. Since highly lipophilic chemicals such as PAHs are known to passively sorb to organic matter such as fungal biomass (Barclay *et al.*, 1995) this is not surprising. Interestingly, comparison of the ICAP formed in the live *Fusarium sp.* with that occurring in metabolically-inactivated SFU403 indicates that the *Fusarium sp.* sorbed ≅1/2 the unreacted pyrene that SFU403 did suggesting that there may be species differences in the ability to sorb unreacted pyrene.

***In vitro* Evidence for One Electron Intracellular Reduction of PQs.** EPR spectroscopy was used to directly monitor the formation of PSQs from incubation of PQs with the intracellular reductant NADPH. 1,6PQ (0.60 mM) and 1,8PQ (0.75 mM) were incubated (separately) with NADPH (20 mM). The experimental spectra obtained (Fig. 2) were identified as 1,6PSQ and 1,8PSQ using computer simulation of theoretical spectra. The signal obtained from 1,8PSQ required a slightly greater concentration of PQ than was required to obtain a signal with 1,6PQ and was less stable than the 1,6PSQ signal. These data clearly demonstrate that 1,6PQ and 1,8PQ can undergo one electron reduction when in the presence of intracellular reductants such as NADPH, and suggest that the 1,6PQ may be more readily reduced than 1,8PQ. Both PSQs were found to redox cycle with oxygen (data not shown).

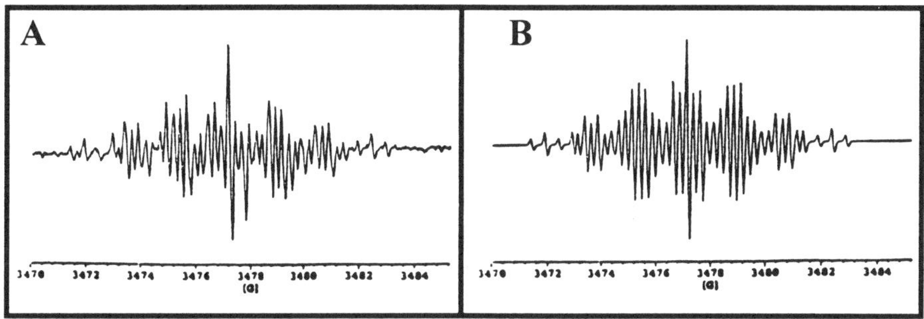

FIGURE 2. Experimental EPR spectrum (A) and the computer simulated spectrum (B) of the 1,8-pyrene semiquinone anion radical. The experimental profile for the 1,6PSQ also matched its simulated spectrum (not shown).

pyrene —(1) P450→ 1-pyrenol —(2) P450→ 1,8-dihydroxypyrene —(3)→ 1,8-pyrenequinone ⇄ (4) O_2 / NADPH +/- NADPH-quinone reductases 1,8-pyrenesemiquinone —(5)→ covalent binding to cell and/or polymerization

FIGURE 3. Pyrene metabolism by *P. janthinellum* SFU403. Steps 1-3 were reported in (Launen *et al.* 1995). Evidence to support steps 4 and 5 are presented in this study. 1,6PQ is expected to proceed through a similar pathway.

CONCLUSIONS

1) *Penicillium janthinellum* SFU403 can oxidize up to 100% of pyrene to pyrene 1,6- and 1,8-quinone (PQs) by three days, after which PQs decline to 26% by 7 days. This indicates that PQs are further metabolized by SFU403.
2) By 21 days of culture, 100% of added ^{14}C-pyrene has formed inextractable cell-associated products (ICAP) in SFU403. Approximately 40% of the ICAP is related to PQs and the balance (≅60%) is due to strong sorption of unreacted pyrene to the fungal mycelia. Metabolism of PAHs to ICAP may represent a means of detoxifying PAHs from contaminated matrices.
3) PQs are reduced by NADPH to their corresponding pyrene semiquinone anionradicals (PSQs). PSQs may then bind to cell material and contribute to the formation of ICAP.

ACKNOWLEDGEMENTS: Funding from Imperial Oil (University Research Grant), the Science Council of British Columbia, and Morrow Environmental Consultants Inc. is gratefully acknowledged.

REFERENCES

Barclay, C. D., G. F. Farquhar, and R. L. Legge. (1995). "Biodegradation and sorption of polyaromatic hydrocarbons by *Phanerochaete chrysosporium.*" *Applied Microbiology and Biotechnology* **42,** 958-963.

Cerniglia, C. E., D. W. Kelly, J. P. Freeman, and W. M. Dwight. (1986). "Microbial metabolism of pyrene." *Chemico-Biological Interactions* **57,** 203-216.

Flowers, L., S. T. Ohnishi, and T. M. Penning. (1997). "DNA strand scission by polycyclic aromatic hydrocarbon *o*-quinones: role of reactive oxygen species, Cu(II)/Cu(I) redox cycling, and *o*-semiquinone anion radicals." *Biochemistry* **36,** 8640-8648.

Lamar, R. T. (1992). "The role of fungal lignin-degrading enzymes in xenobiotic degradation." *Current Opinion in Biotechnology* **3,** 261-266.

Launen, L., L. Pinto, and M. Moore. (in press). "Optimization of pyrene oxidation by *Penicillium janthinellum* using response surface methodology." *Applied Microbiology and Biotechnology* .

Launen, L., L. Pinto, C. Wiebe, E. Kiehlmann, and M. Moore. (1995). "The oxidation of pyrene and benzo[a]pyrene by nonbasidiomycete soil fungi." *Canadian Journal of Microbiology* **41,** 477-488.

van den Brink, H. J., R. F. M. van Gorcom, C. A. M. J. J. van den Hondel, and P. J. Punt. (1998). "Cytochrome P450 enzyme systems in fungi." *Fungal Genetics and Biology* **23,** 1-17.

SURFACTANT ENHANCEMENT OF WHITE-ROT FUNGAL PAH SOIL REMEDIATION

Bill W. Bogan - EarthFax Development Corporation, North Logan, UT, USA
Richard T. Lamar - EarthFax Development Corporation, North Logan, UT, USA

Abstract: Five surfactants were examined for their ability to enhance PAH removal from soils inoculated with *Pleurotus ostreatus* (which performed best in a preliminary screening of six white-rot fungi). In general, non-ionic alkylethoxylate surfactants gave the best results. In soil from a coking facility, PAH removal in 30-day cultures of *P. ostreatus* increased from 23% without surfactants to 65% with surfactant. In an MGP soil, degradation increased from essentially zero to 43%. The amount of surfactant required was less when a short surfactant "pre-soak" of the contaminated soil was incorporated prior to inoculation. Several less-expensive vegetable oils performed nearly as well as the best surfactants. In the MGP soil, for example, 35-45% degradation of PAHs occurred in *P. ostreatus* cultures so supplemented. It was also found that some oil amendments significantly lowered the required inoculation rate, while still supporting high rates of PAH removal (80% for total PAH, >90% for some individual compounds).

Introduction

Our goal was to develop a treatment, based on white-rot fungi (WRF) and surfactants, capable of remediating high-molecular-weight (HMW) polycyclic aromatic hydrocarbons (PAHs) in contaminated soils and sediments. Over 700 sites in the U.S. have sustained creosote contamination, and over 1500 former manufactured gas plant (MGP) sites are contaminated with coal tar and related wastes. Studies of PAH-contaminated sediments (6) have shown that the majority of mutagenic activity in such samples is due to HMW PAHs. Thus, significant reductions in the genotoxicity of PAH-contaminated soils will require removal of these compounds. However, low aqueous solubility and high affinity for soil organic matter together result in very low "bioavailability" of HMW PAHs in soil systems, greatly restricting their biodegradation.

Materials and Methods

Ironton soil was from the vicinity of a waste tar lagoon at a coking facility in Utah. COGEMA soil was from an abandoned MGP site. All fungi were grown (≈5 days) on a 1:1 mixture of alder chips and cottonseed hulls, amended with 6% soy meal. In the first experiment, each of six fungi (*Bjerkandera adusta*, *Irpex lacteus*, *Phanerochaete chrysosporium*, *Pleurotus ostreatus*, *Trametes hirsuta*, and *Trametes versicolor)* was inoculated (20% dry wt basis) into each soil. This provided no-surfactant "baseline" information on the PAH-degrading performances of the different species. After 5 weeks, PAHs were extracted by ultrasonication and quantitated by GC.

Results and Discussion

Although PAH levels were initially very similar in the two soils (Table 1), we observed definite variability in PAH degradability, and distinct differences in the performances of the six fungi. For example, several of the fungi caused significant reductions in the PAH content of the Ironton soil (both total PAHs and carcinogenic PAHs), even without surfactants (Table 1). In COGEMA soil, however, no fungus caused significant PAH depletion, relative to non-inoculated controls, without surfactant addition. This was not due to toxicity, as all of the fungi grew well. Studies indicate that PAHs in MGP soils are frequently bound very tightly to soil constituents, and are particularly resistant to biodegradation (2).

TABLE 1. Removal of PAHs from Ironton and COGEMA soils during 35-day treatment with white-rot fungi (BA = *Bjerkandera adusta*; PO = *Pleurotus ostreatus*), or in non-inoculated controls ("CON" column). All data are given as ppm of PAH in the soil.

	Ironton soil				COGEMA soil			
PAH*	Initial	CON	BA	PO	Initial	CON	BA	PO
ACY	52	40	22	20	28	26	26	30
FLU	41	29	8	7	-	-	-	-
PHE	215	157	55	59	51	46	52	61
ANT	94	70	26	24	58	36	34	36
FLA	190	147	66	65	195	170	176	195
PYR	188	147	68	67	173	150	153	167
BAA	81	66	32	30	88	75	75	82
CHR	69	71	28	27	52	47	44	51
BBF	41	34	19	17	84	75	71	79
BKF	13	11	6	6	15	12	14	14
BAP	81	67	35	33	106	91	88	95
IP	31	27	16	14	46	39	38	41
BP	37	30	16	15	72	58	56	61
Total	1135	900	398	385	968	815	827	912
C-PAH	284	235	124	115	411	350	342	372

* Note: the following abbreviations are used: ACY - Acenaphthylene, FLU - Fluorene, PHE - Phenanthrene, ANT - Anthracene, FLA - Fluoranthene, PYR - Pyrene, CHR - Chrysene, BAA - Benz[*a*]anthracene, BBF - Benzo[*b*]fluoranthene, BKF - Benzo[*k*]fluoranthene, BAP - Benzo[*a*]pyrene, IP - Indeno[1,2,3-*c,d*]pyrene, BP - Benzo[*g,h,i*]perylene. "C-PAH" = carcinogenic PAHs, those with industrial cleanup guidelines of 100 ppm or lower (BAA, BBF, BKF, BAP, IP) and BP.

Evaluation of Surfactants. We evaluated five surfactants; three were non-ionic, including two alkylethoxylates (Brij 30 and Witconol SN-70) and one alkylphenylethoxylate (Trycol 6964). Brij 30 greatly increases solubility of PAHs in soil/water systems, even in soils containing "aged" coal-tar PAHs (5), whereas Witconol SN-70 significantly enhanced bacterial bioavailability of pyrene in soil systems (8). Trycol 6964 is identical to Hyonic NP-90, which also enhances PAH solubility in contaminated soils (5). These three were screened against one cationic surfactant (cetyltrimethylammonium bromide, CTAB), and one anionic surfactant (sodium dodecyl sulfate, SDS). Each was tested, in 1% or 5% doses (w/w surfactant/soil), for its ability to enhance WRF PAH degradation in each soil.

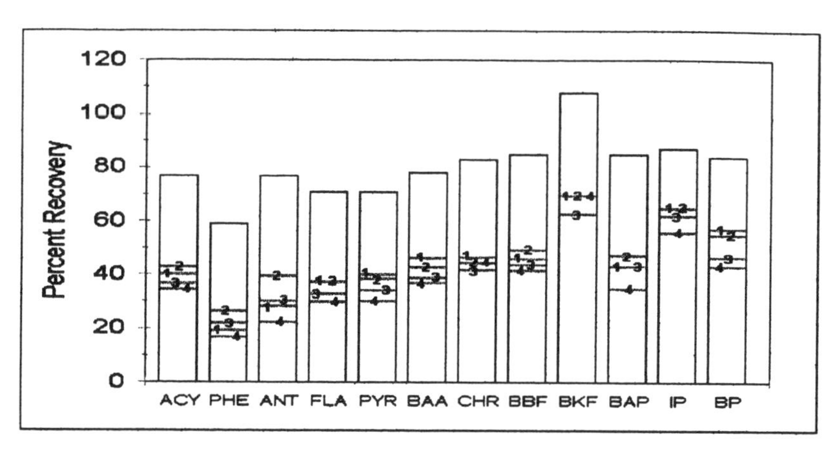

FIGURE 1. Removal of PAHs from Ironton soil during 30-day treatment with *P. ostreatus* with surfactants added (1 = 5% Brij; 2 = 1% Trycol; 3 = 1% Witconol; 4 = 5% Witconol). All percent recoveries are *vs.* initial levels; full bars are recoveries in no-surfactant controls.

Addition of surfactants to the Ironton soil greatly increased PAH removal by *P. ostreatus*. The best treatments were three of the four alkylethoxylates (5% Brij, 1% and 5% Witconol); performance with 1% Trycol 6964 was also good (Figure 1).

As noted above, PAHs in the COGEMA soil were apparently sorbed and/or sequestered more tightly than in the Ironton soil. Nonetheless, several surfactants also significantly enhanced removal of some PAHs from this soil (Figure 2). This is particularly true for the alkylethoxylates Brij 30 and Witconol SN-70. With Brij 30, what had been essentially zero removal of PAH without surfactant increased to over 40% removal with 5% surfactant; this was true for some carcinogenic PAHs (*e.g.* BAP) and for total PAH. Small effects were seen with CTAB (at 1%), Trycol (primarily at 5%), and SDS (both 1% and 5%). As in the "baseline" experiment,

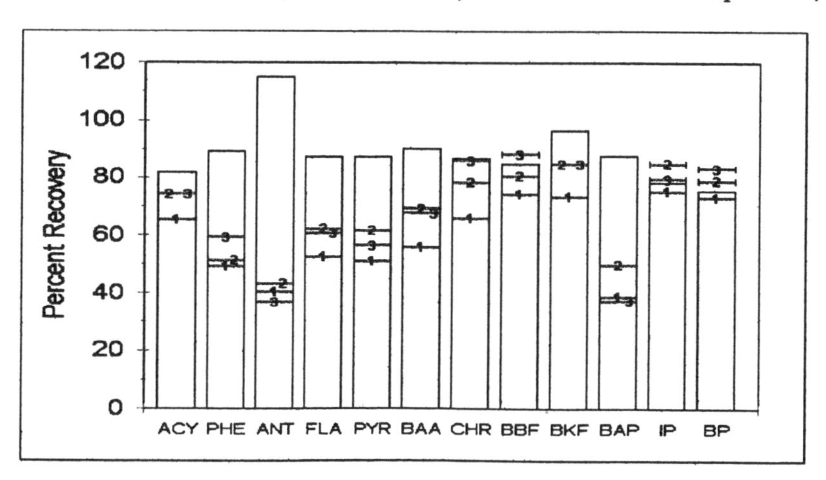

FIGURE 2. Removal of PAHs from COGEMA soil during 30-day treatment with *Pleurotus ostreatus* and various surfactants (1 = 5% Brij; 2 = 1% Witconol; 3 = 5% Witconol).

the PAH levels in the no-surfactant cultures in this experiment were essentially unchanged from initial levels.

The superiority of alkylethoxylates to alkylphenolethoxylates (*i.e.* Trycol) was unexpected; in fact, it is somewhat counterintuitive. The only major difference between the two classes, the presence of a benzene ring in alkylphenyl-ethoxylates (5), might be expected to give alkylphenylethoxylates greater PAH-solubilizing capabilities. This is, however, fortuitous, as this same difference makes the breakdown products of alkylphenylethoxylates considerably more toxic (5).

The fungi undoubtedly also degrade the surfactant, which likely decreases its effectiveness. In one experiment, COGEMA soil was moistened and allowed to "pre-soak" for 6 or 24 days in Brij 30 prior to inoculation. This treatment would hopefully maximize mass transfer of PAHs into surfactant aggregates prior to introduction of the fungi. Results indicate that, for lower Brij 30 concentrations, this was beneficial. Removals of several PAHs met or exceeded those with 5% Brij without a pre-soak (Figure 3). This effect was most pronounced for the highest molecular-weight compounds (IP and BP). Most likely, uptake of these compounds into surfactant aggregates was slowest, and their degradation benefitted most from delaying fungal inoculation until more mass-transfer has occurred. However, when the "soak phase" was longer (24 days), the benefits were lost. In this case, indigenous soil microbes probably degraded some or all of the Brij 30 prior to inoculation.

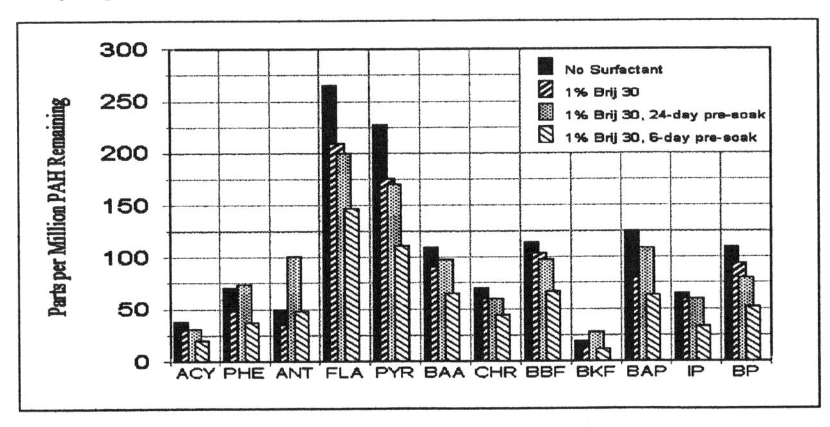

FIGURE 3. Removal of PAHs from COGEMA soil during 35-day treatment with *Pleurotus ostreatus* and 1% Brij 30, with or without "pre-soaks" prior to inoculation.

Use of Lipids. *In vitro* studies have shown that lipid peroxidation, catalyzed by fungal enzymes, is important in PAH oxidation (1, 7). Vegetable oils are also known to solubilize PAHs, sometimes to levels similar to Brij 30 (3, 4). Thus, we examined whether vegetable oils could aid PAH degradation. Several vegetable oils did, in fact, increase PAH removal from COGEMA soil. Oils added at the 5% rate gave PAH removals similar to those with 5% Brij or Witconol (Table 2). Controls (not shown) showed that addition of oil to non-inoculated soils had almost no effect; thus, losses are not due to indigenous microbes, nor to abiotic mechanisms (*i.e.* auto-oxidative lipid peroxidation).

TABLE 2. **Removal of PAHs from COGEMA soil during 35-day treatment with *Pleurotus ostreatus* in the presence of various vegetable oils.**

		Oil #1		Oil #2		Oil #3	
	CON	1%	5%	1%	5%	1%	5%
ACY	31	35	25	31	22	29	22
PHE	63	75	42	61	34	57	35
ANT	58	75	31	77	31	45	25
FLA	230	239	164	213	132	188	130
PYR	198	201	119	179	104	169	104
BAA	99	100	56	88	51	79	54
CHR	61	65	41	64	40	49	37
BBF	96	101	74	89	75	80	64
BKF	18	18	12	16	13	14	8
BAP	110	118	54	105	50	79	50
IP	50	51	36	46	36	40	30
BP	80	84	61	75	63	67	54
Total	1094	1162	705	1046	651	896	613
C-PAH	453	472	283	421	288	359	260

TABLE 3. **Removal of priority PAH compounds from Ironton and COGEMA soils during 30-day treatment with *Pleurotus ostreatus*, inoculated into the soil at either the 5%, 10%, or 20% rate, with or without the addition of 5% oil #3. All data are ppm of PAH remaining.**

	Ironton soil						COGEMA soil					
	5% Inoc		10% Inoc		20% Inoc		5% Inoc		10% Inoc		20% Inoc	
	No Oil	+ Oil	No Oil	+ Oil	No Oil	+ Oil	No Oil	+ Oil	No Oil	+ Oil	No Oil	+ Oil
ACY	25	21	26	14	26	22	27	19	31	25	34	24
PHE	62	48	72	19	67	35	48	30	57	37	92	18
ANT	78	55	89	25	80	46	96	65	114	74	160	62
FLA	75	56	83	26	73	48	183	114	223	140	239	126
PYR	78	52	86	23	74	47	166	94	201	120	211	107
BAA	42	29	46	13	40	27	95	52	110	70	114	59
CHR	32	27	34	15	33	25	52	37	61	44	68	37
BBF	24	21	25	12	23	21	85	66	102	87	104	72
BKF	11	10	12	8	12	12	17	13	22	19	22	17
BAP	43	28	47	11	39	22	106	44	119	62	133	48
IP	20	18	21	12	21	20	49	36	57	49	59	41
BP	23	20	25	12	24	19	69	52	79	69	84	61
Total	513	375	566	190	512	344	993	622	1176	796	1320	672
C-PAH	163	116	176	68	159	121	421	263	489	356	516	298

Effect on Inoculation Rate. To minimize technology costs, inoculation rates should be as low as possible without compromising performance. Besides mediating in peroxidase-catalyzed PAH degradation reactions and enhancing PAH availability, lipids have been shown to greatly stimulate growth of white-rot fungi. These increases, frequently well above those expected if the lipid were merely a carbon source, indicate the presence of fungal growth factors in the oils (9).

P. ostreatus was inoculated into both soils at rates of 5, 10, or 20%, in the presence or absence of 5% vegetable oil (oil #3). The results of this experiment are presented in Table 3. In the Ironton soil, oil increased PAH removal at each inoculation level. Furthermore, inoculation at 10% was clearly superior, as opposed to the 20% rate used previously. Losses of individual PAHs (*e.g.* phenanthrene) were as high as 91 percent; >75% of carcinogenic compounds were removed. In COGEMA soil, very little PAH loss occurred at any inoculation rate in the absence of vegetable oil. Again, however, vegetable oil stimulated PAH removal at each inoculation rate. Here, the best results were with 5% inoculation.

We are continuing to evaluate surfactant and/or oil amendments for their ability to enhance PAH degradation, and are determining whether these beneficial effects can be extended to the degradation of other hydrophobic contaminants (*e.g.* DDT, PCBs, *etc.*).

References

1. **Bogan, B.W.** 1996. "Biochemical and Molecular Biological Aspects of Polycyclic Aromatic Hydrocarbon Degradation by *Phanerochaete chrysosporium*" Ph.D. Thesis. University of Wisconsin, Madison, WI.

2. **Erickson, D.C., R.C. Loehr, and E.F. Neuhauser.** 1993. "PAH Loss during Bioremediation of Manufactured Gas Plant Site Soils" *Water Res.* **27:**911-919.

3. **Grimberg, S.J., J. Nagel, and M.D. Aitken.** 1995. "Kinetics of Phenanthrene Dissolution into Water in the Presence of Nonionic Surfactants" *Environ. Sci. Technol.* **29:**1480-1487.

4. **Klevens, H.B.** 1950. "Solubilization of Polycyclic Aromatic Hydrocarbons" *J. Phys. Coll. Chem.* **54:**283-298.

5. **Liu, Z., S. Laha, and R.G. Luthy.** 1991. "Surfactant Solubilization of Polycyclic Aromatic Hydrocarbon Compounds in Soil-Water Suspensions" *Water Sci. Technol.* **23:**475-485.

6. **Marvin, C.H., J.A. Lundrigan, B.E. McCarry, and D.W. Bryant.** 1995. "Determination and Genotoxicity of High Molecular Mass Polycyclic Aromatic Hydrocarbons Isolated from Coal-Tar-Contaminated Sediment" *Environ. Toxicol. Chem.* **14:**2059-2066.

7. **Moen, M.A. and K.E. Hammel.** 1994. "Lipid Peroxidation by the Manganese Peroxidase of Phanerochaete chrysosporium Is the Basis for Phenanthrene Oxidation by the Intact Fungus". *Appl. Environ. Microbiol.* **60:**1956-1961.

8. **Thibault, S. L., M. Anderson, and W. T. Frankenberger, Jr.** 1996. "Influence of Surfactants on Pyrene Desorption and Degradation in Soils". *Appl. Environ. Microbiol.* **62:**283-287.

9. **Wardle, K.S., and L.C. Schisler.** 1969. "The Effects of Various Lipids on Growth of Mycelium of *Agaricus bisporus*". *Mycologia.* **61:**305-314.

USE OF RESIDUAL SUBSTRATE FROM MUSHROOM FARMS TO STIMULATE BIODEGRADATION OF POORLY AVAILABLE PAH

Joop Harmsen, Antonie van den Toorn (SC-DLO, Wageningen, The Netherlands)
Jan Heersche, Durk Riedstra (BION, Almelo, The Netherlands)
Aldert van der Kooij (DHV, Amersfoort, The Netherlands)

ABSTRACT: Commercial used fungi like *Agaricus bisporus* (common mushroom) and *Pleurotus spp.* (oyster mushroom) produce extracellular enzymes that may increase the degradation of contaminants. We have investigated if the application of these free available substrates can be used to increase the degradation of poorly available PAH. Research under laboratory and field conditions shows that the applied fungi only develop further under controlled circumstances. The substrate itself has a positive effect on the degradation, although it is shown that the fungi are not responsible for the degradation.

INTRODUCTION

White-rot fungi have the ability to metabolize contaminants by their production of extracellular ligninolytic enzymes (Bar and Aust, 1994). *Phanerochaete chrysosperium* and *Bjerkandera sp. BOS55* are well known for degradation of contaminants. Other commercial used fungi like *Agaricus bisporus* (common mushroom) and *Pleurotus spp.* (oyster mushroom) also have these enzymes although they are less active (see table 1).

TABLE 1 Production of extracellular oxidative enzymes

Fungi	Ligninolytische enzymen					
	Ligine peroxidase	Mn peroxidase	Peroxidase	Laccase	Glyoxal oxidase	Arylalcohol oxidase
Agaricus bisporus	(+)	+		+		
Pleurotus	-	+		+		+
Phanerochaete chrysosperium	+	+	-	-	+	-
Bjerkandera sp. BOS55	+	+	+	+	+	+

De Jong et al.,1994, values for *Agaricus* Bonnen et al., 1994

Most fungi grow on substrate. For further growth and survival it is necessary to inoculate them and add them to the soil on a substrate. The production of an active substrate makes the use of fungi expensive. In view of the major importance attached to the economical feasibility of the method, commercially grown residual fungi substrates from *Agaricus bisporus* and

Pleurotus spp. are favorable. The major mushroom production countries are the USA, France, the Netherlands, England and Italy. In 1993, 190 million kg of mushrooms were produced in the Netherlands (Van Horen, 1994). This gives finally a large amount of residual substrate, 5 kg for every kg of mushrooms. The fungi in the residual substrate are still active and the substrate is available for free and could be used to stimulate biodegradation.

Objective. In aerobic soils the bioavailable PAH and mineral oil are well degradable by the endogenous microbiological population. Fungi must have a function for the degradation of the poorly available PAH. Sediments are not aerobic from origin. They are completely saturated with water. Fungi cannot survive under these circumstances and it is necessary to dewater the sediment. Figure 1 gives the period in which adding of fungi may have a positive effect on the degradation. In this period the endogeneous population has already degraded the bioavailable PAH. To be successful, adding of fungi after dewatering must lead to lower residual concentrations compared to the original sediment and adding of inactive substrate. This paper describes the research to this hypothesis, which is also illustrated in figure 1.

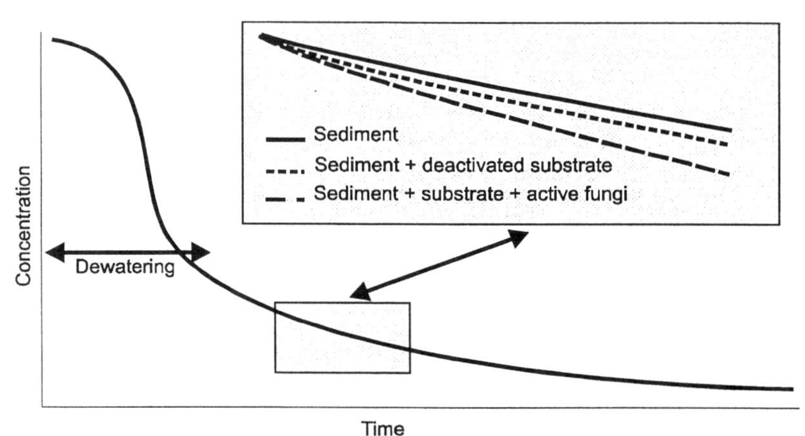

FIGURE 1. Decrease of contaminant concentration and the possible effects of fungi substrates during bioremediation

MATERIALS AND METHODS

The experiments were designed to be applicable for future use. Mixing of the sediment should be easy and not leading to an unacceptable dilution. Therefor the mixing ratio was 1:1 on the basis of volume; 28% and 20% (on weight) for respectively *Agaricus* and *Pleurotus*. The hypothesis has been tested on three levels:

- *Controlled system in containers on a research laboratory*. These systems on the Mushroom Experimental Station in Horst were designed to produce mushroom substrate. The containers were placed in a tunnel during 24 days and temperature and aeration were controlled. Each container was 105*105*180 cm (l*w*h). They were filled with 100 cm sediment or sediment substrate mixture. The following experimental set up has been used:

- Sediment only, the sediment was mixed comparable to the sediments in the other containers (blank)
- Sediment mixed with substrate, *Agaricus* or *Pleurotus*
- Sediment mixed with deactivated substrate. This was achieved by steaming the substrate during 12 h. at 70°C.

- *Test system in a soil-cleaning hall.* Circumstances in the hall were simulated in small containers during four months. It was possible to aerate and to irrigate the soil (Riedstra et al.,1997). Experiments had the same design as in the containers. An extra blank has been used in which the nutrients mixture normally applied in the hall has been added (Bion blank).

- *Outside experiment on a landfarm.* The outside landfarm was situated near the Kreekraksluizen in the southwest part of the Netherlands. On this site several sediments have been treated by landfarming. On this site only the *Agaricus* has been tested in different layer thicknesses, 30, 50 and 100 cm during 4 months. Again necessary blanks have been used.

Dewatered and ripened sediment from the Petroleum harbor has been used. The PAH and mineral oil concentrations in this sediment were original 550 and 12,000 mg/kg d.m. The sediment taken for this experiment had already been dewaterd and landfarmed during 2 years and the residual concentrations were ca 30 mg/kg d.m. for the PAH and 4000 mg/kg d.m. for mineral oil. In the Bion hall also a fresh sediment from a Water Counsil and a soil from a former manufactured gas plant has been tested.

All figures presented are the average off 5 analyzed samples. PAH have been measured after extraction with aceton/petroleumether using HPLC and fluorescence detection. Mineral oil has been measured in the same extract after clean up with florisil by GC-FID. Results have been corrected for the added amount of substrate.

RESULTS AND DISCUSSION

Container experiments. The results of the container experiments showed that adding of the *agaricus* substrate leads to a lower residual concentration (see table 2). However, the effect of the active and deactivated substrate was the same. The fungi activity did not give extra degradation. The extra degradation by the substrate could be caused by the improved soil structure, nutrients in the substrate or presence of extra dissolved organic molecules, but this was not investigated further. The active and deactivated *Pleurotus* substrate did not give extra degradation. Both fungi did grow in the containers. Especially the *agaricus* developed well, even within lumbs of sediment. The extra degradation potential of fungi for high molecular PAH and mineral oil by the *Agaricus* could also not be established (see Table 3)

TABLE 2. Degradation and 95% confidence level of PAH concentrations in dewatered Petroleumharbour sediment in the container experiments (corrected for added substrate)

Substrate	Concentration T_0 (mg/kg^{-1} d.m.)	Concentration T_{24} (mg/kg^{-1} d.m.)	Decrease (mg/kg^{-1} d.m.)
Only sediment	24,1 ± 3,2	24,4 ± 1,0	0 ± 1,5
Only sediment (duplo)	24,2 ± 3,2	22,1 ± 2,2	1,8 ± 1,7
Agaricus	32,4 ± 10,8	25,7 ± 11,0	6,7 ± 6,8
Agaricus (duplo)	33,2 ± 12	19,4 ± 2,9	13,8 ± 5,5
Agaracus deactivated	32,8 ± 12	23,3 ± 3,6	9,5 ± 5,5
Pleurotus	32,7 ± 6,7	33,6 ± 8,5	-0,9 ± 4,9
Pleurotus (duplo)	30,3 ± 4,9	29,5 ± 8,5	0,8 ± 2,8
Pleurotus deactivated	28,0 ± 1,2	29,5 ± 11	-1,5 ± 4,9

TABLE 3. Relative degradation (%) of different ringsystem PAH and mineral oil in a mixture of dewaterd Petroleum harbor sediment and *Agaricus* substrate

Group of compounds	*Agaricus* + Petroleumharbour sediment		
	Active fungi		Deactivated
2 + 3 ring PAH	19,9	49,5	36,9
4 ring PAH	32,1	46,2	28,5
5 + 6 ring PAH	18,8	24,2	14,5
Sum PAH	20,6	41,7	29,0
Mineral oil	2,2	4,0	1,8

Hall experiments. In the BION-hall it was shown that both fungi could survive and be active under the circumstances in the hall. The substrate concentrations in the Petroleumharbour sediment were 15 and 26% (w/w). The *Agaricus* in the sediment was well developed. Development with *Pleurotus* could not visually be observed. Mushrooms were growing after three weeks on both sediment/substrate mixtures.

Adding the *Agaricus* substrate to the gasfactory soil again showed that in a period of three months no extra degradation occurs (no data shown). After a fast dewatering period the fresh sediment was mixed with the *Agaricus* substrate. The results in figure 2 are a confirmation of the hypothesis shown in figure 1. During dewatering of the substrate (X=1 to X=2) the bioavailable part has already been degraded. Mixing with substrate also leads to a decrease. Adding of substrate has a positive effect on the bioremediation, but fungi do not give an extra decrease. It should however been mentioned that in the sediment as well as in the gas factory soil the fungi development was poor, probably caused by the bad soil structure.

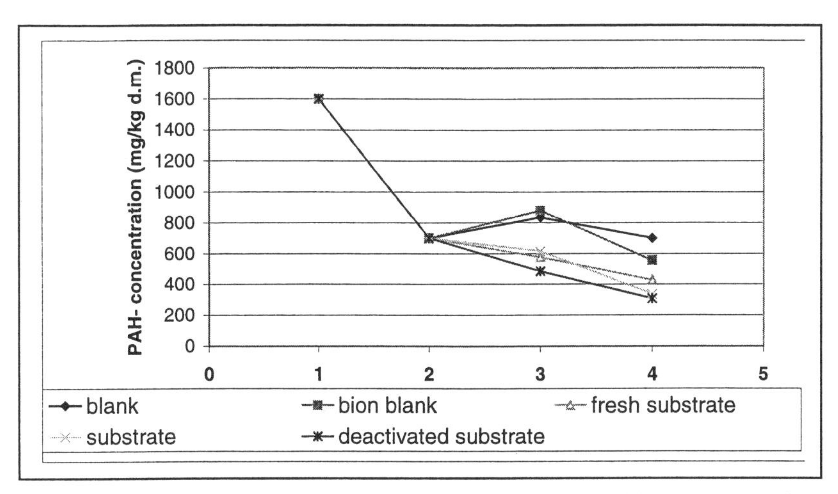

FIGURE 2. Decrease of PAH concentrations during different steps of the bioremediation. X=0, original sediment: X=1, after dewatering: X=2, after adding the substrate; X=3, at the end of the experiment (four months)

Field experiments. The field experiments were started in june 1997 (summer). The temperature and moisture contents were not limiting factors till September. This was a very dry month leading to a very dry soil, about 17% moisture at the end of September compared to about 30% in august. Oxygen was also not a limiting factor with this sediment, although there was a remarkable decrease of the oxygen in the first days (5% oxygen). In the first month the growth of mushrooms was observed. After two and four months it was not possible to detect *Agaricus* anymore. *Agaricus* has lost the competition with the endogenous microbiological population. All treatments have lead to comparable end concentrations, the fungi were not able to degrade the poor available PAH (table 4).

TABLE 4. PAH and mineral oil concentrations (mg/kg d.m.) and the 95% confidence level measured on the landfarm. (corrected for added substrate)

	2+3 ring	4 ring	5+6 ring	Mineral oil
***T = 0*, average all fields**	12.3 ± 1.9	16.4 ± 2.1	13.6 ± 0.9	2400 ± 160
T = 4 month				
100 cm active	14.2 1.7	15.5 ± 4.6	13.5 ± 2.0	2410 ± 530
50 cm active	11.2 ± 2.6	11.6 ± 2.6	13.0 ± 1.5	2710 ± 100
30 cm active	10.4 ± 2.8	12.0 ± 2.0	12.4 ± 1.2	1900 ± 230
30 cm deactivated	10.1 ± 3.0	10.5 ± 2.3	11.1 ± 1.1	2100 ± 220
Only sediment	9.8 ± 2.2	10.3 ± 4.0	11.0 ± 1.2	2230 ± 140
Sediment with Bion nutrients	10.7 ± 1.7	11.4 ± 3.2	12.1 ± 2.1	2120 ± 100

CONCLUSIONS. Adding of a substrate with fungi in order to achieve fungi activity in polluted soils, asks for development of the fungi. Well-controlled circumstances are necessary. With less optimal conditions the desired fungi may disappear due to competition with the endogenous population. This research shows that under controlled hall conditions fungi technology will be favourable compared to outside conditions in a landfarm.

Adding of a residual substrate from commercial grown fungi substrates *Agaricus bisporus* and *Pleurotus spp.* could be of economical benefit. It is shown that this substrate may have a positive effect on the degradation of PAH. However, the fungi are not responsible for the observed degradation. The same results could be obtained with substrates with deactivated fungi.

ACKNOWLEDGEMENTS

This study was part of the Research Progamm of the Ministry of Agriculture, Nature and Fishery and part of the Dutch Research Programme Bio-Technological In-Situ Bioremediation (NOBIS)

REFERENCES

Bar, D.P., and S.D. Aust, 1994. "Pollutant Degradation by White Rot Fungi". In G. Ware (ed), *Revieuws of Environmental Contamonation and Technology.* Vol 138, pp. 49-72 Springer-Verlag, NY.

Bonnen, A.M., L.H. Anton, A.B. Orth, 1994. "Lignin Degrading Enzymes of the Commercial Button Mushroom, Agaricus bisporus". *Appl Environ Microbiol 60: 960-965.*

Van Horen, L.G.J., 1994. "Landbouwtelling 1993, Cijfers over de champignonteelt". *De Champignoncultuur 38, 35.*

De Jong, E., J.A. Field, T.W Joyce, 1994. "Aryl alcohols in the physiology of lignilolytic fungi". *FEMS Microbiol Rev 13: 153-158.*

NOBIS, 1998. *Biodegradation of Micro Pollutants with Fungi.* (in Dutch) NOBIS-report 96-1-08, Gouda, The Netherlands.

Riedstra, D., J.A.N.M. Heersche, K. Westenterp, H. van den Beld, Th. Van Dam, P.H.J. Hamers and M. Bos, 1997. "Neural Network-facilitated Prediction of the Bioremediation Time". *In Situ and On-Site Bioremediation*: Volume 5, pp. 565-570. Battelle Press, Columbus.

DEGRADATION OF BENZO (A)PYRENE IN SOIL BY WHITE ROT FUNGI

Refugio Rodríguez V., Claudia Montalvo P., Luc Dendooven, Fernando Esparza G. (CINVESTAV, México City, México City) and Luis Fernández L. (IMP, México City, México)

ABSTRACT: The effect of the inocula obtained from solid culture on CO_2 production and the removal of benzo(a)pyrene from a contaminated soil was investigated. A Latin square was used and effects of type of fungus (*P. ostreatus*, *T. versicolor* and *P. chrysosporium*), the inoculum time on sugar cane bagasse pith (4, 9 and 18 days) and culture medium (medium A with 2.40 g C and 0.08 g N, C:N ratio 28; medium B with 4.02 g C and 0.03 g N, C:N ratio 121; and medium C with 12.08 g C and 0.09 g N, 3 times C:N ratio 121). Statistical analysis showed that the largest CO_2 production was obtained with an inoculum of *T. versicolor*. The greatest amount of benzo(a)pyrene that could not be accounted for (ca. 40 %) was obtained in soil added with an inoculum conditioned for 9 days in a medium with C:N ratio of 121 and incubated for 10 days.

INTRODUCTION

Benzo(a)pyrene, is high molecular weight Polycyclic Aromatic Hydrocarbon (PAH) (5 rings), with mutagenic and carcinogenic effects (Cavalieri and Rogan, 1985). White rot fungi are able to degrade an extremely diverse range of pollutants; pesticides, dyes, chlorinated aromatic compounds, and PAH's (Barr and Aust, 1994).

The abilities of white rot fungi to degrade aromatic compounds had been extensively studied in liquid culture (Wunch et al.. 1997). PAH's degradation has been found under ligninolytic (N limited) (Sanglard et al., 1986; Haemmerli et al., 1986) and non-ligninolytic (Bezalel et al., 1997).

Bioremediation is a technique with a lot of potential to clean up contaminated site. Factors, such as, pH, nutrients and water content, effecting the efficiency of bioremediation have been studied intensively (Kästner et al., 1998) but we are not aware of reports on the effect of inoculum conditions on solid culture in the degradation of PAH's.

Objective. The objective of this study was to investigate whether an inoculum produced under variable solid culture conditions on sugarcane bagasse pith, such as, incubation time, culture medium, and type of fungus (*T. versicolor*, *P. ostreatus* and *P. chrysosporium)* and added to a benzo(a)pyrene contaminated soil had an effect on its removal and production of CO_2 .

MATERIALS AND METHODS

Inoculum. Strains *Phanerochaete chrysosporium* h-298, *Pleurotus ostreatus* IE8 and *Trametes versicolor* CDBB h-1051, were obtained from the micro-organisms collection of CINVESTAV-IPN. *P. ostreatus* and *T. versicolor* were grown on malt extract 2 % agar at 28°C, for 8 and 6 days respectively, and *P. chrysosporium*, at 39°C for 26 h.

Culture conditions

Propagation on wheat seed. Twenty-five grams of moistured wheat seeds (40%) were added to serum bottles (124 ml) and inoculated with 4 plugs of fungi grown on 2 % malt extract. The incubation was for 7 days for *T. versicolor* and *P. ostreatus* at 28°C, and for *P. chrysosporium* at 39 °C. The wheat seeds were adjusted to water holding capacity and sterilized for 15 min at 121 °C.

Sugarcane bagasse pith culture. Eighteen tubular columns (20 cm length) were packed with 15 g pith (particle size 0.86 to 0.56 mm) with a moisture content of 60 % and inoculated with 3 g of wheat seeds with fungi. Six columns were amended with medium A containing 2.40 g of C and 0.08 g of N (C:N ratio 28), six with medium B containing 4.02 g of C and 0.03 g of N (Kirk medium, a C:N ratio of 121) and medium C containing 12.08 g of C and 0.09 g of N (3 times Kirk medium, and a C:N ratio of 121).

The columns were incubated for 4,9 and 18 days at temperature specific for each fungus as mentioned earlier. Columns were aerated with air at a flow rate of 0.38 ml $min^{-1}g^{-1}$ dry matter.

Experimental Design. The Latin square experimental design contained three independent factors; the type of fungus, the inoculum incubation time on solid culture, and the culture medium (TABLE 1).

TABLE 1. Latin square for the inoculum on solid substrate.

Treatment	Fungus	Culture medium	Incubation time
1	*P. ostreatus*	C	4
2	*P. ostreatus*	B	9
3	*P. ostreatus*	A	18
4	*T. versicolor*	A	4
5	*T. versicolor*	C	9
6	*T. versicolor*	B	18
7	*P. chrysosporium*	B	4
8	*P. chrysosporium*	A	9
9	*P. chrysosporium*	C	18

The response variables were CO_2 producton and the benzo(a)pyrene concentration in soil after 10 days of incubation with *P. chrysosporium* at 39 °C , *P. ostreatus* and *T. versicolor* at 28 °C.

Statistical Analysis. All analysis were done with SAS statistical package (1988).

Preparation of soil samples with benzo(a)pyrene. The soil was sieved (1mm) and sterilized with dry heat at 140 °C for 48 h. Hundred and sixty six ml of a benzo(a)pyrene acetone solution (3 g benzo(a)pyrene L^{-1}) was added to the soil. A concentration of approximately 500 mg benzo(a)pyrene kg^{-1} dry soil was obtained.

Application of fungi to soil. Six g of soil adjusted to 60 % of field capacity was mixed with bagasse inoculated with or without fungi at a 85:15 ratio and added to 120 ml serologic bottles. Triplicates of each treatment were incubated for 10 days at temperature specific for each fungus and then stored at –20°C pending analysis.

Metabolic activity. The headspace volume of each serologic bottle was sampled and analyzed for CO_2 using a Gow Mac 580 GC fitted with a thermal conductivity detector. A CTR1 column maintained at 40°C was used to separate CO_2 from the other gasses and helium was used as carrier gas flowing at a rate of 55 ml min^{-1}.

Extraction of benzo(a)pyrene. Soil samples were extracted with acetone, in a Soxhlet apparatus for 8 h. Acetone extract were amalyzed for benzo(a)pyrene by high performance liquid chromatogragy (HPLC), in a Varian 9012 chromatograph. A C18analytic column (15 cm) was used for separation and detection of benzo(a)pyrene was performed at 254 nm. The mobile phase was acetonitrile:water in a 75:25 ratio.

RESULTS AND DISCUSSION

A significant greater production of CO_2 was found for *T. versicolor* and *P. chrysosporium* than for *P. ostreatus* ($P<0.05$). The media had a different effect on the CO2 production of the different fungi. The smallest CO2 production for *P. ostreatus* was found when grown on a medium B, *T. versicolor* on a medium A and for *P. chrysosporium* on a medium C. The production of CO2 was significantly greater for fungi grown on a medium B ($P<0.05$). (FIGURE 1). Eggen and Majcherczyc (1998) observed benzo(a)pyrene removal from soil is not only depending on the metabolic activity of the fungi but also on the initial bioavailability of the toxic compound.

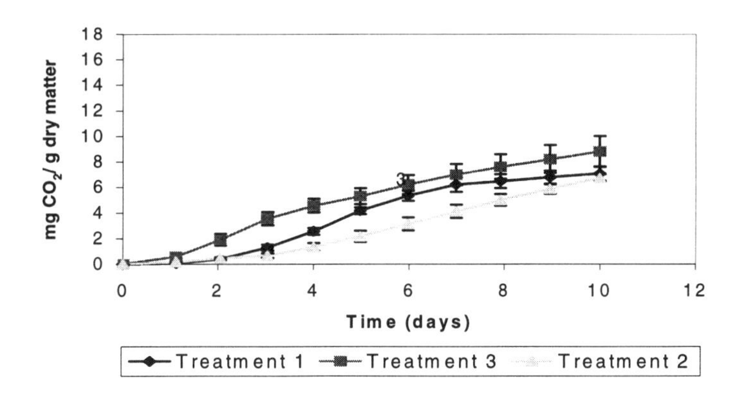

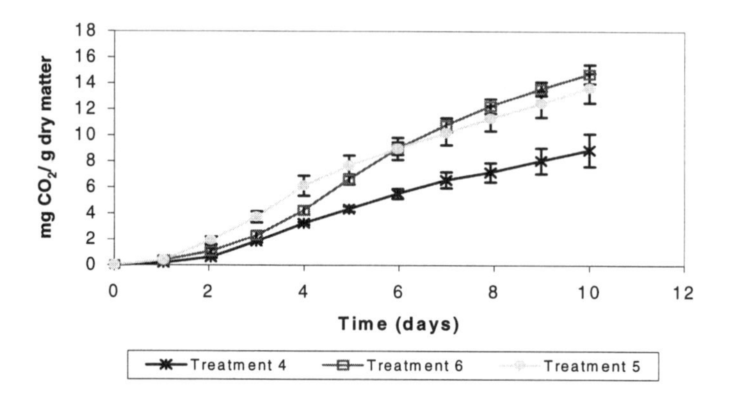

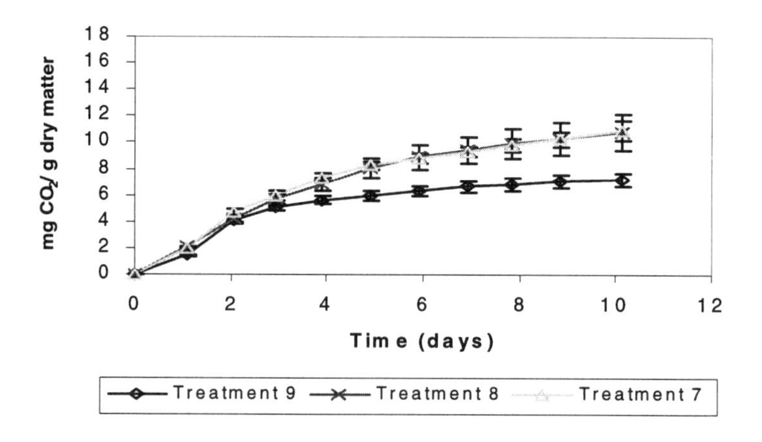

FIGURE 1. CO_2 production during the incubation with *P. ostreatus* (A), *T. versicolor* (B) and *P. chrysosporium* (C) in a contaminated soil with 500 mg kg^{-1} of benzo(a)pyrene. References to the treatments can be found in TABLE 1.

The fungi were able to remove up to 40 % of benzo(a)pyrene and the rate of removal depended on culture media and incubation time of the inoculum. Inocula of *T. versicolor* and *P. ostreatus* incubated for 4 days, independent of the culture media, removed only approximately 17 % benzo(a)pyrene from soil (FIGURE 2). Statistical analysis showed that culture medium and period of incubation had a significant positive effect ($P<0.05$) on the benzo(a)pyrene removal, but not the type of fungi. The greatest removal was obtained with fungi inoculated for 9 days and on the culture medium B. Storing samples, however, could had an effect on benzo(a)pyrene concentrations due to adsorption.

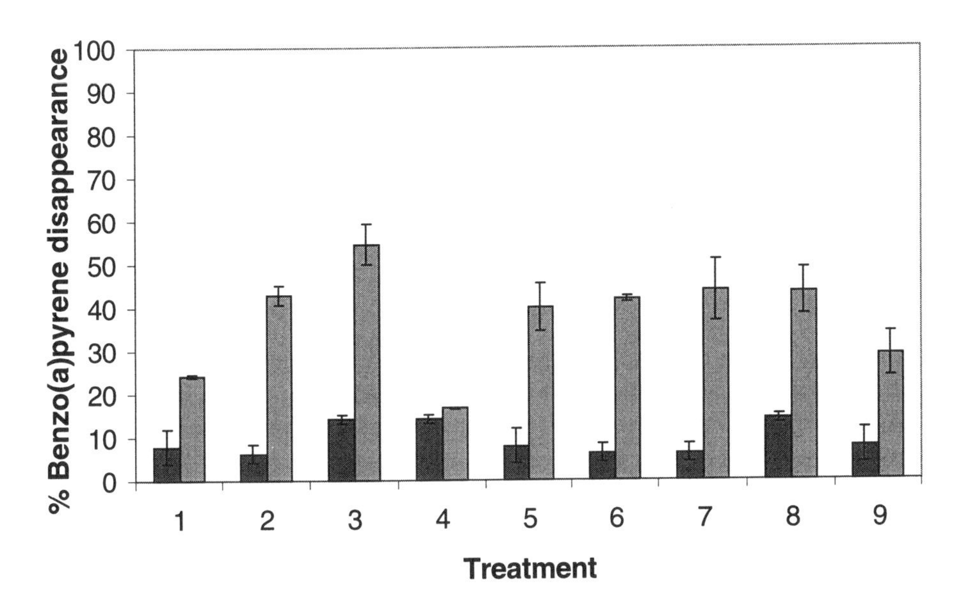

FIGURE 2. Removal of benzo(a)pyrene in soil by the inoculum obtained. The incubation temperature for *P.chrysosporium* was 39 °C and for *T. versicolor* and *P. ostreatus* was 28 °C (control,   treatments with fungus)

Even when fungi were used without a period of adaptation to benzo(a)pyrene they coul growth in soil added with 500 mg kg^{-1} of benzo(a)pyrene. Other studies have shown that in soils contaminated with 60 mg kg-1 benzo(a)pyrene 90 % was removed in five months (Kanaly et al., 1997).

The culture medium B, limited in N, and the time of the inoculum on sugarcane basse, presumably favors the production of ligninolytic enzymes when fungi are added to soil.

REFERENCES

Barr, D. P., and Aust, S. D. 1994. "Mechanism White Rot Fungi Use to Degrade Pollutants." *Env. Sci. Technol.*. 28:78A-87A.

Bezalel, L., Hadar, Y., and Cerniglia C. E. 1997. "Enzymatic Mechanims Involved in Phenanthrene Degradation by the White Rot Fungus *P. ostreatus*. "*Appl. Environ. Microbiol.* 63(7):2495-2501.

Cavalieri, E., and Rogan, E. 1985. "Role of Radical Cations in Aromatic Hydrocarbon Carcinogenesis." *Env. Health Perspectives*. 64: 69-84.

Eggen, T., and Majcherzyk, A. 1998. "Removal of Polycyclic Aromatic Hydrocarbons (PAH) in Contaminated Soil by White Rot Fungus *Pleurotus ostreatus*." *Inter. Biodeter. and Biodegr.* 41:111-117.

Haemmerli. S. D., Leisola, S. A., Sanglard, D., and Fiechter, A. 1986. " Oxidation of Benzo(a)pyrene by Extracellular Ligninases of *P. chrysosporium*." *J. Biol. Chem.*. 261(15):6900-6908.

Kanaly, R., Bartha, R., Fogel, S., and Rinday, M. 1997. "Biodegradation of (^{14}C) Benzo(a)pyrene added in Crude Oil to Uncontaminated Soil." *Appl. Environ. Microbiol.* 63(11):4511-4515.

Kästner, M., Brever-Jammali, M., and Mahro, N. 1998. "Impact of Inoculation Protocols, Salinity, and pH on the Degradation of Polycyclic Aromatic Hydrocarbons (PAHs) and Survival of PAH Degrading Bacteria Introduced into Soil." *Appl. Environ. Microbiol.* 64(1):359-362.

Sanglard, D., Leisola, M. S., and Fiechter, A. 1986. "Role of Extracellular Ligninases in Biodegradation of Benzo(a)pyrene by *Phanerochaete chrysosporium.*" *Enzyme Microbiol. Technol.* 8:209-212.

Wunch, K., Geibelman, T., and Bennett, J., 1997 "Screening for Fungi Capable of Removing Benzo(a)pyrene in Culture." *Appl. Microbiol. Biotechn*ol. 47:620-624.

DEGRADATION OF AGED CREOSOTE-CONTAMINATED SOIL BY *Pleurotus ostreatus*

Trine Eggen (Jordforsk, Norway)
Edgardo Araneda (Jordforsk, Norway)
Øistein Vethe (Jordforsk, Norway[1])
Per Sveum (Statskog Miljø og anlegg AS, Norway[2])

ABSTRACT: Static composting was simulated in medium sized polyethylene columns with aged creosote contaminated soil. In the study three variables associated with white rot fungi in bioremediation are studied: i) spent mushroom compost (fungal substrate) from two commercial mushroom suppliers are compared, ii) for one mushroom compost supply the difference in using the substrate before and after mushroom production is studied, iii) pile layout, i.e. organizing soil and fungal substrate in layers or a homogenized mix is evaluated. The experiment was run over 7 weeks at room temperature. Start concentration was 1990 mg/kg (16 US.EPA). Pile layout and source of fungi inoculum were major factors for PAHs degradation. With the most optimal fungi inoculum, mixed soil and substrate and amended fish oil, the following removal was obtained (expressed as percentage of total input concentration): 75-99% Acenaphtalene, Fluorene, Phenanthrene, Anthracene, Fluoranthene and Pyrene,86.7% Benzo(a)anthrancene, 79%Chrysene/Triphenylene, 51% Benzo(b+k)fluaranthene, 42%Benzo(a)pyrene, and 86% PAH (16 US.EPA).

INTRODUCTION

Polycyclic aromatic hydrocarbons (PAH) are pollutants typically found at wood preservation plants and gas work sites. With remediation of polluted soil during the last decades, biodegradation of PAHs has been in focus. Particularly, degradation of high molecular weight PAHs have gained much attention due to their recalcitrant properties (Mueller et al., 1991). Another important factor connected to remediation of PAHs is the difference in bioavailability of contaminants in artificially and newly polluted soil (Erickson et al., 1993). Field et al. (1995) found a doubling in the refractory fraction of Benzo(a)pyrene, with an increase in the pollution age from zero to only three months. This reduced biodegradation over time can be explained by reduced bioavailability (immobilization in micropores or changes in binding forms, e.g. oxidative coupling reactions) (Bollag et al., 1992).

White rot fungi secrete non-specific extracellular enzymes that are involved in the degradation of lignin (Barr and Aust, 1994). The same mechanisms that give these fungi the ability to degrade lignin are also used to degrade a wide range of environmental pollutants, such as DDT, TNT, PCBs and PAHs (Higson, 1991; Barr and Aust, 1994). The most extensively investigated

---

[1] new adress: Norwegian Standards Association, Norway
[2] new adress: Deconterra as, Norway

fungi, *Phanerochaete chrysosporium*, degrades lignin under lignolytic conditions. For use of white rot fungi in bioremediation there is particular interest in using fungi that produce extracellular enzymes in excess of nitrogen. Therefore, species as *Bjerkandera adusta*, *Trametes vericolor*, *Lentinus edodes* and *Pleurotus ostreatus* have been recently studied (Field et al., 1995; Okeke et al., 1996; Bezalel et al., 1996; Morgan et al., 1991).

When using white rot fungi for bioremediation, availability of fungal inoculum is an economical consideration. Therefore, it would be advantageous to use spent mushroom culture (Buswell, 1994). In this study the following concerns associated with white rot fungi in bioremediation are examined:

i) spent mushroom compost (fungal substrate) from two commercial mushroom suppliers are compared,
ii) for one mushroom compost supply the difference in using the substrate before and after mushroom production is studied,
iii) pile layout, i.e. organizing soil and fungal substrate in layers or a homogenized mix is evaluated.

MATERIALS AND METHODS

Inoculum. The fungal inoculum used, *Pleurotus ostreatus,* was obtained from two commercial oyster mushroom producers, Frillesås Swamp (Frillesås, Sweden) and Funginova (Halmstad, Sweden). From Frillesås, colonized mushroom substrate before mushroom production (fruitbody production) was also tested as a fungal inoculum.

Soil. Creosote contaminated soil came from an abandoned wood preservation site in southern Norway (Lillestrøm). Distribution of PAH in start soil is shown in Table 1. Soil characteristics were: pH 7.5, Tot.N 0.1g/100g, Tot.P 658 mg/kg, TOC 2.4 g/100g, TC 2.9g/100g.

Experimental design. The soil was sieved (< 1cm) and mixed thoroughly. Water was added to adjust soil humidity to 70% of the water holding capacity. The soil was contained in nine polyethylene columns (d=12.5 cm, h=47 cm) and aerated daily. Each column was a unique pile design (Table 2). In the first row (substrate source) the letters F and H indicate the source of the inoculum. For substrate F the subscripts A and B identify if the substrate was after or before mushroom production. The second row identifies if the substrate and soil were mixed (M) or layered (L). Fish oil was expected to have a positive effect on bioavailability of hydrophobe PAH, therefor fish oil (Norwegian Herring Oil and Meal Industry Research Institute) was added (1% solution) to most of the columns. Fish oil was omitted from two columns that are marked O in subscript.

The columns were incubated for 7 weeks at room temperature (22-25°C). Fish oil or water was added periodically to maintain soil humidity. The columns were covered with perforated parafilm. Five soil samples from each column were mixed in a blender and sieved (<2 mm) then analyzed for PAHs, laccase activity and FDA hydrolysis. The interaction between PAH degradation and microbial activity will be discussed in a forthcoming paper.

Chemical analysis. Soil samples (10 g) were extracted with dichloromethane (DCM) and water (3:1) for 30 h in an ultrasound bath and 2 h on shaking table (150 rpm). The extraction was performed twice before the DCM phases were purified in a NH_2 fast phase column. Samples were analyzed by gas chromatography with flame ionization detector (FID) (Hewlett Packard 5890 Serie II) (DB-5 (5% diphenyl, 95% dimethyl polysiloxan, 30m x 0.32mm x 1 µm capillary column,). After 1 min at 35°C, the temperature was increased to 200°C at a rate of 10°C/min, and then to 300°C (5°C/min) and maintained for 20 min.

TABLE 1. Initial concentration and distribution of 16 PAHs (n=3).

Compound	Percentage of Total PAH	Avg. mg/kg	std	%std
Naphthalene	0.8	13.7	4.9	36.0
Acenaphtylene	1.0	19.9	4.0	20.3
Acenaphthene	9.4	179.4	32.5	18.1
Fluorene	9.5	182.6	48.9	26.8
Phenanthrene	20.3	394.0	140.7	35.7
Anthracene	19.6	369.4	109.1	29.5
Fluoranthene	17.2	327.9	54.3	16.6
Pyrene	12.7	242.5	40.8	16.8
Benz(a)anthracene	3.2	61.5	10.2	16.6
Chrysene	2.7	52.0	8.2	15.7
Benzo(b,k)fluoranthene	1.9	36.0	5.6	15.6
Benzo(a)pyrene	0.7	13.5	2.0	15.1
Indeno(1,2,3-cd)pyrene	0.1	1.3	1.2	89.5
Dibenzo(ah)anthracene	0.5	10.0	5.2	52.0
Benzo(ghi)perylene	0.3	5.5	0.7	13.2
Total 16 PAH		1909.2	357.9	18.7
Σ3-ring PAHs	59	1113.6		
Σ4-ring PAHs	36	696.6		
Σ5-ring PAHs	3	59.3		

TABLE 2. Experimental design.

Column	1	2	3	4	5	6	7	8	9
Substrate	H_A	H_A	F_A	F_A	F_A	F_B	F_B	F_A	F_A
Soil:substrate	M	L	M	L	L	M	L	M_o	L_o

Substrate: F= Frillesås, H= Halmstad , A and B= after and before mushroom production, respectiviely
Soil:substrate: M=mixed, L= in layers, o= fish oil omitted

RESULTS AND DISCUSSION

All treatments degraded Σ3-ring PAHs well (> 73%) (Fig. 1a). For the single 3-ring compounds (Acenaphthene, Fluorene and Phenanthrene) the removal rates were > 80%, and in the most optimal treatment the removals were 94-99% (Fig. 1b). The removal of Anthracene was consistently lower and more varied than the other 3-ring compounds.

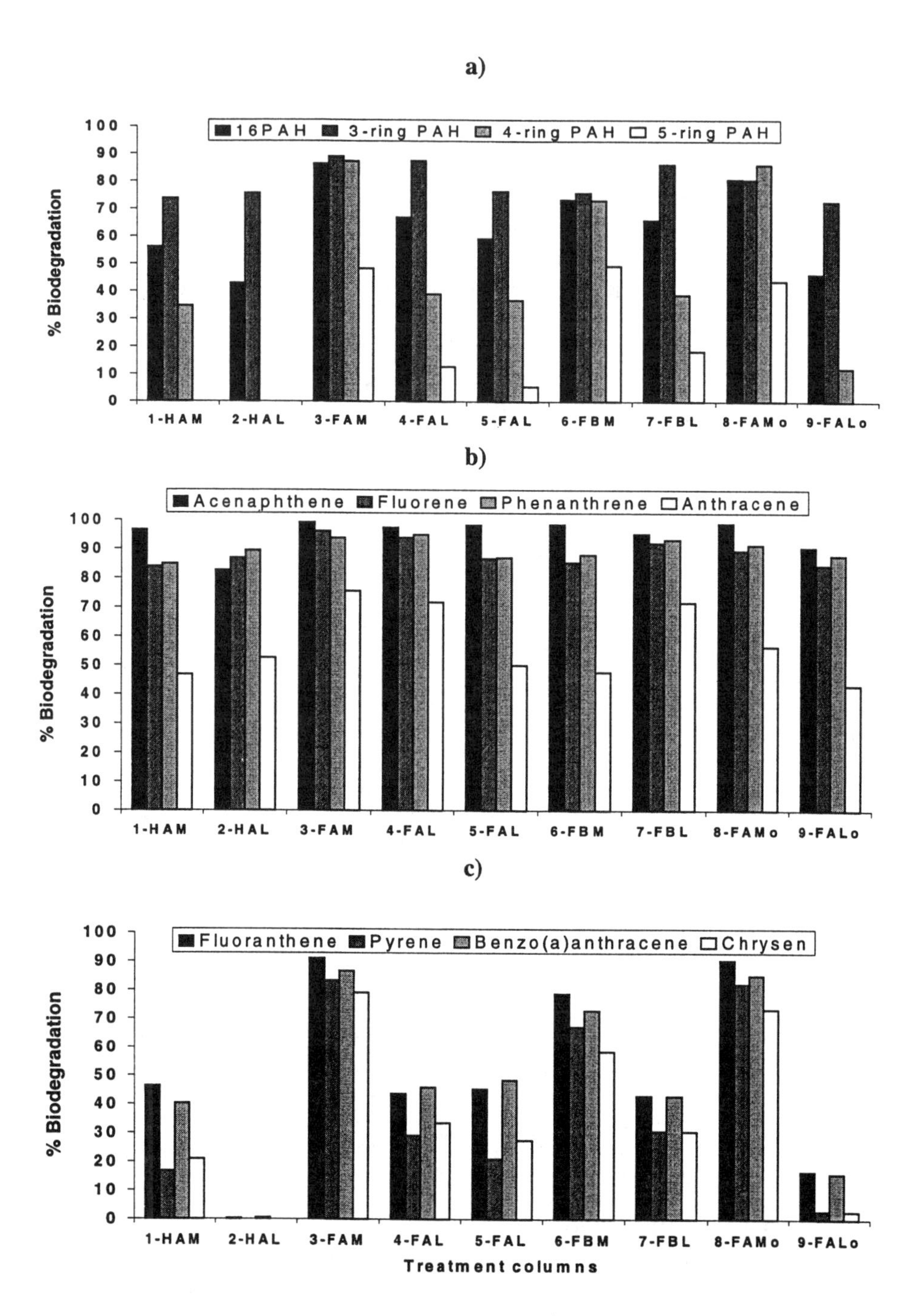

FIGURE 1. Percent biodegradation after 7 weeks incubation with *Pleurotus ostreatus*, a) Σ16PAHs, Σ3-ring-, 4-ring- and 5-ring PAHs, b) single 3-ring PAHs, c) single 4-ring PAHs.

For removal of compounds that are more difficult to degrade, 4-ring- and 5-ring PAHs, spent mushroom substrate F showed a greater potential than spent substrate H (Fig. 1a). There was also a clear improvement in mixing compared to layering the soil-substrate. When soil-substrate was mixed, there was a slightly lower degradation rate for substrate before mushroom production (column 6) than after mushroom production (columns 3 and 8). This was not evident when soil-substrate was layered.

The degradation of the Σ4-ring PAHs were more varied than the Σ3-ring PAHs. Substrate F, mixed rather than layered with soil, proved the most effective treatment medium (column 3): Fluoranthene 91%, Pyrene 83%, Benzo(a)anthracene 87%, and Chrysene/Thriphenylene 79% (Fig. 1c). Highest degradation of Σ5-ring PAHs (Fig. 1a) was also found in columns with mixed soil-substrate and substrate F (column 3, 6 and 8); however, the removal rate (44-49%) was consistently less than for 16 PAH, 3-ring and 4-ring compounds.

In contrast to removal of 3-ring PAHs, which showed relatively low variation in the different treatments, removal of 4- and 5-ring PAHs were highly dependent of treatment methods. Homogenizing soil and fungal substrate, combined with use of substrate A was the optimal treatment for the compounds that are difficult to degrade. Columns 4 and 5 were duplicates of the treatment substrate F after mushroom production and soil and substrate in layers. Comparison between these columns showed a low variation. Although fish oil enhanced degradation, only two of the nine columns did not have the amendment; hence one must consider this positive effect with caution. Comparing column 3 and 8, both substrate F, after mushroom production, mixed with soil, there was a slightly lower degradation in the column without fish oil (column 8). Columns 4, 5 and 7, all inoculated with substrate F, organized in layers, and added fish oil, do have comparable degradation results, and obtained a better result than obtained in column 9 with same treatment except no fish oil addition.

CONCLUSION

The study investigated different factors associated with the use of white rot fungi in bioremediation. Apparently, both the source of inoculum and the type of PAHs one wishes to treat are equally important. The two commercial sources of inoculum treated the easier to degrade 3-ring compounds similarly.
However, for compounds that are more difficult to degrade (4- and 5-ring), there was a significant difference. Spent mushroom compost tends to be more effective than fungal substrate before mushroom production. Moreover, the 4- and 5-ring compounds degrade better if the soil and substrate are mixed rather than layered. Adding fish oil enhances the degradation process, but the lack of replication in this study makes it difficult to determine the significance of this amendment. Nevertheless, when fish oil was added to spent mushroom compost, mixed with creosote contaminated soil, the following removal occurred after 7 weeks incubation: 86% ΣPAHs, 89% Σ3-ring PAHs, 87% Σ4-ring PAHs and 48% Σ5-ring PAHs.

REFERENCES

Barr, D. P., and S.D. Aust. 1994. Mechanisms White Rot Fungi use to Degrade Pollutants. *Environ. Sci. Technol.* 28(2): 79-87.

Bezalel, L., Y. Hadar, and C.E. Cerniglia. 1996. Mineralization of Polycyclic Aromatic Hydrocarbons by the White Rot Fungus *Pleurotus ostreatus*. *Appl. Environ. Microbiol.* 62: 292-295.

Bollag, J.-M., C.J. Myers, and R.D. Minard. 1992. Biological and Chemical Interactions of Pesticides with Soil Organic Matter. The Science of the Total Environment (123/124), 205-217.

Buswell, J.A. 1994. Potential of Spent Mushroom Substrate for Bioremediation Purposes. *Compost Sci. Utiliz.* 2(3): 31-36.

Erickson, D. C., R.C. Loehr, and E.F. Neuhauser. 1993. PAH Loss During Bioremediation of Manufactured Gas Plant Site Soils. *Water research,* 27(5): 911-919.

Field, J. A., H. Feiken, A. Hage, and M.J.J. Kotterman. 1995. "Application of a White-Rot Fungus to Biodegrade Benzo(a)pyrene in Soil." In R.E. Hinchee, J. Fredrickson and B.C. Alleman (Eds.) *Bioaugmentation for Site Remediation.* pp. 165- 171, Battelle Memorial Institute, Ohio.

Higson, F. K. 1991. "Degradation of Xenobiotics by White Rot Fungi". In G. W. Ware (Ed.). *Reviews of Environmental Contamination and Toxicology*, Vol. 122. Springer Verlag, New York, Berlin.

Morgan, P., S.T. Lewis, and R.J. Watkinson. 1991. Comparison of Abilities of White-rot Fungi to Mineralize Selected Xenobiotic Compounds. *Appl. Microbiol. Biotechnol.* 34: 693-696.

Mueller, J. G., D.P. Middaugh, S.E. Lantz, and P.J. Chapman 1991. Biodegradation of Creosote and Pentachlorophenol in Contaminated Groundwater: Chemical and Biological Assissment. *Appl. Environ. Microbiol.* 57(5): 1277-1285.

Okeke, B.C., J.E. Smith, A. Paterson, and I.A. Watson-Craik. 1996. Influence of Environmental Parameters on Pentachlorophenol Biotransformation in Soil by *Lentinula edodes* and *Phanerochaete chrysosporium*. *Appl. Microbiol. Biotechnol.* 45: 263-266.

DEGRADATION OF AROMATIC POLLUTANTS BY A NON-BASIDIOMYCETE LIGNINOLYTIC FUNGUS

CLEMENTE, A. R.; FALCONI, F. A.; ANASAWA, T. A. & ***DURRANT, L. R.***
DCA/FEA- UNICAMP (Campinas State University), CP 6121 - 13081-970
Campinas – SP, Brazil.

Abstract. A ligninolytic fungal strain, F 898, was grown in media containing either naphthalene, phenanthrene, anthracene, tannic acid, or phenol as the carbon source. Various levels of lignin peroxidase (LiP), manganese peroxidase (MnP) and laccase activities were detected following growth for three, six, and ten days. Regardless of the carbon source, however, MnP activity levels were higher than LiP and laccase activities. Degradation (D) of the five aromatic compounds, measured by reversed-phase high-performance liquid chromatography (HPLC) on a C_{18} column, varied with each carbon source. Phenol and naphthalene were the least degraded compounds (D<20%). About 27% of the anthracene present in the medium was degraded, whereas for phenanthrene and tannic acid degradations of 90% and 100% were obtained, respectively.

INTRODUCTION

Toxic aromatic compounds concentrated in industrial wastes and contaminated sites potentially can be eliminated by bioremediation systems. Ligninolytic fungi produce an unusual highly nonspecific battery of enzymes which may be used to breakdown the lignin macromolecule. These lignin-degrading systems oxidize a very large variety of compounds in addition to lignin, among which are numerous environmental pollutants.

In nature, different groups of microorganisms are involved in lignin degradation. The white-rot fungi are the most important, and of these, *Phanerochaete chrysosporium* has attracted the greatest interest from researchers. Brown-rot and soft-rot fungi, although they modify and degrade lignin only to a limited extent (Arora, 1995), also have potential as bioremediation agents.

The irregular and recalcitrant nature of lignin and the fact that it contains substructures found in primary pollutants (such as phenols, anisoles, biphenyls, and diarylethers) led researchers to postulate that the same nonspecific ligninolytic system produced by the ligninolytic fungi may be used for the degradation of recalcitrant xenobiotics such as polycyclic aromatic compounds, chlorinated aromatics, polycyclic chlorinated aromatics and non-aromatic chlorinated compounds (Sayadi & Ellouz, 1995; Kay-Shoemake & Watwood, 1996).

Polyphenolic compounds are commonly found in wastewaters derived from coal conversion, petroleum refining, and the resin, plastic and textile industries (Lee et al., 1996). These compounds also are major constituents of some coffee-processing, pharmaceutical, petrochemical, and other chemical effluents (Ambujon & Manila, 1995). Phenolic and nitrophenolic compounds also are present in wastewaters from, for example, the dye, pesticide, explosive, solvent industries (Abu-Salah et al, 1996), and from olive mill (Benitez et al., 1997), and wine distilleries (Borja et al., 1993). They can be toxic to microorganisms, plants, and animals.

Polycyclic aromatic hydrocarbons (PAHs) are a class of fused-ring aromatic compounds which occur as ubiquitous environmental pollutants. Besides being present in petroleum, PAHs are formed during the incomplete combustion of almost any organic material, and they have been isolated from air, soil, water, river, and marine sediments. These compounds are of great environmental and human health concern because of their potential trophic biomagnification and because some low-molecular-weight PAHs are acutely toxic, and most higher-molecular-weight compounds have mutagenic, teratogenic, and carcinogenic effects (Cerniglia, 1992).

OBJECTIVES

The purpose of this work was to investigate the ability of a non-basidiomycete ligninolytic fungus to grow and degrade phenolic/aromatic compounds, the ligninolytic enzymes produced during growth and the extent of the degradation caused.

MATERIALS AND METHODS

Microorganisms: The fungal strain 898 was isolated from soil samples collected along the Una do Prelado river (Jureia-Itatins Ecological Reserve, SP - Brazil), and was maintained on PDA (potato dextrose agar - DIFCO) slants.
Media: The medium used contained: $(NH_4)_2HPO_4$ 0.5 $g.l^{-1}$; KH_2PO_4 0.8 $g.l^{-1}$; $MgSO_4$ 0.3 $g.l^{-1}$; $ZnSO_4$ 0.04 $g.L^{-1}$; $MnSO_4$ 0.05 $g.L^{-1}$; CaCl 0.1 $g.L^{-1}$; CoCl 0.02 $g.L^{-1}$; $FeSO_4.7H_2O$ 0.01 $g.L^{-1}$; yeast extract 0.2 $g.L^{-1}$ and a carbon source at a final concentration of 0.05% (w/v) for the PAHs and of 0.1% (w/v) for tannic acid and phenol.
Culture conditions: The fungus was inoculated in Erlenmeyer flasks (30 mL medium/ 250 mL) under stationary conditions at 30ºC. On the 3rd, 6th and 10th day of growth the supernatant fluids were collected and used for enzyme assays and high-performance liquid chromatography (HPLC) analysis.
Enzyme assays: Lignin peroxidase (LiP), Manganese peroxidase (MnP), and Laccase (Lac.) activities were assayed spectrophotometrically in the extracellular fluid of the culture supernatants. LiP was determined by measuring veratryl alcohol oxidation as described previously (Tien & Kirk, 1984). MnP determinations were based on the oxidation of phenol red, according to Kuwahara et al. (1984). Laccase was determined via the H_2O_2-independent oxidation of syringaldazine (ethanol solution) to its quinone form (Szklarz et al., 1989). All enzyme activities are expressed in units per liter, with one unit equal to 1 μmol of substrate oxidized per minute.

High Performance Liquid Chromatography: All HPLC analyses described herein were performed with a Zorbax ODS (0,46 x 15 cm) C_{18} reverse-phase column (SUPELCO Chromatography Products). Separation of PAHs was achieved by isocratic elution in acetonitrile:water (70:30), at a flow rate of 0.8 mL/min., and the retention times of the products were determined using a UV detector (279 nm). For tannic acid, separation was achieved by isocratic elution in water:methanol (70:30), at a flow rate of 0.5 mL/min., and the retention times of the products were determined using a UV detector (254 nm); for phenol, separation was achieved by isocratic elution in ammoniun acetate: acetonitrile: methanol (56:34:10), at a flow rate of 0.8 mL/min., and the retention times of the products were determined using a UV detector (280 nm).

RESULTS AND DISCUSSION

As shown in Figures 1 through 3, various levels of LiP, MnP, and laccase activities were detected in the culture supernatants following its growth in phenolic/aromatic compounds. The lowest levels of LiP and MnP activities were produced in a medium containing naphthalene or phenol. The highest levels of laccase activities were observed in phenol medium and reasonable levels were observed in naphthalene medium. No degradation and negligible levels (D< 6%) were observed for naphthalene and phenol, respectively (Figure 4). The absence of LiP and MnP activities and the low degradation obtained indicate that these two enzymes are important in the breakdown of naphthalene and phenol. The presence of laccase is probably related to a detoxification mechanism of this fungus. A longer period of incubation might have produced higher degradation.

The highest levels of LiP were detected in tannic acid medium. MnP and laccase were produced, and the highest levels of degradation were also obtained (D= 100%) in this medium. Although the highest levels of MnP activities were produced in anthracene medium, only 27% was degraded. High degradation of

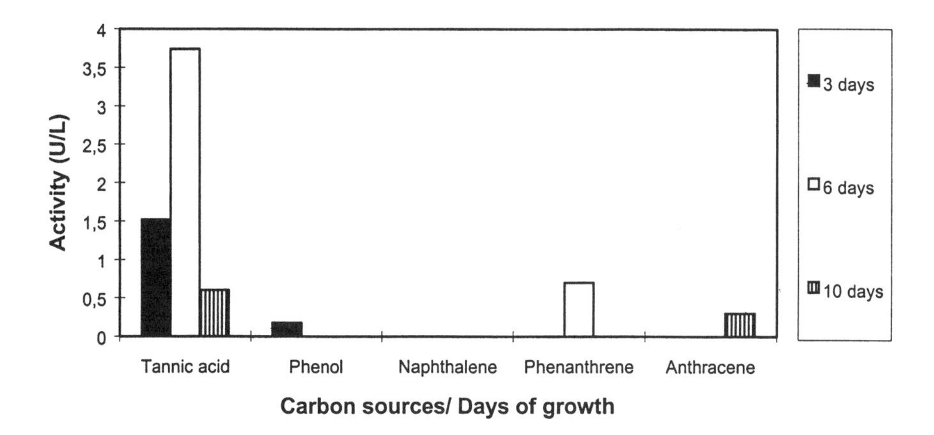

FIGURE 1: Lignin Peroxidase activities produced by the fungal strain 898 following its growth in various carbon sources.

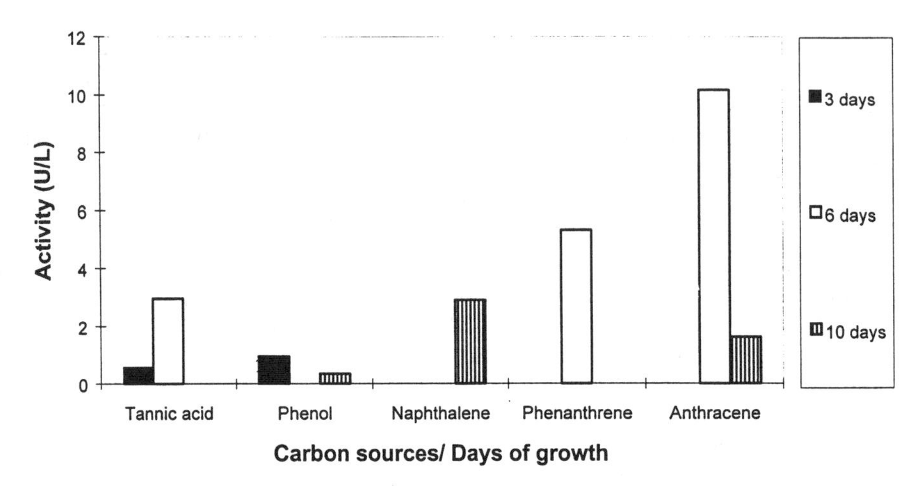

FIGURE 2: Manganese Peroxidase activities produced by the fungal strain 898 following its growth in various carbon sources.

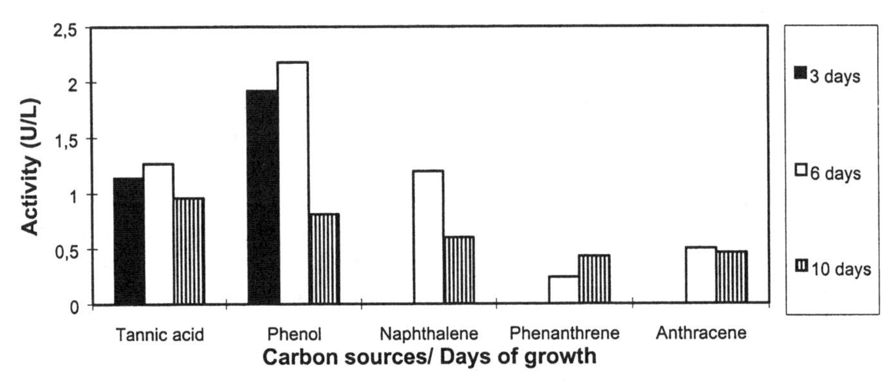

FIGURE 3: Laccase activities produced by the fungal strain 898 following its growth in various carbon sources.

phenanthrene (D> 90%) was observed. Reasonable levels of LiP and MnP and low levels of laccase activities were produced in this medium. The results obtained suggest that LiP followed by MnP are the two most important enzymes for degradation of the compounds studied here. It is also possible that a certain combination in the levels of activities of these two enzymes is necessary to produce a high percentage of degradation.

Although the degradation of naphthalene and phenol was very low, the strain 898 was able to grow in their presence and to produce LiP, MnP, and laccase. Because of its ability to grow in all the carbon sources used here, to degrade phenol and anthracene to some extent, and to greatly degrade phenanthrene and tannic acid, this fungus shows considerable promise as a bioremediation agent for use in the restoration of contaminated environments.

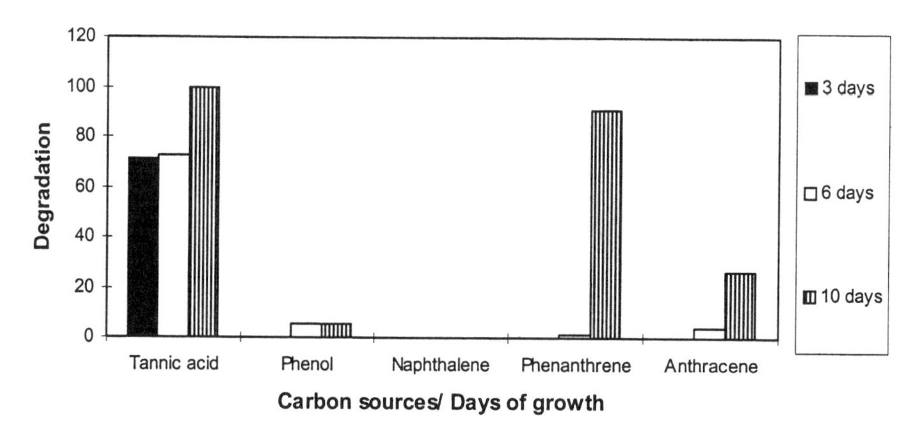

FIGURE 4: Degradation (%) of the carbon sources by the fungal strain 898.

ACKNOWLEGMENTS

We thank FAPESP and CNPq for financial support.

REFERENCES

Abu-Salah, K.; Shelef, G.; Levanon, D.; Armon, R.; Dosoretz, C.G. 1996. Journal of Biotechnology, 51: 265-272.

Ambujon, S. andManila, V.B. 1995. Biotechnology letters, 17 (4): 443-448.

Arora, D.S. 1995. *Biodegradation*, 6: 57-60.

Benitez, J.; Beltran-Heredia, J.; Torregosa, J. Acero, J.L.; Cerca, V. 1997. Applied Microbiology Biotechnology, 47: 185-188.

Borja, R.; Martín, A.; Maestro, R., Luque, M.; Murán, M.M. 1993. Biotechnology Letters, 15(3): 311-316.

Cerniglia, C. E. 1992. *Biodegradation*, 3: 351-368.

Kay-Shoemake, J.L. and Watwood, M.E. 1996. Applied Microbiology Biotechnology, 46: 438-442.

Kuwahara, M.; Glen, J.K.; Morgan, M.A.; Gold, M.H. 1984. FEBS Letters, 169: 247-250.

Sayadi, S. and Ellouz, R. 1995. Applied and Environmental Microbiology, 1098-1103.

Szklarz, G. D.; Antibus, R. K.; Sinsabaugh, R. L.; Linkins, A. E. 1989 Mycology, 81: 234-240.

Tien, M. and Kirk, T.K. 1988. In: Biomass, part B, pp 238-249. Editors: Wood, W.A. & Kelloj, S.T.

INVESTIGATIONS INTO THE FACTORS LIMITING BIOREMEDIATION OF PAH CONTAMINATED SITES

ESMAEIL S. ALSALEH (King's College London, UK)
STEVE SMITH and JEREMY MASON (King's College London)

ABSTRACT: The biodegradative activity towards polycyclic aromatic hydrocarbons (PAH) present in soil samples from an uncontaminated and a disused gas-works site were investigated by means of ^{14}C- naphthalene mineralization assays. Higher rates of activity were observed in soil from the uncontaminated site. Augmentation of these soils with a previously characterized PAH degrading organism resulted in stimulation of activity in both soils. However, the stimulation in activity was significantly less in the case of contaminated soil indicating that low rates of PAH degradation in the contaminated soil were due to constraints and not to an inadequate indigenous microflora. Physical and chemical characterization of both soils revealed similar properties except for the elevated levels of some metals in the contaminated soil. A bioluminescence toxicity assay using *lux*-marked *Escherichia coli* demonstrated that the primary toxicity was due to non-volatile organics and that this effect was compounded by the inorganic fraction.

INTRODUCTION

The application of a bioremediation technology may be assisted by the demonstration of enhanced biodegradation under controlled laboratory conditions where typical parameters such as microbial counts, isolation of microbes capable of degrading specific pollutant, and determination of degradation rates of individual or total pollutants can be measured. This investigation should be complemented with toxicity tests to elucidate whether the determined biodegradative potential is constrained by organic or inorganic material, and which would affect the choice of a remediation technology (Burlage *et al.*, 1994). One of the best approaches for the estimation of toxicity is the use of *lux*-marked biosensors (Paton *et al.*, 1997).

Objective. The objective of this study was to assess the biodegradative potential of soil from a PAH-contaminated coal gas manufacturing site. The feasibility of bioaugmentation to remediate the site was demonstrated. In addition, the presence of toxic materials was demonstrated and the extent of their inhibitory effects estimated.

MATERIALS AND METHODS

Total viable and PAH degrading bacterial counts (cfu g^{-1}) were determined using the plate-dilution technique. Naphthalene mineralization rates were measured using ^{14}C-naphthalene (Fleming *et al.*, 1993) and $^{14}CO_2$ was quantified using a liquid scintillation counter (Packard Tri Carb). Bioluminescence toxicity testing was conducted according to the method of Paton *et al.*, (1997). The effects of aqueous extract from soil were tested directly, after N_2 sparging, muffle furnacing and pH adjustment on the bioluminescence of *E. coli* HB101 *pUCD607* measured using a Bio-orbit 1251 luminometer.

RESULTS AND DISCUSSION

Isolation and Counting of Bacteria. Aqueous extracts of uncontaminated and contaminated soil samples showed similar numbers of heterotrophic bacteria ($4.50 \pm 0.03 \times 10^6$ and $3.50 \pm 0.02 \times 10^6$ cfu g^{-1}, respectively). However, the contaminated soil showed significantly higher numbers of naphthalene degrading organisms ($5.00 \pm 0.03 \times 10^4$ cfu g^{-1}) than the uncontaminated soil ($1.20 \pm 0.01 \times 10^3$ cfu g^{-1}).

Mineralization Rates. The rate of ^{14}C-naphthalene mineralization by the two soil samples was estimated. Lower rates of mineralization were measured in contaminated soil (1.3 ng/min/g), than in contaminated soil (9.5 ng/min/g) Fig 1. Thus despite having higher numbers of naphthalene degrading organisms, the contaminated soil showed substantially lower biodegradative activity (by a factor of 400 on a per cell basis).

In order to determine whether this lower activity in contaminated soil was due to an inadequate microflora, augmentation experiments were performed using a strain of *Pseudomonas cepacia* EA100 isolated previously in our laboratory and demonstrated to have a broad substrate specificity and high PAH degradation activity in liquid culture. This strain showed rates of phenanthrene, fluorene, pyrene and fluoranthene degradation of 21.7 ± 0.20, 5.90 ± 0.04, 4.20 ± 0.03, and 2.83 ± 0.02 ng/min/10^9 cells respectively.

Soils augmented with *Pseudomonas cepacia*EA100 (1.35×10^9 cells/g soil) showed increased naphthalene mineralization rates of 11.3 ng/min/g soil and 87.8 ng/min/g soil for contaminated and uncontaminated soils respectively. Thus although both soils showed stimulation in biodegradative activity upon augmentation, the relative percentage inhibition remained the same in both augmented and non-augmented soils. This would indicate that the low rate of PAH degradation in contaminated soil was due to constraints other than an inadequate microflora.

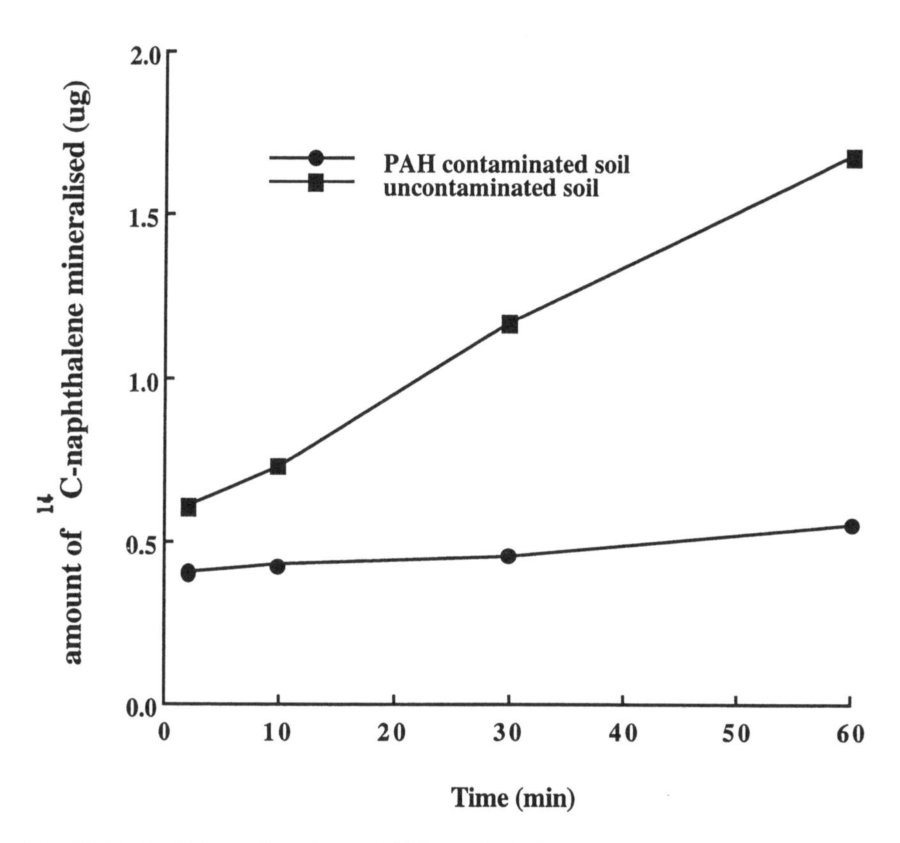

FIGURE 1. Mineralization of ^{14}C-naphthalene (0.5μmol/g soil) by the indigenous microflora of contaminated (●) and uncontaminated (■) soils at 30 °C. Values are the average of two determinations, error bars lie within symbols.

Soil Characterization. Physical and chemical characterization of both soils revealed similar properties (data not shown) except for a five and a three fold increase in the amount of copper and lead respectively in the contaminated soil (table 1).

TABLE 1. Concentration of some metals (μg/g soil) in uncontaminated and contaminated soil measured by atomic absorption spectroscopy

Metals	Pristine soil (μg/g soil)	Contaminated soil (μg/g soil)
Fe	3999±26.0	20540±33.0
Mg	658 ±9.0	732 ±9.0
Cu	87.0±5.0	473 ±7.0
Mn	114 ±6.0	188 ±7.0
Zn	59.0±5.0	131 ±5.0
Pb	50.0±7.0	169 ±6.0
Cr	26.0±5.0	34.0±4.0
Ni	2.76±1.0	14.0±3.0

Toxicity Assay. In order to determine the nature of the constraints to bioremediation, a bioluminescence toxicity assay was employed using *lux*-marked *E. coli* HB101*pUCD607*. This assay demonstrated that the primary toxicity was due to non-volatile organics that caused a 13% inhibition in bioluminescence. When the pH of the soil extract was adjusted to 5.5 the inhibition of bioluminescence was increased to 25% relative to a similarly adjusted control (fig 2). This would indicate the presence of an inhibitory fraction that is more bioavailable at acidic pH. Metals are potential candidate for this inhibition but a reconstituted muffle furnaced fraction resulted in only an 8% inhibition in bioluminescence indicating that the inorganic component (e.g. metals) was not the primary constraint to activity. Another potential inhibitory fraction that is usually associated with coal gas work sites may be cyanide (Morris *et al.*, 1988). However, reconstitution experiments in which cyanide is added to uncontaminated soil to a level comparable to that in contaminated soil show no appreciable inhibition of mineralization activity (data not shown). Work is in progress to ascertain the nature of the non-volatile organics inhibitory component.

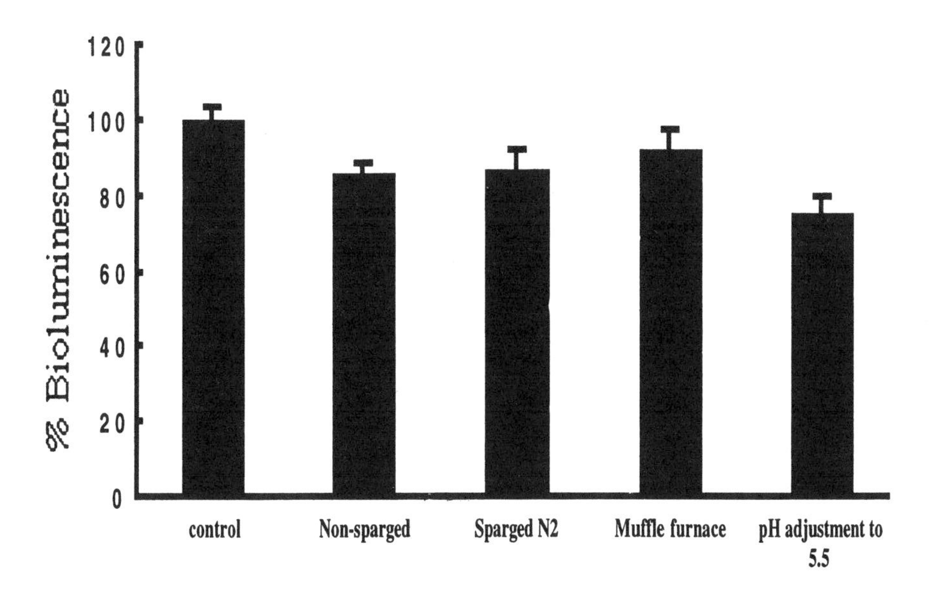

FIGURE 2. The effects of four treatments of an aqueous extract of contaminated soil on the percentage bioluminescence of *E. coli* HB101 *pUCD607*

ACKNOWLEDGEMENT

This project is sponsored by a grant from Kuwait University

REFERENCES

Burlage, R. S., A. V. Palumbo, A. Heitzer, and G. Sayler. 1994. "Bioluminescent Reporter Bacteria Detect Contaminants in Soil Samples." *Appl. Biochem. Biotech.* 45/46: 731-740.

Fleming, J. T., J. Sanseverino, and G. Sayler 1993. "Quantitative relationship between naphthalene catabolic gene frequency and expression in predicting PAH degradation in soils at town gas manufacturing sites". *Environ Sci Technol* 27: 1068-1074.

Morris, R., Hunt, L., S. Jones, and A. Hart. 1988. "Landfarm Bioremediation of Polycyclic Aromatic Hydrocarbons". *Land Contam Reclam* 6: 17-25.

Paton, G. I., E. A. S. Rattary, C. D. Campell, M. S. Cresser, L. A. Glover, J. C. L. Meeussen, and K. Killham. 1997, "Use of Modified Microbial Biosensors for Soil Ecotoxicity Testing." In Biol Indicators Soil Health Sustainable Productivity, pp. 379-418. Edited by C, Pankhurst.

CHARACTERIZATION OF MICROFLORAE DEGRADING POLYCYCLIC AROMATIC HYDROCARBONS OF CONTAMINATED SOILS

Frank Haeseler, Denis Blanchet (Institut Français du Pétrole, Rueil Malmaison, France)
Vincent Druelle (Gaz De France, La Plaine-Saint-Denis, France)
Peter Werner (Technische Universität, Dresden, Germany)
Jean-Paul Vandecasteele (Institut Français du Pétrole, Rueil Malmaison, France)

ABSTRACT

The influence of environmental conditions on the capacities of microflorae of former manufactured gas plants sites (MGP) to degrade polycyclic aromatic hydrocarbons (PAH) was investigated. A series of soil samples from different MGP sites was collected and quantitative evaluation of PAH degraders was carried out. Two parameters strongly influencing soil population levels, water content and oxygen availability, were identified. Population levels of PAH degraders appear to be indicators of metabolic activity taking place in the sample in a given set of conditions, including pollutant availability, rather than of the degradative potential (intrinsic capacities) of the microflora itself. Besides PAH degraders, the development of other populations was observed in cultures in reactors using PAH mixtures as sole carbon sources, in particular of *Pseudomonas aeruginosa* strains unable to degrade PAH. Evidence that such strains could grow on metabolites resulting from incomplete PAH degradation and improved the overall degradative performance of the soil consortia was obtained.

INTRODUCTION

Soils from contaminated former MGP sites have been widely used for the isolation of bacterial strains capable of growing on 2-4 cycle PAH (Boldrin et al. 1993; Bouchez et al. 1995a; Weissenfels et al. 1991).

Experimentation on biological treatability of these soils showed that they contained microflorae capable of degrading a large spectrum of PAH (Stieber et al, 1990).

However the relationships between the nature and the population levels of PAH degraders, the environmental conditions of the sites and the intrinsic capacities for PAH degradation of the indigenous microflorae have been little studied. The results of an approach to these two aspects are presented below.

MATERIALS AND METHODS

All soil samples originated from contaminated former MGP sites.

Biodegradation experiments of PAH supplied as crystals were carried out in slurry reactors at 30°C and performed in a Sapromat respirometer (type D, Voith, Ravensburg, Germany) as described by Bouchez et al 1995b. The PAH mixture was added as an acetone solution, which was evaporated from the flask, before mineral salt medium and inoculum addition.

PAH were extracted with cyclohexane and quantified by gas chromatography with a flame ionization detector (FID).

For bacterial enumeration, microflorae were extracted from 10 g of soil by shaking in 100 mL sterile physiological saline solution. Enumeration of microorganisms was performed on the supernatant. The most probable number (MPN) of specific PAH-degrading microorganisms was determined on 96-well microplates according to Stieber et al. (1994) with the following PAH used as sole energy and carbon source: naphthalene (NAP), acenaphthene (ACE), acenaphthylene (ACY), fluorene (FLU), phenanthrene (PHE), anthracene (ANT), fluoranthene (FLA) and pyrene (PYR). Bacteria degrading PAH metabolites were enumerated with the same method using appropriate carbon sources. For total microflora the growth substrate was meat broth.

RESULTS

The capacities of microflorae of three former MGP sites (SA, LH and G) to use PAH as sole carbon and energy source are shown in figure 1. Positive responses reached up to MPN per g of soil for naphthalene degraders and up to 10^4 MPN for 4-cycle PAH degraders but the population levels in different soils was found quite variable (by a 10^3 factor). No direct correlation could be noticed between the population levels of PAH degraders of soil microflorae and the PAH concentrations in the soils.

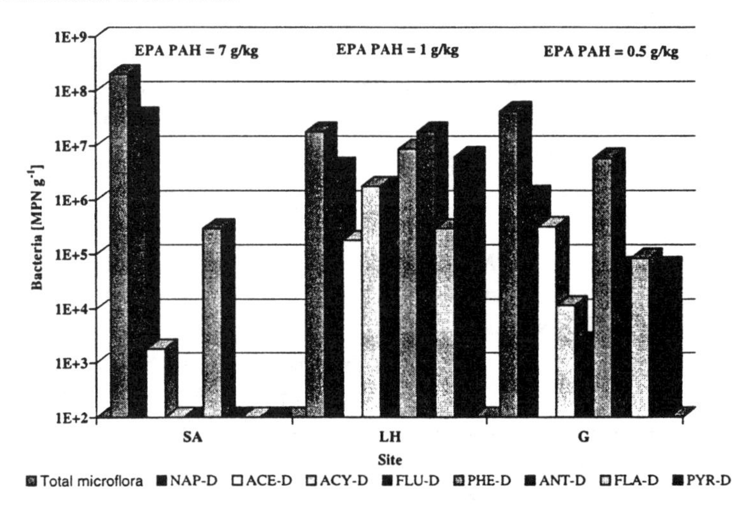

Figure 1. Population levels of PAH-degraders (PAH-D) of three MGP soils presenting various PAH concentrations.

Microbiological characterization of PAH-degrading bacteria in soil stored under different environmental conditions lead to the identification of two parameters strongly influencing soil population levels, i.e. water content and oxygen availability. Figure 2 shows that soil SA presented initially low numbers of PAH degraders. The population level of this soil remained very low when kept with a low water content (<5%) under aerobic conditions at 20°C (Figure 2a). However the low population levels of soil samples with low water contents kept at room temperature remained measurable for several years. The initially low number of PAH degraders of soil SA increased by a factor up to 10^4 over a period of three months when incubated with a higher water content (18%) under aerobic conditions at 20°C (Figure 2b). This increase of PAH degraders did not occur in this soil when stored in anoxic conditions.

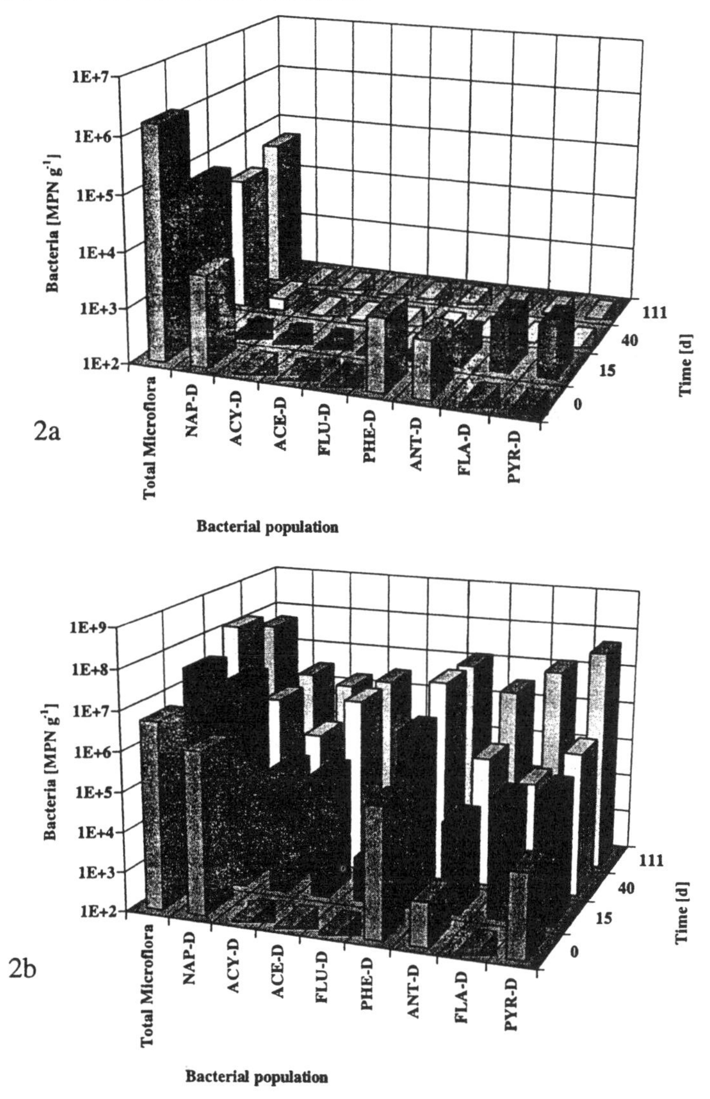

Figure 2. Evolution of population levels in soil SA incubated under aerobic conditions at 20°C. 2a: 3% water content; 2b: 18% water content.

The intrinsic capacities of the microflora present in soil G were investigated with a mixture of ten 2 to 5-cycle pure PAH. The inoculum used was 1 g of contaminated soil. Sizable amounts (20 to 80 mgL^{-1}) of each PAH (including benzo(e)pyrene) were degraded (Table 1). The final degradation performances obtained with soil SA were quite similar (data not shown).

Table 1. Biodegradation of a mixture of pure PAH in slurry reactor using MGP soil G as inoculum.

	Concentration [mgL^{-1}]	Experiment 1 Biodegradation [%]	Experiment 2 Biodegradation [%]
Naphthalene	28	100,0	100,0
2 cycles	**28**	**100,0**	**100,0**
Acenaphtylene	57	100,0	100,0
Acenaphtene	80	100,0	100,0
Fluorene	36	99,9	99,7
Phenanthrene	36	99,9	99,8
Anthracene	35	97,2	94,9
3 cycles	**244**	**99,5**	**99,2**
Fluoranthene	37	99,5	98,4
Pyrene	36	98,7	97,8
Benzo(a)Anthracene	54	92,5	88,0
Chrysene	36	85,2	82,2
4 cycles	**163**	**93,9**	**91,4**
Benzo(e)Pyrene	34	47,3	56,0
5 cycles	**34**	**47,3**	**56,0**
Sum of PAH	468	93,8	93,4

Figure 3 presents the oxygen uptake during two biodegradation experiments of the mixture of pure PAH (shown above) with soil G and soil SA. It appears that in spite of different kinetics the inocula lead to nearly the same final oxygen uptake. So we can conclude that in spite of different population levels present in soils from the different former MGP sites studied (Figure 1) the intrinsic capacities of the microflora were found to be very similar.

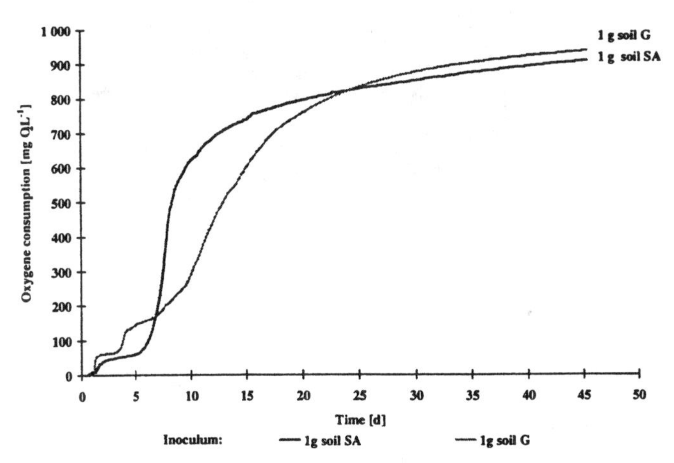

Figure 1. Oxygen consumption during the biodegradation of a mixture of pure PAH in slurry reactors using different MGP soils as inocula

During the PAH biodegradation experiment with 1g of soil G as inoculum a microbiological characterization was done. As shown in Figure 4 a sequential growth, first of naphthalene degraders (NAP-D), then of fluoranthene degraders (FLA-D) and finally of *Pseudomonas aeruginosa* was observed. The strain *P. aeruginosa* GL1, unable to use PAH as sole carbon and energy sources and isolated during one of these biodegradation experiments, remained present in the microbial community after ten successive enrichment cultures on mineral salt medium containing the mixture of 2-, 3- and 4-cycles PAH as sole growth substrate. We found that *P. aeruginosa* GL1 was able to grow on compounds such as salicylic acid and o-phthalic acid (data not shown), two known PAH metabolites (Cerniglia and Heitkamp 1989). Furthermore *P. aeruginosa* GL1 was able to excrete high amounts ($3.5 g.L^{-1}$) of rhamnolipids (Arino et al. 1998). These two points, susceptible to promote overall PAH assimilation, can also provide an explanation for the persistence of *P. aeruginosa* GL 1 in this bacterial community.

In fact, in the reactor culture inoculated with soil G, significant numbers of bacteria capable to grow on known PAH metabolites were found: up to 10^7 $MPN.L^{-1}$ on salicylic acid, 10^7 $MPN.L^{-1}$ on o-phthalic acid and 10^3 $MPN.L^{-1}$ on 1-hydroxy-2-naphthoic acid.

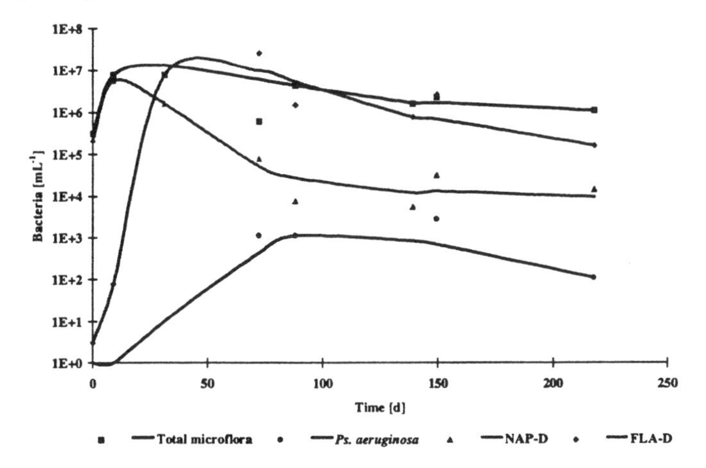

Figure 3. Evolution of the bacterial population of the MGP site G during biodegradation of a mixture of pure PAH in a slurry reactor.

DISCUSSION

Population levels of degraders appear to be indicators of metabolic activity taking place in the sample in a given set of conditions, including pollutant availability, rather than of the degradative potential (intrinsic capacities) of the microflora itself. Two environmental parameters strongly influencing soil population levels, water content and oxygen availability, were identified.

In the biodegradation experiments with a mixture of pure PAH as sole carbon and energy source, the development of naphthalene degraders was more rapid and higher than the development of other degraders such as those using fluoranthene or pyrene but the final population levels reached with different inocula were comparable (data not shown). Understandably, the isolation of individual strains using as carbon source one of the eight 2 to 4-cycle PAH listed above was found most efficient when starting from samples that had been subjected to amplification of the population of PAH degraders.

Besides PAH degraders, the development of other strains was observed in several cultures in reactors using PAH mixtures as sole carbon sources, in particular *Pseudomonas aeruginosa* strains unable to degrade PAH. Evidence that such strains could grow on metabolites resulting from incomplete PAH degradation was obtained. This suggests that the bacteria unable to degrade PAH were susceptible to improve the overall degradative performance of the soil consortia.

REFERENCES

Arino, S. Marchal, R. Vandecasteele, J.-P. 1998. "Involvement of a rhamnolipid-producing strain of *Pseudomonas aeruginosa* in the degradation of polycyclic aromatic hydrocarbons by a bacterial community", *Appl. Microbiol. Biotechnol.*, 84, 769-776.

Boldrin, B. Tiehm, A. Fritzsche, C. 1993. "Degradation of phenanthrene, fluorene, fluoranthene, and pyrene by a *Mycobacterium sp.*", *Appl. Environ. Microbiol.*, 59, 1927-1930.

Bouchez, M. Blanchet, B. Vandecasteele, J.-.P 1995a. "Degradation of polycyclic aromatic hydrocarbons by pure strains and by defined strain associations: inhibition phenomena and cometabolism", *Appl. Microbiol. Biotechnol.*, 43, 156-164.

Bouchez, M. Blanchet, D. Vandecasteele, J.-P. 1995b. "Substrate availability in phenanthrene biodegradation: transfer mechanism and influence on metabolism", *Appl. Microbiol. Biotechnol.*, 43, 952-960.

Cerniglia, C.E. Heitkamp, M.A. 1989. "Microbial degradation of polycyclic compounds (PAH) in the aquatic environment", *In* Varanasi V (ed) Metabolism of PAHs in the aquatic Environment, CRC Press Inc. Boca Raton, Florida, pp. 41-68.

Stieber, M. Böckle, K. Werner, P. Frimmel, F.H. 1990. "Biodegradation of polycyclic aromatic hydrocarbons (PAH) in the subsurface", *In* Arendt F Hinsenveld M and Van den Brink W.J. (eds), Contaminated Soil '90, pp. 473-479.

Stieber, M. Haeseler, F. Werner, P. Frimmel, F. 1994. "A rapid screening method for micro-organisms degrading polycyclic aromatic hydrocarbons in microplates", *Appl. Microbiol. Biotechnol.*, 40, 753-755.

Weissenfels, W.D. Beyer, M. Klein, J. Rehm, H.J. 1991. "Microbial metabolism of fluoranthene isolation and identification of ring fission products", *Appl. Microbiol. Biotechnol.*, 34, 528-535.

HYDRO-BIOLOGICAL CONTROLS ON TRANSPORT AND REMEDIATION OF ORGANIC POLLUTANTS.

A.D.G. Jones (King's College London)
J. Mason and S. Smith (King's College London)
H. S. Wheater, A. B. Butler, J. N. B. Bell and H. Gao (Imperial College London)
A.Shields and P. E. Hardisty (Komex Clarke Bond, Bristol)
S. Wallace (BG plc Property Division, UK)

ABSTRACT: The processes controlling the fate of polynuclear aromatic hydrocarbons (PAHs) at a former coal-gas manufacturing site are being investigated, with the aim of developing appropriate modeling tools for risk-based site remediation. The site is being characterized physically, chemically, and geochemically. Aerobic microbiological activity has been demonstrated from sediment samples assessed through naphthalene radiolabelled mineralisation assays, with rates of up to 30μg h^{-1} kg^{-1} soil. These have been shown to be independent of temperature over a range of 8-30°C, but dependent on concentration. Residual sediment PAHs have been indicated to be bioavailable. The results are being used to identify biodegradation rate controls for a multi-species contaminant transport model, which is currently under development.

INTRODUCTION

The characterization and enhancement of *in situ* biodegradative activity towards PAHs is a necessary target for many former gas work sites. These sites can be contaminated as a result of their long industrial history and combine problems of complex contaminant chemistry with man-made and physical characteristics. Microbial degradation of PAHs in the environment is often limited by a large number of potential controls, including the partial pressure of oxygen, bioavailability and toxicity of the contaminants, pH, temperature and the ability of endogenous microorganisms to degrade the target contaminants.

In this field-based study the influence of hydrobiological, bioavailability and toxicity controls on the *in situ* bioremediation process are being investigated, with the aim of incorporating these parameters into a model utilizable for risk assessment and remedial design. A combination of microbiological *ex situ* mineralisation rates, and molecular *in situ* techniques have been adopted to establish *in situ* biodegradation rates.

In this paper quantified *ex situ* PAH mineralisation rates are presented with site characterisation and initial results from the modeling process.

Site Description. The site chosen for this study was a former gas works in the north-west of England, covering an area of approximately 0.7ha. Town gas production started on site in the 1850s and continued in operation for the following 100 years until production ceased in the 1950s, for the past 40 years the site has remained derelict. Site investigations have identified BTEX, PAH and complex cyanide contamination, frequent pollutants at town gas sites. Two hydrologically distinct regions of PAH and BTEX contamination have been found, and the corresponding source material for these areas removed as part of the remediation programme. The top 1-2m of the site is made-ground, beneath which fine sand extends to a depth of 7 to 10m. Underlying the sand aquifer is a layer of very low permeability clay with a thickness greater than 3m. Limited evidence has been observed of clay lenses within the sand layer. The groundwater table at the site varies between 1 to 2m below ground surface (bgs) with a groundwater flow rate of 20-50m $year^{-1}$.

MATERIALS AND METHODS Prior to the research presented in this paper, extensive surveys on the site had been conducted on behalf of BG plc Property Division. On the basis of the results in the previous surveys, five further boreholes (MW 98-4 -8) were drilled (Figure 1).

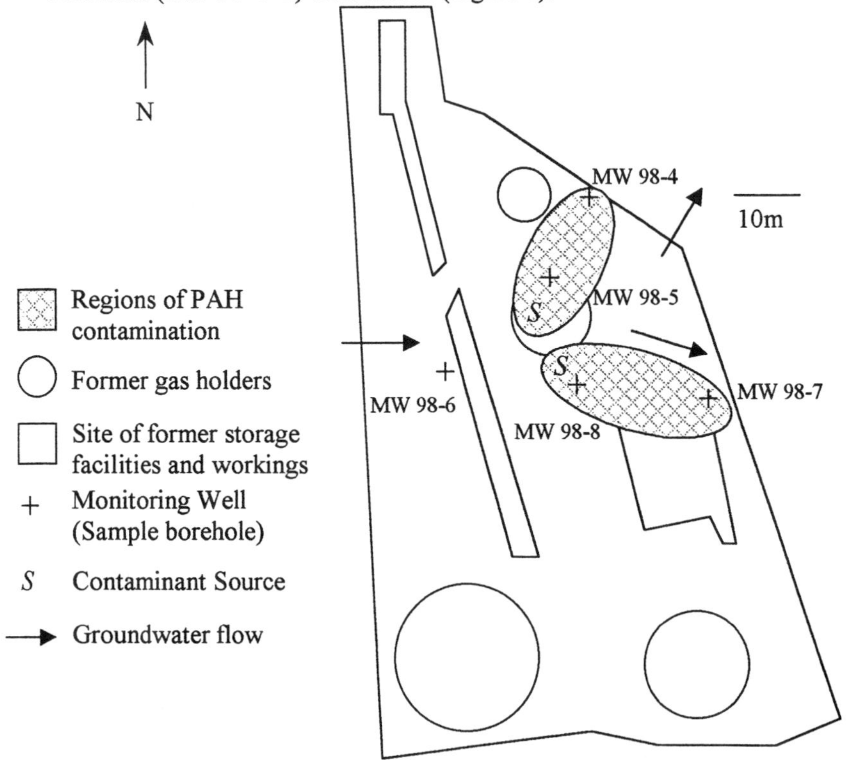

Figure 1 Schematic plan view of the former coal gas manufacturing site, illustrating regions of PAH contamination, former structures and groundwater flow.

Two boreholes were placed in each of the two regions of PAH contamination, separated by a distance of between 10 to 20m. One borehole was drilled in an uncontaminated region of the site, to act as a control. Each borehole was drilled to a depth of 9-10m using a hollow-stem auger drilling rig, producing continuous sediment cores of 1.5m. Boreholes were screened and maintained as monitoring wells for subsequent groundwater analysis. The high degree of homogeneity in the sediment allowed for each 1.5m core to be divided further into three sections of 0.5m with satisfactory reproducibility, one section was analysed by an accredited contract laboratory (TES Bretby Ltd.,) using standardised methodology (Wild and Jones, 1995). The analysis included the 16 PAHs on the U.S. EPA list of priority pollutants, metals, cyanide, phenols and BTEX compounds as well as other basic parameters. Another section was analyzed for hydrogeochemical and physical characteristics and the third section was used for microbiological analysis.

Microbiological analysis. 30 core sections were available for microbiological analysis. Samples (approximately 100g) were extracted and flash-frozen in liquid nitrogen and stored at –70°C for subsequent molecular analysis. Two further samples of 50g were extracted for mineralisation assays before the core section was returned to storage at 4°C. Mineralisation assays were conducted by combining soil sediment and minimal medium with radiolabelled naphthalene (^{14}C-1-naphthalene) according to the method of Fleming *et al.*, (1993). Quadruplicate samples were used at each time point and mercuric chloride (200 mg l^{-1}) and no-sediment controls were included. The average rate of biodegradation was calculated and displayed as an unadjusted value.

Although *in situ* degradation rates can be determined *ex situ*, for a soil sample, by measuring the mineralisation rates of spiked radiolabelled substrates, cross-correlation requires further experimentation. Specifically, the (dilution) effect of endogenous PAHs in the soil complex and the temperature sensitivity of the assay system. To this extent two further experiments were conducted. Two soil samples were selected, one with marginal to zero PAH contamination detectable (MW 98-4 2.8m bgs <1 mg kg^{-1} naphthalene) and another with significant PAH contamination (MW 98-5 1.5m bgs 9mg kg^{-1} naphthalene). A standard mineralisation assay was performed as outlined above, with the further modification of dual isotope activities. Both sets of soil samples (98-4 and 98-5) had 1.2µg of total added naphthalene present in 5ml of slurry as well as endogenous PAHs, and were exposed to both diluted and un-diluted radiolabels, the diluted samples had 1 part radiolabel added to 3 parts cold naphthalene, retaining the same total naphthalene concentration whilst reducing the specific activity 4 fold. The experiment was run as before in quadruplicate with three time points and the rates determined from a mean of the four values at each time point. Temperature sensitivity was assessed with a sample of MW98-4 (2.1-3.8m bgs) soil to maintain consistency with previous assays. The mineralisation assay was performed as before with a range of temperatures of 8, 20, and 30°C using radiolabelled naphthalene.

Modeling

Modeling the transport of dissolved organic contaminants in the near-surface environment is controlled by a generation term, in this case representing the dissolution from residual phase sources present at the site, coupled with migration due to groundwater flow. Owing to differential flow velocities arising from the porous structure of the geological material, the flow transport process is conceptualized using the advection-dispersion equation. Where advection arises from the mean bulk groundwater flow and dispersion is related to flow variations about this mean value. Interactions of the organic contaminants with the solid medium, particularly organic fractions present can lead to sorption partitioning, which can be beneficial in retarding contaminant movement. However, in terms of the eventual risk to a target receptor, a key consideration is, as already described, the ability of microorganisms present within the subsurface to degrade the contaminant. Thus, a basic mathematical formulation for the fate and transport of organic compounds is the variably saturated groundwater flow equation (1) coupled with the contaminant mass transport equation (2).

$$\frac{\partial \theta}{\partial t} = \nabla . q \qquad (1)$$

$$\frac{\partial(\theta c_i + \rho_b S_i)}{\partial t} = \nabla.(\theta D_i \nabla c_i - q c_i) + G_i - D_i \qquad (2)$$

Where θ is the volumetric water content (m^3m^{-3}), q is the groundwater flux ($m^3m^{-2}s^{-1}$), c is the dissolved phase concentration ($kg.m^{-3}$), ρ_b is the dry bulk density ($kg.m^{-3}$), S the sorbed concentration ($kg.kg^{-1}$), D is the dispersion tensor (m^2s^{-1}) and t is time (s). The final two terms represent the source term generation G and the contaminant decay D ($kg.m^{-3}s^{-1}$). The subscript i denotes the ith chemical component and allows for multiple-species to be simulated simultaneously.

The experimental work described earlier has demonstrated the ability of the microbial activity at the contaminated land site to degrade PAHs, in particular naphthalene. As part of the ongoing research work we are seeking to use the above model to represent PAH fate and transport. However, it is important to recognize that, owing to the complex nature of the contamination, this takes place within the context of a wide range of contaminants present (e.g. BTEX compounds) along with controls on the efficiency of the degradation process (e.g. dissolved oxygen availability, potential toxicological effects etc.). Therefore, the experimental work is being used to determine site-based empirical relationships for degradation processes. Thus the degradation term D_{PAH} is related to contaminant concentration c_{PAH} though a Michaelis-Menten relationship, which is, in turn, dependent on the level of dissolved oxygen (c_{oxy}), as this determines whether the system is aerobic or anaerobic. This is an important dimension to the model's performance as anaerobic degradation rates for PAH compounds are typically (1-2 orders of magnitude lower) and hence this can greatly affect the extent of the contaminant plume.

RESULTS AND DISCUSSION

Chemical characterization.

The surface and subsurface substrate is composed of loose sandy material, low in organic matter (generally <1.0% LOI), although the surface section in one of the contaminated cores was as high as 11% LOI. Site pH ranges from 6-7.5, with a few acidic pockets. MW 98-8, the most contaminated core recorded a pH of 8. Redox values were <0 below the water table.

MW 98-8 (Figure 1) had 10 000 mg kg^{-1} ΣPAHs and > 3 000 mg kg^{-1} Σphenols at 1m bgs with comparatively low BTEX concentrations (< 70 mg kg^{-1}). The high organic content of 11% in this section contrasts to 2% for the rest of the core and <1% for the remainder of the site, indicating the heavy organic contamination of this section. Whilst contamination decreased bgs, even at 6.7m bgs ΣPAH concentrations were > 300mg kg^{-1} and Σphenols are > 100mg kg^{-1}.

Microbiological characterization.

The mineralisation experiments have produced several important results from the perspectives of both a site remediation programme and PAH degradation. Most significantly, endogenous bacteria in the sediment are able to mineralise naphthalene (Figure 2 S.E. bars have been omitted, but accounted for up to 30% of actual value) and phenanthrene (data not shown). No bioremediation programme would be possible without such activity.

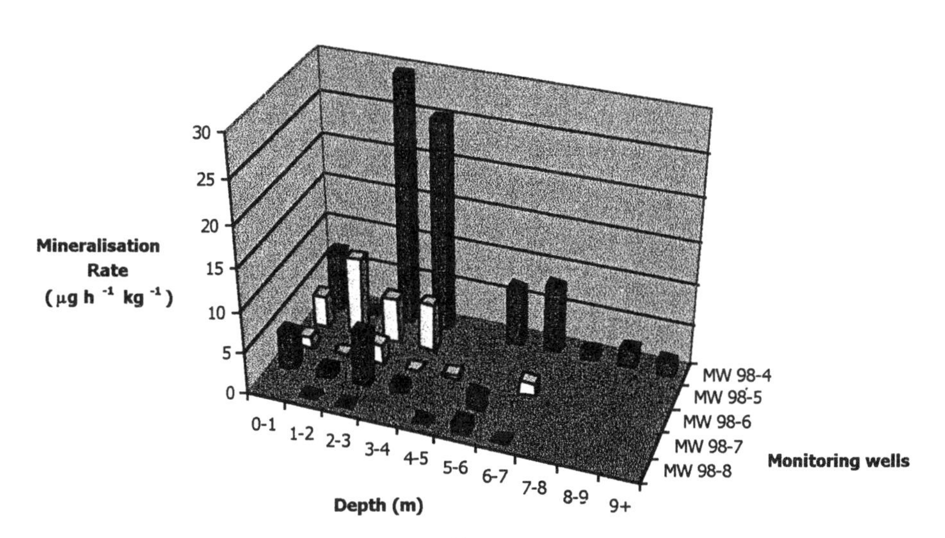

Figure 2. Mineralisation rates of ^{14}C-naphthalene derived from site sediment samples.

Mineralisation rates in the upper 4m of the sediment (Average value 5.7 µg h^{-1} kg^{-1} soil) are considerably higher than below 4m (Average rate 2.2 µg h^{-1}

kg^{-1} soil), possibly corresponding to a reduction in partial pressure of oxygen. Interpretation of mineralisation rates requires care, the presence of endogenous PAHs may or may not act to dilute the radioisotope (dilution effect) depending on their bioavailability.

A dilution experiment was conducted with two sediments as outlined earlier, sediment with no endogenous PAHs produced a corresponding 72% reduction in mineralisation rate with a 75% diluted isotope, whilst sediment with endogenous PAHs showed a reduction in mineralistion rate of only 2.5% after a 75% isotope dilution (Table 1). Consequently, the actual mineralisation rates may be double the observed values, but more significntly, the endogenous PAHs in MW 98-5 appeared bioavailable, despite having been present in the soil for between 50-150 years.

Table 1. Dilution effect of endogenous naphthalene on ^{14}C-naphthalene mineralisation rates ($\mu g\ h^{-1}\ kg^{-1}$ soil, % of maxima).

Sample	Undiluted mineralisation rates	Diluted mineralisation rates
MW 98-4 <1mg kg^{-1} naphthalene	11.5 (100%)	3.2 (28%)
MW 98-5 9mg kg^{-1} naphthalene	4.0 (100%)	3.9 (98%)

The effect of temperature was investigated and demonstrated no significant difference over a 22°C temperature range (8-30°C data not shown), suggesting PAH degradation in this sediment may be conducted by a microbial consortia with a broad temperature affinity range.

Ks (affinity constants) for both naphthalene and phenanthrene substrates in soil-based systems are currently being derived, to allow for greater model sophistication in definition of D_{PAH} and c_{PAH} terms.

REFERENCES.

Fleming, J. T., J. Sansevenio, and G. Sayler. 1993. Quantitative relationship between naphthalene catabolic gene frequency and expression in predicting PAH degradation in soils at town gas manufacturing sites. Environmental Science and Technology. **27:**1068-1074.

Wild, S. R., and K. C. Jones. 1995. Polynuclear aromatic hydrocarbons in the UK environment: A preliminary source inventory and budget. Environmental Pollution. **88**:91-108.

CASE HISTORY: RECLAMATION OF A FORMER MANUFACTURING GAS PLANT

C. Di Leo, A. Barbaro, A. Bittoni, P. Colapietro, F. Fabiani, A. Robertiello
(EniTecnologie, Monterotondo, Italy)

ABSTRACT: Site assessment was performed on 36 hectares of a former MGP. Evidences of mono- and polycyclic aromatic hydrocarbons, phenols and cyanides were found. About 30% of the total surface of the site was also contaminated by pyrite ash containing lead and arsenic. Laboratory tests were performed to assess the feasibility of organic pollutants biodegradation. Leaching tests, acute toxicity test and ecotoxicological studies were performed on surface soils containing arsenic and lead. On the basis of the laboratory test results a reclamation plan was designed and realized. Reclamation plan was based on three different kinds of actions: controlled dumping of small quantities of unsaturated soil, bioremediation of aromatics contaminated areas, using indigenous bacterial flora and securing of arsenic and lead contaminated areas, by surface decortication and transferring the soil to a smaller area; soil was then covered with a layer of asphalt.

SITE ASSESSMENT

The site, about 36 hectares, is the former MGP of the city of Avenza, located in the industrial area of Massa (Tuscany). The activity of the plants was discontinued in the seventies.

The soil is a highly permeable sand (permeability: 10^{-2}-10^{-3} cm/sec), the water table was about at 3 m below the surface.

For site assessment about 650 cores, to a maximum depth of 10 m, for a total of 1100 soil samples were realized. Samples were analized for organics and heavy metals to map sites topological distribution. On contaminated samples, pH, redox potential, conductivity, phosphate and ammonium availability were measured and microbiological analysis (total eterotrophic and hydrocarbon oxidizing microorganisms counts) were carried out.

PAH (polycyclic aromatic hydrocarbons) and TPH (total petroleum hydrocarbons) were found above limits (according to the guidelines of the Tuscany province) in both vadose and saturated zones. VOCs (volatile organic carbon), phenols and cyanides were found above limits only in some vadose areas, easily removable by scraper.

30 % of the total area was contaminated with lead and arsenic, from pyrite ash, used as filler material during the plant construction.

Microorganisms found in the contaminated areas ranged from 10^3 to 10^6 UFC/g of soil. The ratio: total eterotrophic/hydrocarbon oxidizing microorganisms was about 1, showing that almost all the microorganisms of the contaminated areas belonged to the hydrocarbon oxidizing group.

Subsoil environmental conditions (pH, redox potential and conductivity) were suitable for microbial life; the availability of phosphorous and oxygen appears to be the growth limiting factor.

LABORATORY FEASIBILITY STUDIES

Laboratory tests were performed on some samples to assess the feasibility of bioremediation approach for organic pollutants degradation in saturated soil. Tests were based on the degradative activity of autochthonous microorganisms mesured with respirometers, microcosms and simulation apparata.

The respiration tests were performed in 500 ml glass bottles, containing 50 g of contaminated soil, with 75% of moisture content and adding nitrogen and phosphorous sources. The bottles were connected to the Micro-Oxymax respirometer. The experiment was carried out at room temperature for about 1 month.

Glass pots of 1 l volume, containing 100 g of contaminated soil and 200 ml of colture medium (either water or one of two different saline colture media: in one case tripolyphosphate as a slow release phosphorous source was used) were used. Microcosms were hold at room temperature in orbital shaker.

Biosparging was simulated in 1 l glass cylinders, filled with about 2 kg of contaminated soil. Air spargers were placed at the bottom of the cylinders. The soil was saturated using a saline medium containing a suitable phosphorous source. The cylinders were mantained at room temperature.

Soils samples were taken at the beginning and at the end of the experiment for the respiration test. Samples were also taken periodically to perform chemical and microbiological analysis in microcosms and in simulation apparata.

On arsenic and lead contaminated soils chemical-physical leaching tests, with CO_2 saturated solutions and acute toxicity test, utilizing the Microtox, were carried out. Plant uptake studies were performed to evaluate metal bioaccumulation in plants, through the roots activity.

RESULTS

The results of the tests showed a very good biological abatement of the organic contaminants in the soil indicating the possibility of biological reclamation of the polluted areas.

Respiration tests. The respirometric data showed the presence, in the polluted soils, of an hydrocarbon oxidizing autochthonous microflora having the capability of utilizing the contaminants present in the soil as carbon and energy source. Moreover, the toxicity of the contaminants on the microorganisms was negligible. In fact, the kinetics of contaminants degradation for two soil samples, having one PAH contamination double than the other, was similar and without any lag period (Figure 1).

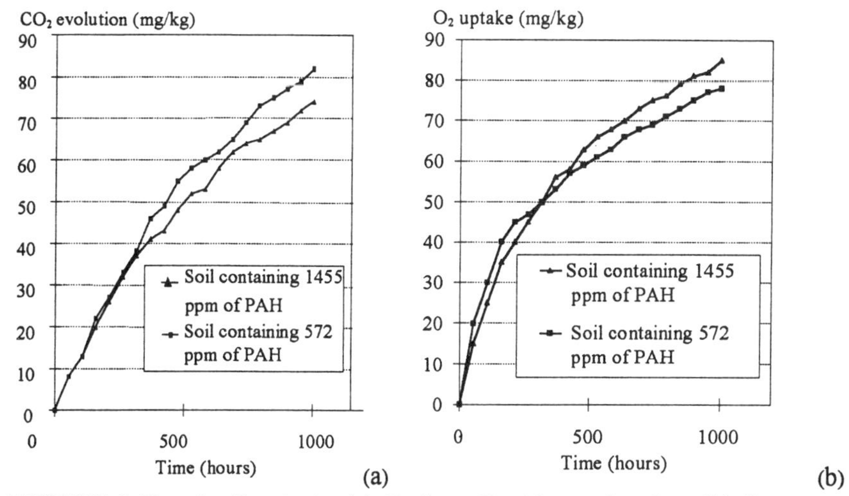

FIGURE 1. Respiration tests. (a) Carbon dioxide production. (b) Oxygen consumption.

Microcosms. The experiments with the microcosms confirmed the degradative capabilities of the autochthonous microflora. After 14 days, total abatement (final concentration less than 1 ppm) of PAH was found when the initial concentration was of 325 ppm. 70-80% of abatement was measured when PAH were initially present at 1648 and 1221 ppm. The addition of nutrients and, in particular, of tripolyphosphate as additional phosphorous source was proved to be very useful and suitable to be used in field work.

Simulation of the biosparging. The experiment showed the abatement of PAH down to and below 200 ppm (the law limit in Tuscany for industrial soils) in about one month for light contaminated soils (572 ppm) and in 40 days for more heavily contaminated soils (1455 ppm).

Arsenic and lead contaminated soils. The results of chemical-physical leaching tests have shown that the slags are very stable; no leached metal was measured in the test. Hydrogeological investigation has verified that pyrite ashes don't contaminate the soils and the waters below.

No acute toxicity effects were shown by samples in the toxicological tests.

The ecotoxicological studies showed lack of bioaccumulation of metals in plants grown in presence of the pyrite ash.

THE REMEDIATION PLAN

Based on the evidence gathered, a reclamation plan was designed and implemented, based on three different kinds of actions:

- controlled dumping of small quantities of unsaturated soil (about 3000 m^3)
- bioremediation (biosparging), using autochthonous bacteria, of 4 areas (indicated as A, C, D and F, respectively) contaminated mainly by aromatics

for a total of about 28.000 m^3 of saturated sandy soil (A: 225 m^2, C: 400 m^2, D: 100 m^2 and F: 2112 m^2; depht 10 m)

- securing of the arsenic and lead contaminated area, by surface decortication and transferring of the polluted soil (about 30000 m^3) on a smaller area. The area was then covered with a layer of asphalt to eliminate the risk of wind transportation.

THE BIOSPARGING PLANT

Bioremediation action consisted of building four plants for injecting oxygen into the soil, together with the nutrients necessary for the aerobic microbial flora to accelerate biodegradation of the contaminants.

The biosparging plants were built taking into account the hydrogeological features of the site, the distribution of pollutants and the results of laboratory experiments.

Numerous air and nutrient injection points were digged in the areas to be treated. Nutrients were added at two different levels (Figure 2).

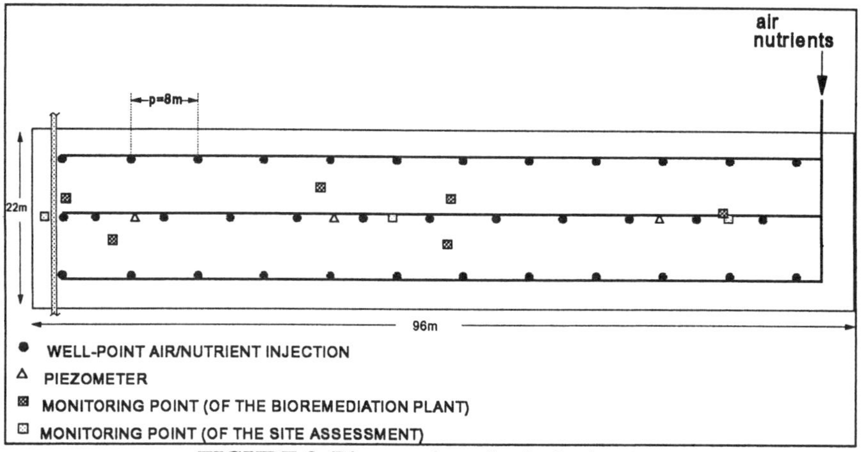

FIGURE 2. Biosparging plant planimetry

Soil sampling and piezometer monitoring points were established in each of the four treated areas. During treatment, soil and water were sampled to evaluate the decontamination process.

Figure 3 shows the PAH abatement in the soil in one of the four treated areas.

Moreover, the VOC present in the ground water, below the contaminated zones, where also reduced by bioremediation activities (Table 1).

CONCLUSIONS

Site assessment of a former MGP and laboratory feasibility studies showed the possibility of biological reclamation of the saturated soils contaminated essentially by PAH. Reclamation was performed building four biosparging plants.

The bioremediation activities were followed in each plant by soil and water monitoring points; bioremediation yielded complete recovery of the treated zones.

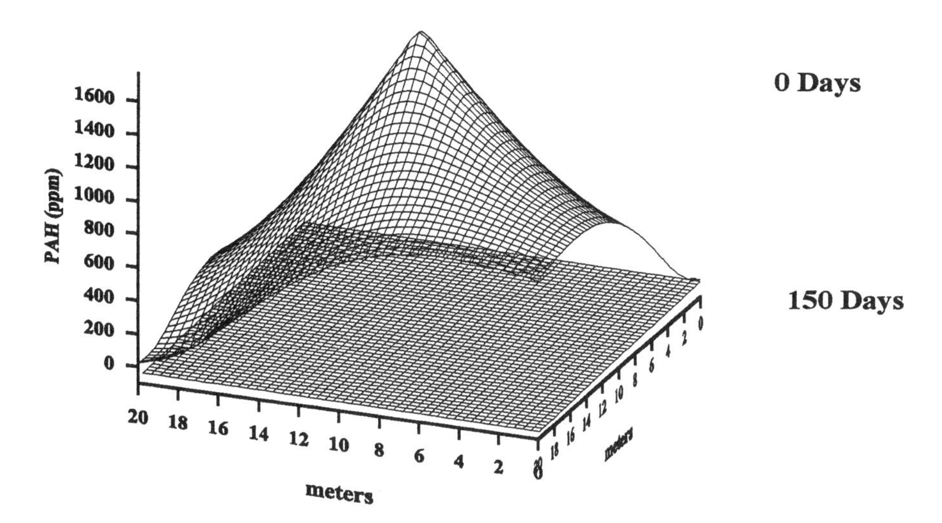

FIGURE 3. Area C biosparging plant: PAH abatement.

TABLE 1. Area F biosparging plant: groundwater remediation (ppm)

Time (days) and Piezometer	Benzene	Toluene	Ethyl-benzene	p- and m-Xylene	o-Xylene
0 Pz1	4,6	1,6	<0,05	0,1	<0,05
30 Pz1	<0,05	<0,05	<0,05	<0,05	<0,05
70 Pz1	0,1	<0,05	<0,05	<0,05	<0,05
115 Pz1	0,2	<0,05	<0,05	<0,05	<0,05
0 Pz2	0,1	<0,05	<0,05	0,2	0,1
30 Pz2	<0,05	<0,05	<0,05	<0,05	<0,05
70 Pz2	<0,05	<0,05	<0,05	0,4	<0,05
115 Pz2	<0,05	<0,05	<0,05	<0,05	<0,05
0 Pz3	0,4	0,3	0,2	2,4	0,9
30 Pz3	<0,05	1,2	<0,05	<0,05	<0,05
70 Pz3	0,5	0,1	0,1	0,8	<0,05
115 Pz3	0,3	<0,05	0,1	0,2	<0,05

Soil remediation activities began at the end of 1995 and ended in May 1998 with a total cost of about 2.25 millions of dollars, excluded the cost of securing the metal contaminated area.

ANALYSIS OF PAHs AND METABOLITES DURING TREATMENT OF CONTAMINATED SOIL

Staffan Lundstedt, Peter Haglund, Bert van Bavel and Lars Öberg
(Umeå University, Umeå, Sweden)

ABSTRACT: Heavily contaminated gasworks soil was treated in a pilot-scale bio-slurry reactor. Samples were taken and analyzed for more than 30 PAHs and several alkylated and heterocyclic PAHs in order to monitor the degradation process. In addition, some PAH-ketons and PAH-quinones were analyzed as possible degradation products. The total concentration of the monitored contaminants in the soil was reduced to 50% of the initial concentration during the treatment. However, the concentration of some of the low-molecular weight PAHs was reduced to 95%, while the concentration of the high-molecular weight PAHs (5 and 6 rings) was not reduced at all. The concentrations of the methylated and heterocyclic PAHs decreased to the same extent as the corresponding PAHs of the same size. No build up of metabolites was observed during the treatment. However, three of the studied metabolites, anthrone, 7H-benz(de)anthrone and 1,2-benzanthraquinone, were found at a rather constant concentration. The concentrations of two other, 9-fluorenone and 9,10-anthraquinone, were decreasing significantly during the treatment. In order to evaluate if any of the contaminants were partitioned from the soil into the water-phase instead of being degraded, the water from two slurry samples was analyzed. The fraction of the contaminants found in the water was, however, very low compared to the fraction found in the soil. Moreover, the concentration in the water-phase was higher early during the process than during the final stage of the process, indicating that the contaminants are more easily degraded once they have partitioned into the water-phase.

INTRODUCTION

During remediation of contaminated soil it is always necessary with a careful monitoring, to verify that the concentration of the contaminants decreases. The more contaminants that are analyzed the more accurate is the monitoring. However, to save time and money, often only a limited number of compounds are analyzed. The remediation is considered to be successful when the concentration of these compounds has reached a desired level. Sometimes, however, the remediation may lead only to partial transformation of the contaminants, and metabolites are formed that can be as toxic as or even more toxic than the parent compounds (Dubourguier *et al.* 1997). This potentially may create a situation worse than the original. It is known, for example, that PAHs are microbially degraded to phenols, diols and quinones, which in some cases may lead to increased toxicity. In addition some microorganisms are able to methylate PAHs, which drastically increases their carcinogenicity (Cerniglia 1984). The problem is also complicated by the altered polarity of the metabolites, which may lead to increased mobility and risk for elevated exposure. The metabolites may also undergo polymerization, or conjugation to humic matter, leading to decreased bioavailability (Wischmann *et al.*1997).

Objective. The objective of this study was to identify any changes in the contaminant pattern during biological treatment of PAH-contaminated soil. The two main questions to be answered were; 1) if the contaminants behaved differently during the biological treatment, 2) if metabolites were formed and accumulated during the process. In order to answer these questions, two soil samples from a bio-slurry process were taken for characterization, which included tentative identification of a large number compounds. Their contaminant patterns were then compared. Second, five soil samples from the process were taken for quantitative determination of several of the identified compounds. Differences in degradation rate were studied. Third, to study if contaminants were partitioned from the soil into the water-phase of the slurry during the treatment, the water from two samples were also investigated.

MATERIALS AND METHODS

The soil in this study was taken from a gasworks site in Stockholm, Sweden. One cubic meter of this soil was treated for 29 days in a pilot-scale bio-slurry reactor in cooperation with a Swedish bioremediation company. Daily samples were taken during the process for the monitoring of PAH-degradation. A small number of these samples were chosen for this study. One soil sample from the first day (Day 0) and one soil sample from after three days treatment (Day 3) were taken for characterization. Five soil samples, Day 0, Day 3, Day 8, Day 24 and Day 29, were taken for contaminant quantification. In two of these samples, Day 0 and Day 24, also the water-phase of the slurry was analyzed.

Sample pre-treatment. The soil was separated from the water by simultaneous filtration through a 0.45 μm polyamide filter (Sartolon, Sartorius) and a solid-phase extraction disk (Envi C-18 Disk, Supelco). The dissolved contaminants were in this way adsorbed to the adsorbent disk. The filtrated soil and the disk were both air-dried. The dry soil was then ground in a mortar before extraction.

Soil extraction. The soil was extracted by using Accelerated Solvent Extraction (ASE 200, Dionex, UK). One-gram dry soil was weighed into the extraction cell, which was then filled with anhydrous sodium sulfate. Each sample was extracted with two solvents, toluene and toluene/methanol (1:1). The extractions were carried out at 120°C and 136 atm, with 6 min dynamic and 5 min static extraction, two cycles for each solvent. The extracts were combined and approximately one tenth was taken out for further clean up. This fraction was spiked with internal standard consisting of perdeuterated PAHs and then evaporated to a small volume.

Disk extraction. The solid phase extraction disks were also extracted using ASE. The procedure was the same as described for the soil samples, with a few minor exceptions. The solvents used were 100 % methanol followed by 100% toluene, only one cycle for each solvent. Half of the extract was taken out and spiked for further clean up.

Fractionation. The spiked extracts were fractionated on a column consisting of 10% deactivated silica gel. Five fractions (1. hexane, 2. hexane, 3. hexane/methylene chloride (3:1), 4. methylene chloride, 5. methanol) were collected during the initial characterization. For the quantification the three first fractions were collected and combined. This combined fraction was evaporated and toluene was added before analysis.

Analysis. The samples were analyzed by HRGC/LRMS. The GC (Fisons GC 8000, 60 m DB-5 capillary column, 0.32 mm i.d., 0.25 µm film from J&W Scientific, CA) was operated in splitless mode. The MS (Fisons MD 800) was operated in fullscan mode during the characterization and in SIR-mode during the quantification. The different compounds were identified according to retention time and mass spectra. The quantification was carried out by using a reference standard solution containing all the compounds listed as quantified in Table 1, and a internal standard solution containing perdeuterated PAHs.

RESULTS AND DISCUSSION

Soil characterization. Several PAHs, alkylated PAHs, heterocyclic PAHs, PAH-ketons, PAH-quinones and cyano-PAHs were tentatively identified in the soil (Table 1). No significant differences in the contaminant pattern between Day 0 and Day 3 could be observed, except for some cyano-PAHs that were found in the Day 0 sample but that could not be found in the Day 3 sample.

Quantification study. Of the identified substances 19 PAHs (including biphenyl), five methylated PAHs, three PAH heterocyclics and five PAH-ketons/quinones were quantitatively determined (Table 1). The original concentration of the 19 PAHs was 2090 mg/kg totally. This was reduced to 1340 mg/kg during the 29 days treatment. The degradation of the 2-3 ringed PAHs started immediately, and after 29 days only 5-40% could be found in the soil (Fig 1 and 2). The concentration of the 4-ringed PAHs was unaffected the first week, but was then reduced to 40-70% of the initial concentration (Fig 2). The concentration of the 5-6 ringed PAHs was slightly increasing during the process (Fig 3). This could be due to an increased availability for extraction during the treatment, as a consequence of the grinding effect and degradation of matrix substances, or due to fluctuations during sampling and analysis.

The concentrations of the five methylated PAHs and the three PAH-heterocyclics were reduced to 10-35% of their initial concentrations during the process. They seem to behave just like their corresponding PAHs of the same size (Fig 4 and 5). It should be noted, though, that only 2- and 3-ringed methylated PAHs were quantified in this study. Larger methylated PAHs are expected to be more complicated to degrade (Cerniglia 1984).

The five PAH-metabolites showed different behavior in the bio-slurry reactor. 9-Fluorenone and 9,10-anthraquinone behaved like the other 2- and 3- ringed compounds, and their concentration decreased to 25% of the initial concentration during the 29 days. However, the concentration of the other three metabolites, anthrone, 7H-benz(de)anthrone and 1,2-benzanthraquinone, were unaffected of the treatment. The formation and degradation of these compounds seemed to be in equilibrium. As can be seen, no build up of the studied metabolites was observed during the process. However, the risk for metabolite accumulation may be greater when the high molecular weight PAHs are degraded, which was not the case during this bio-slurry treatment.

TABLE 1. Quantified (mg/kg d.w.) and tentatively identified compounds in the PAH-contaminated soil that was treated in a pilot-scale bio-slurry reactor.

Substance		Conc. Day 0	Substance		Conc. Day 0
PAHs	**Quantified**	**mg/kg**	**Alkyl-PAHs**	**Quantified**	**mg/kg**
Naphthalene		32	2-Methylnaphthalene		14
Biphenyl		5.8	1-Methylnaphthalene		8.8
Acenaphthylene		23	2,6-Dimethylnaphthalene		11
Acenaphthene		3.2	2,3,5-Trimethylnaphthalene		2.4
Fluorene		58	1-Methylphenanthrene		6.1
Phenanthrene		303		**Identified**	
Anthracene		78	C2-naphthalenes		
Fluoranthene		381	C3-naphthalenes		
Pyrene		257	Methylbiphenyls		
Benzo(a)anthracene		180	C2-biphenyls		
Chrysene		175	C3-biphenyls		
Benzo(b)fluoranthene		119	Methylfluorenes		
Benzo(k)fluoranthene		110	Methylphenanthrenes		
Benzo(e)pyrene		83	Methylanthracenes		
Benzo(a)pyrene		96	C2-Phenanthrene		
Perylene		27	C2-Anthracene		
Indeno(cd)pyrene		74	Methylfluoranthene		
Dibenz(ac)anthracene		24	Methylpyrene		
Benzo(ghi)perylene		64	Methylchrysenes		
	Identified		Methylbenzo(a)anthracenes		
1-Phenylnaphthalene			C2-Chrysenes		
2-Phenylnaphthalene			C2-Benzo(a)anthracenes		
Dihydrocyclopenta(def)phenanthrene			Cyanonaphthalenes		
Dihydrofluoranthene			Cyanoanthracene		
Benzo(c)phenanthrene			Cyanofluorene		
Triphenylene					
Naphthacene			**PAH-heterocyclics**	**Quantified**	**mg/kg**
Acepyrene			Dibenzofuran		40
Benzo(ghi)fluoranthene			Dibenzothiophene		20
Cyclopenta(cd)pyrene			Carbazole		43
Binaphthalene				**Identified**	
Benzo(j)fluoranthene			Methyldibenzofurans		
			Xanthene		
Metabolites	**Quantified**	**mg/kg**	Methyldibenzothiophene		
9-Fluorenone		25	Benzo(b)naphtho(2,1-D)thiophene		
Anthrone		0.7	Benzo(b)naphtho(1,2-D)thiophene		
9,10-Anthraquinone		22	Benzo(b)naphtho(2,3-D)thiophene		
7H-Benz(de)anthracene-7-one		6.2	Methylcarbazoles		
Benz(a)anthracene-7,12-dione		7.7	Benzo(b)naphtho(2,3-D)furan		
	Identified		Benzo(c)carbazole		
Methylfluorenones			Benzo(a)carbazole		
Benzofluorenone			Benzoquinoline		
Cyclopenta(def)phenanthrene-4-one			Acridine		
5,12-Naphthacenedione			Indenoquinoline		
Dibenzanthracenequinone			Benzo(a)acridine		
Fluorene-9-imine			Naphthalimide		

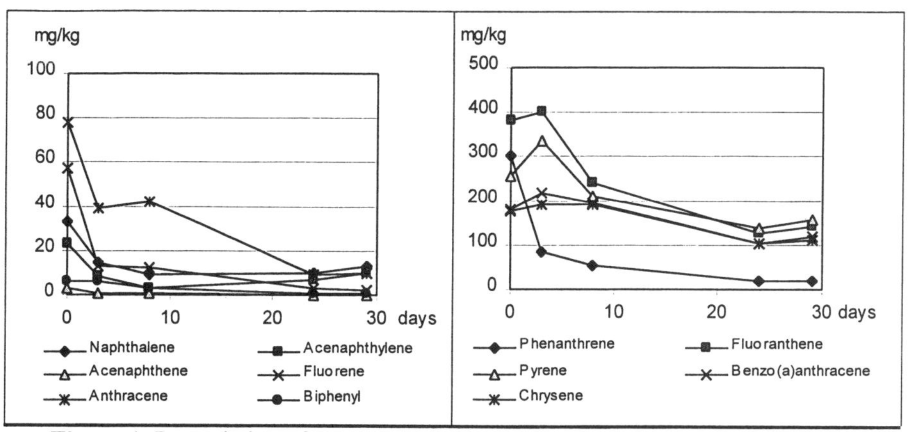

Figure 1. Degradation of 2- and 3-ringed PAHs

Figure 2. Degradation of 3- and 4-ringed PAHs

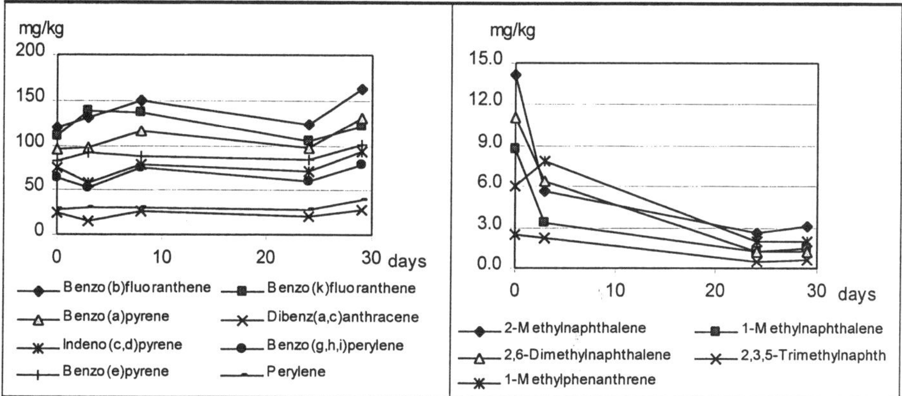

Figure 3. Degradation of the 5- and 6-ringed PAHs

Figure 4. Degradation of the methylated PAHs

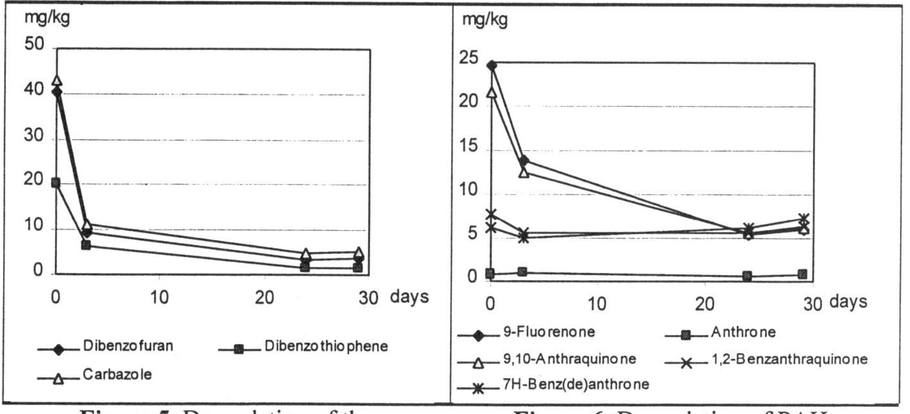

Figure 5. Degradation of the heterocyclic PAHs

Figure 6. Degradation of PAH-metabolites

Analysis of the water-phase. The fraction of the contaminants that were found in the water-phase was very small compared to the fraction found in the soil, as presented in Table 2. Moreover, the concentration in the water-phase was higher when the reactor was started up than it was at the final stage of the treatment. It is obvious that the studied soil contaminants, including metabolites, are more easily degraded once they have partitioned into the water phase.

TABLE 2. Dissolved fraction of the contaminants in the soil-slurry reactor, calculated as water concentration / soil concentration.

Substance	Fraction in water phase compared to the soil	
	Day 0	Day 24
PAHs (2-4 rings)	0.05 %	0.008 %
PAHs (5-6 rings)	0.001 %	0.001 %
Methyl-PAHs	0.09 %	0.01 %
PAH-heterocyclics	0.04 %	0.004 %
PAH-metabolites	0.04 %	0.005 %

CONCLUSIONS

Several PAHs, methylated PAHs, heterocyclic PAHs and PAH-metabolites were identified in the slurry samples, taken during the biological treatment. The concentration of the low-molecular weight PAHs, the methylated PAHs, the heterocyclic PAHs and two metabolites decreased to 5-70 % of their initial concentration during the treatment. The concentration of the high molecular weight PAHs and three metabolites were, however, unaffected. Small amounts of all contaminants were found in the water-phase of the slurry, compared to the amounts found in the soil. Moreover, the concentration in the water-phase was higher when the reactor was started up than during the final stage of the treatment.

ACKNOWLEDGEMENTS

This study was carried out within the Swedish national research program COLDREM (Soil remediation in a cold climate). The authors wish to acknowledge MISTRA (Foundation for Strategic Environmental Research) for the financial support. EkoTec is acknowledged for providing facilities and technical expertise.

REFERENCES

Cerniglia, C. E. 1984. "Microbial Metabolism of Polycyclic Aromatic Hydrocarbons" *Adv. App. Microbiol.* 30:31-71.

Dubourguier H. C., Duval M. N., Cazier F., Wijjfels P., 1997. "Bioremediation of Polycyclic Aromatic Hydrocarbons from Former Coal Industries." *In situ and On-site Bioremediation,* Vol. 2, 4(2): 79-84.

Wischmann H., Steinhart H., 1997. "The Formation of PAH Oxidation Products in Soil and Soil/Compost Mixtures. *Chemosphere* 35(8): 1681-1698.

TRANSFORMATIONS IN THE SAND FILTRATION STEP OF A GROUNDWATER TREATMENT PLANT SITUATED ON A FORMER GASWORKS SITE

Michael Reiners, Joachim Fettig, Hans-Peter Rohns, Jürgen Schubert, Paul Eckert, Stefan Kamphausen, and ***Ralf Schramedei***

Besides the desired retention of iron and manganese also further reactions such as the microbial degradation of aromatics and a nitrification occur in the sand filtration step of a groundwater treatment plant. This could be shown during the biotechnological on-site groundwater treatment at a former gasworks site. After an almost 100% microbial degradation of the aromatic pollutants the re-infiltrated water is of perfect chemical quality - even before it passes the activated carbon filtration. The results presented here, allow for a further optimization of the plant - also with regard to economical aspects.

The sand filtration step is employed in the treatment of contaminated groundwater because it allows for the elimination of iron oxides and bacteria. This step ensures the continuous operation of the entire treatment plant, especially of the subsequent activated carbon filtration in which the remaining pollutants are absorbed. Under specific circumstances, manganese ions will also be retained in these sandfilters.

Investigations carried out within the scope of a diploma thesis on groundwater treatment showed that the sand filtration step of a biotechnological on site remediation plant did not only meet the requirements regarding iron and biomass retention but also enhanced the purification efficiency [1]. It could be shown that sand filtration does not only meet the requirements concerning the iron and biomass separation but also enhances the purification efficiency by means of different – at least microbially supported – processes. In this context, the manganese oxidation and its interactions with other processes taking place in the sandfilter is of particular importance and will be more closely examined below.

During 1890 and 1968 city gas was generated on the site of the former gaswork Düsseldorf-Flingern. Due to improper handling and destructions owing to war BTEX (destruction of the plant during WWII) aromatics and polycyclic aromatic hydrocarbons (PAH) were introduced into the groundwater forming a circa 500m long and 100m wide plume in the area of the former benzene plant [2,3,4]. Since December 1995, Stadtwerke Düsseldorf AG operates a groundwater remediation plant on this former gasworks site (fig.1).

The polluted groundwater is pumped out of two wells into a bio-reactor where the major part of the aromatic pollutants are eliminated by aerobic microorganisms. The biologically treated water is then fed via an intermediate reservoir into the sandfilter. In the final two-stage activated carbon filtration process, the remaining pollutants are eliminated. The purified water is infiltrated into the contaminated aquifer and external electron acceptors are added to enhance microbial degradation(nitrate, oxygen).

Purification capacity of the remediation/treatment plant

Fig 2 shows the mean concentrations of mono and polycyclic aromatics during the treatment process. On the average, 90% of the BTEX and 80% of the PAH (mostly naphthalene and acenaphthene) are transformed in the bio-reactor under aerobic conditions. During the subsequent sand filtration the iron and manganese compounds are retained and the residual aromatic concentrations are almost reduced down to the limit of detection (LOD). Consequently the activated carbon filters are only loaded with very low pollutant concentrations. This finding led to further investigations into the transformations occurring in the sand filtration step.

Fig. 1 + 2

Fig. 3 + Tab. 1

Transformations in the sandfilter step

The sandfilter stage of the groundwater treatment plant consists of two pressure filters operated in parallel with a volume of $20m^3$ each. The maximum filtration rate at the current pump rate (ca. $12m^3/h$) is ca. 1.7m/h. Thus the current filtration rate falls somewhere in between the range of slow sand filtration (v = 0,05 to 0.5m/h) and fast filtration (v = 3 to 25m/h).

Focusing on the most important concentrations in the filter influent and effluent Fig. 3 gives a first impression of the manifold processes taking place in the sandfilter.

Deferrization

By means of pressure aeration in the bio-reactor the soluble iron (II) ions are oxidated to ferric iron. The following hydrolysis and flocculation forming hydrated ferric oxide also occurs in the bio-reactor so that the iron flocs only have to be mechanically filtered out in the sandfilter step. Experience shows that the iron concentrations in the sandfilter are reduced to the detection limit (0.07mg/l) when using a filter sand of 2 to 3.15mm in diameter.

Demanganisation

In contrast to the chemical oxidation of iron, manganese ions can only be oxidised microbially in the treatment plant [1,5]. Fig. 4 graphically represents the manganese concentrations found over the entire treatment process.

During the remediation phase described the manganese concentration amounts to ca. 1.5mg/l. In March 1996, manganese concentrations in the range of the LOD (0.03mg/l) were found in the sandfilter effluent for the first time. At the beginning of the groundwater treatment the hydrated ferric oxide flocs mostly remained in the bio-reactor system and the intermediate reservoir. Consequently, the sandfilter was only slightly loaded and did not require backwashing.

Until the first backwashing in April 1996, the manganese (II) ions in the filter-bed were increasingly oxidated microbially and retained as hydrated oxide. The high shear forces arising during the backwashing of the filter impaired the microbial activity in the sandfilter. So, it was not until October 1996 that manganese could be retained in the filter again. This means that the demanganizing microorganisms needed six months to adapt to the prevailing conditions (i.e. high shear forces during the filter flushings, pH value, oxygen supply, redox potential (ORP)). Once the adaptation had taken place, the demanganisation reached a satisfactory level.

Aromatics

The load of mono and polycyclic aromatic compounds in the bio-reactor effluent (ca. 200µg/l BTEX and 100mg/l PAH) was mostly eliminated in the sandfilter. Analyses carried out on the sandfilter back-wash sludge and water showed no significant, detectable contamination. Therefore, it is to be assumed that the pollutants in the sandfilter are microbially mineralised or transformed into aliphatic metabolites.

The correlation between the microbial degradation of the contaminants and manganese oxidation was particularly significant/obvious during the period from July 10th until July 23rd, 1997, during which heavily loaded groundwater stemming/originating from a tracer test was temporarily fed directly into the intermediate reservoir (fig. 5). This feeding of pollutants into the intermediate reservoir caused an abrupt rise of the BTEX concentration in the sandfilter influent to 2500 mg/L.

As the microbial consortium in the sandfilter could not process the heavy load of pollutants, the concentration in the effluent increased as well.

(The microorganisms in the sandfilter could not fully receive this shock load so that the concentrations in the filter's efflux also rose) Within a few days the microorganisms adapted to the new conditions and transformed the BTEX loads.

However, the rise in microbial degradation activity resulted in a shortage of dissolved oxygen required for the microbial manganese oxidation. Due to this, no manganese ions could be retained in the filtration step.

Phosphorus and Nitrogen

The sandfilter influent normally shows a phosphate concentration of ca. 0.7mg/l. The major share of this phosphate amount is held back by the iron flocs sedimenting in the filter-bed so that the concentrations in the sandfilter effluent fall significantly below 0.1mg/l. When the filter is flushed the phosphates adsorbed are washed out of the sandfilter system.

The inorganic nitrogen concentrations in the sandfilter influent mainly consist of 1.5mg/l ammonia and 1.5mg/l nitrate. During the filter passage, the major part of the ammonia compounds is transformed into nitrates so that the filter effluent shows concentrations of 0.1mg/l ammonia and 6.5mg/l nitrates. The overall nitrification may be described as follows:

$$NH_4^+ + 2O_2 \rightarrow NO_3^- + H_2O + 2H^+$$

The oxidation of 1 mg ammonia requires 3.3mg oxygen. The relation between the amount of nitrified ammonia and the oxygen or manganese concentration respectively may be taken from fig. 6.

Until December 1996, the difference between the ammonia concentrations in the sandfilter influent and effluent was about 1.2mg/l on the average. Up to then, relatively low oxygen concentrations were measured in the sandfilter effluent (ca. 1.5mg/l). From February 1997 on, the concentrations recorded during the sandfilter passage fell to just 1mg/l. During the same period, the oxygen concentrations in the filter effluent reached significantly higher levels (ca. 3.5mg/l). This illustrates the direct correlation of the oxygen concentrations measured to the ammonia turnover in the filter. In this context, the fact that the manganese oxidation only sets in at a low nitrification capacity underlines the assumption that an adequate oxygen level (> 1mg/l) is crucial for the microbial demanaganisation.

Oxygen

In the sandfilter numerous oxygen consuming processes occur. The oxidation of divalent manganese is only successful if all readily oxidizable substances contained in the water have already been transformed. This particularly refers to ferrous/ ferric iron and ammonia ions as well as all organic compounds [6]. Therefore, the oxygen concentration in the influent must be high enough to allow for all those redox reactions mentioned above.

The following formula may be used to describe the overall oxygen balance in the sandfilter. All concentrations are given in [mg/l].

$c(O2)$ sandfilter effluent = $c(O2)$ sandfilter influent
$-0.14 \cdot c(\text{Iron})$
$-3.08 \cdot c(\text{DOC + benzene})$
$-3.13 \cdot c(\text{toluene})$
$-3.47 \cdot c(\text{ethylbenzene + xylene})$
$-2.38 \cdot c(\text{phenole})$
$-3.00 \cdot c(\text{naphthalene + acenaphthene})$
$-3.31 \cdot c(\text{ammonia})$
$-0.29 \cdot c(\text{manganese})$

The factors given here have been derived from the stoichiometry of the respective redox reaction under application of the same oxidation factor for organic carbons (DOC) as for benzene.

Table 1 shows an exemplary oxygen balance for the period of September 1997. The mean flow rate was $10.8 m^3/h$.

The measured and calculated oxygen consumption correspond with good approximation so that these parameters provide for a sufficiently accurate description of the oxygen depletion taking place in the sandfilter. Furthermore, the balance indicates that the higher oxygen consumption is caused by the nitrification.

In order to ensure the oxidation of the reduced substances the oxygen content of the effluent should be above 1mg/l [5]. This means that during September 1997 ca. 2.5mg oxygen/l remained for further oxide formations oxidation reactions, i.e. without impairing the treatment targets, the sandfilter could have been theoretically loaded with further 1mg/l BTEX aromatics, 0.2mg/l PAH and 0.05mg/l phenols provided an otherwise stable environment.

Conclusion

The description of the chemical, physical and biological transformations in the sandfilter of the groundwater treatment plant showed that in addition to the retention of iron and manganese also further reactions such as the microbial degradation of aromatics and a nitrification occur.

With respect to the filtration step under examination, the microbial degradation of contaminants is of particular practical interest. At low concentrations of aromatics in the filter influent the biotechnological treatment should be modified in a way allowing for a reduction of the specific overall costs.

References

[1] Reiners, M. (1997): Untersuchungen zur Optimierung der Verfahrensstufe Kiesfiltration bei der Grundwassersanierung auf einem ehemaligen Gaswerksgelände; Diplomarbeit an der Universität-Gesamthochschule Paderborn, Abteilung Höxter, Fachbereich 8 Technischer Umweltschutz.

[2] Rohns, H. P.; Dörk, B; Eckert, P.; Raphael, T.(1995): Zur Wirtschaftlichkeit einer biotechnologischen On-site-Sanierung von Grundwasser; Terra Tech 4, Nr. 4, S. 57-59

[3] Wisotzky, Eckert, P.(1997): Sulfat-dominierter BTEX-Abbau im Grundwasser eines ehemaligen Gaswerksstandortes; Grundwasser - Zeitschrift der Fachsektion Hydrogeologie 1/1997

[4] Jütte, Tillmanns: Gefährdungsabschätzung für den Altstandort 12.03 „Ehem. Gaskokerei Flingern"; Internes Gutachten der Stadtwerke Düsseldorf AG

[5] Bohm, L.(1992): Optimierung der chemikalienlosen Entmanganungsfiltration; Dissertation an der Technischen Universität Dresden.

[6] Bernhard, H.(1996): Vorlesungsskript Trinkwasseraufbereitung 1995/96; Eigenverlag im Wahnbachtalsperrenverband, Siegburg.

PILOT SCALE DESIGN OF IGT'S MGP-REM PROCESS FOR SOIL REMEDIATION

Diane Saber (Institute of Gas Technology, Des Plaines, Illinois)
Nancy Huston (Nicor Gas, Naperville, Illinois)
Larry Milner (Black and Veatch, Chicago, Illinois)
Tom Hayes (Gas Research Institute, Chicago, Illinois)

ABSTRACT: The Institute of Gas Technology developed a pilot scale design of their MGP-REM Process for remediation of Nicor Gas' Bloomington site soils. The MGP-REM Process sequentially combines chemical oxidative treatment with aerobic biological treatment. Development of the design was based upon parameters determined through a three-phase treatability protocol for technology evaluation. Each phase of testing yielded specific design parameters required for optimal technology performance. Phases I and II laboratory testing answered basic questions regarding the effectiveness of the technology under ideal conditions and focused on specific engineering parameters necessary for MGP-REM installation by simulating field conditions. A Phase III MGP-REM pilot scale program was then designed with particular goals in mind. Development of the pilot scale design resulted from integration of site specific considerations necessary for successful application of the Process. Phase III will evaluate and determine full-scale remedial design parameters, overall cost-effectiveness and potential applicability of the technology to other sites.

INTRODUCTION

Remediation of soils and groundwater contaminated with organic chemicals from current and past gas industry operations is becoming increasingly important to the U.S. natural gas industry. With co-funding from Nicor Gas, the Gas Research Institute (GRI) and IGT's Sustaining Membership Program (SMP), The Institute of Gas Technology (IGT), in cooperation with Black and Veatch LLP, developed a pilot-scale design for remediation of the Nicor Gas Bloomington site using IGT's MGP-REM Process. The pilot-scale design consisted of a three-phase treatability protocol for determining the applicability and efficacy of the technology for site remediation. Phase I rapidly and reliably characterized the soil matrix and contaminants of interest and identified achievable treatment endpoints using the specific technology, Phase II determined critical engineering parameters for applying the technology and Phase III will demonstrate technology performance at field-scale, thereby providing full-scale remedial design parameters, economics and potential applicability to other sites.

IGT developed the MGP-REM Process, a chemically-accelerated, biological process, to remediate soil, sludge and water contaminated by former manufactured gas plants (MGP). This treatment system sequentially combines two powerful and complementary remedial techniques: 1) chemical oxidative treatment using a modified Fenton's Reagent, and, 2) an aerobic biological treatment process. The steps can be applied in different sequences, depending upon the nature and degree of contamination. The MGP-REM Process is believed to an effective and cost efficient remedial process for use at MGP sites.

The technical goal of this remediation project was to render the Bloomington site soil contaminants non-leachable. This required destruction of, as much as possible, the 2-3 ring polyaromatic hydrocarbons (PAHs); solubility of the 4-6 ring PAH's is inherently very low and, therefore, highly non-leachable. Results of the Phase I and II testing showed that the MGP-REM Process was very effective in treating PAH-contaminated soils from the site. The Phase III pilot is expected to efficiently remove the leachable portion of contaminants within Bloomington site soils.

SITE DESCRIPTION

The Bloomington site is a former MGP property in the upper Midwest. It occupies approximately 4 acres of land in an area of mixed residential and industrial uses. Most of the site is covered with asphalt although some grassy areas are also present. Manufactured gas was produced at the site between 1883 and 1951 using coal carbonization and carbureted water gas processes. The soil at this site is contaminated with coal tar, contained in the sand and gravel which underlie the site and offsite properties.

Based upon data obtained during the remedial investigation, depths greater that approximately 30 feet below land surface (bls) encounter a thick clay till layer, creating an impermeable boundary across the site. A sand and gravel layer, 5 to 15 feet thick, with discontinuous silty-clay lenses was found directly above the clay till layer. The sand and gravel layer is 4 and 21 feet bls and is contaminated with varying degrees of coal tar across the entire site. The permeability of the sand and gravel layer is approximately 5.8×10^{-3} cm/sec. The clay till layer was determined to have a permeability of less than 10^{-7} cm/sec.

RESULTS OF PHASE I AND II TESTING

Phase I testing was conducted in accordance with IGT's established Standard Operating Protocols. Using Bloomington site soils, the MGP-REM Process was evaluated by small-scale laboratory tests under optimal conditions. A series of basic questions were asked, decision trees aided in the interpretation of results and the following answers were noted:

- Physical, chemical and biological characterization of Bloomington site soils indicate that the MGP-REM Process is applicable for the site.
- Desorption studies indicate that contaminants in the sand and gravel layer are not tightly bound to the soil particles and, therefore, may be relatively easy to extract and degrade (i.e., *in situ* design possible).
- Biological and chemical activity tests indicate that biological treatment is effective on two and three ring PAHs, and chemical treatment breaks down leachable four to six ring PAHs.
- Total PAHs were reduced by 76 percent using the MGP-REM Process. Two to three ring PAHs were reduced 82 percent and four to six ring PAHs were reduced 53 percent.

Based on Phase I results, IGT performed a Phase II study to further investigate the degradation of PAHs in soil columns that simulated actual field conditions. Columns were packed with soil collected from the sand and gravel layer. Free product was removed by sieving prior to initiation of the study.

The Phase II study illustrated the advantage of MGP-REM over "chemical only" or "biological only" treatment. Biological treatment alone indicated little reduction in total PAHs. Chemical treatment alone achieved only 45 percent reduction of total PAHs, with about 35 percent reduction in difficult-to-degrade four to six ring PAHs. The MGP-REM treatment columns, however, indicated up to 65 percent reduction of total PAHs, up to 60 percent reduction of four to six ring PAHs and up to 73 percent reduction in two to four ring PAHs.

Other observations applicable to pilot-scale design of the MGP-REM Process were noted during Phase II testing. These include the following:

- Site soils have a notable buffering capacity.
- Slow delivery of iron citrate is necessary to lower soil pH for chemical treatment.
- Hydraulic conductivity was not significantly reduced during iron citrate addition.
- A selected medium is suggested as a nutrient amendment during the biological treatment, and compressed air should be used as the oxygen source.
- Bridging and flow channeling due to biological growth were not observed.

The Phase II study also identified a range of operational parameters for the pilot study design.

PHASE III PILOT DESIGN

Based upon information gathered during Phase I and Phase II treatability studies, a Phase III pilot design was prepared for the Bloomington site. Phase III pilot study testing models full-scale design, with increased emphasis on parameters which will most likely affect technology performance.

Careful consideration was paid to selection of the pilot study test area, as this is a critical parameter to the success of the full-scale remediation effort. A variety of considerations were applied in selecting a suitable pilot-study area. The selected area was chosen because of the confirmed presence of two coal tar contaminated sand and gravel layers (upper and lower layer) separated by a thin silty-clay lense, and PAH and BTEX concentrations similar to those detected in the sand and gravel layers identified at other areas. The presence of underground piping at the Test Case site also restricts placement of the test plot. The pilot study will be performed within a 25 x 50 foot area. The test plot will be divided into two 25 x 25-foot test cells.

The test cells will be defined by either a physical or hydrodynamic barrier, in order to isolate the test area from outside ground water flow or other influences to the pilot study. Prior to the installation of pilot study components, hydrogeological testing will be performed in the study area. These tests are important for the application of the MGP-REM Process to the Bloomington site. The testing will include flow tracer studies, a dissolved oxygen test, a slug test, a free product recovery optimization and a hydrogen peroxide injectivity test.

The Phase III pilot scale system will consist of six (6) recovery wells, drilled in two (2) rows of three (3) wells. This provides two (2) contiguous square patterns. An injection well will be drilled at the center of each square to form a "5-spot well pattern" in each of the two squares, as shown in Figure 1. An additional infill well is drilled in the center of the two-square pattern. The infill well (In-1) located between the test plots will be utilized during the biological treatment phase and during the last phase of the MGP-REM Process test. Using this arrangement of injection and recovery wells, fluid flow is maximized through the contaminated, porous subsurface media (Figure 2). If this well arrangement is found to be effective in removing the contaminants, it can easily be scaled up for full scale remediation.

In addition to the injection and recovery wells, 12 air sparging wells will be drilled for use during the biological treatment (Figure 3). Vapor recovery wells will be installed with the air sparging well set in order to recovery any fugitive emissions from the treatment zone.

Upon completion and qualification of all pre-pilot scale testing and installation of subsurface equipment, a skid-mounted surface groundwater augmentation facility will be connected. The main functions of the surface facility will be to separate and remove free product from the extracted groundwater, mix nutrients and other chemicals into the injection water and to circulate treatment solutions and air through the treatment cell. The main components of the surface groundwater augmentation facility are as follows: oil/water separator, additive preparation tank, hydrogen peroxide tote, pH adjustment tote, blower, transfer pumps and data acquisition and system control instrumentation. A flow diagram of the surface facilities, including tanks, injection/recovery wells piping connections, metering and regulator systems, is shown in Figure 4. The surface facility will consist of one 300-gallon and three large capacity (1000-, 2000-, and 5000-gallon) tanks. The tanks will be used for preparation of the treatment fluids for subsurface injection, holding recovered water and pumping of recovered water to the treatment plant. As shown in Figure 4, one pump will be used for injecting and one pump will be used for recovering treatment fluids from the recovery wells, and a blower will be used to supply air for sparging. An additional pump will be used to deliver the recovered liquid to the treatment plant. After passing through the settling tank, separator and air stripper systems components of the treatment facility, the treated groundwater will then be augmented for biological or chemical treatment of the subsurface. The augmented groundwater will then be re-injected into the contaminated zone.

Once all components necessary for the Phase III testing have been installed, the MGP-REM Process will begin. As much as possible, mobile contaminants (i.e., free product) will be removed prior to treatment of soil. This is necessary to provide direct contact between the injected fluids and the leachable contaminants adhering to the pore surfaces of the soil.

The nutrient solution will then be injected at the same rate as the extraction of groundwater in order to maintain a steady-state flow condition in the study cell. All extracted groundwater will be treated in the surface facility and returned to the test cell.

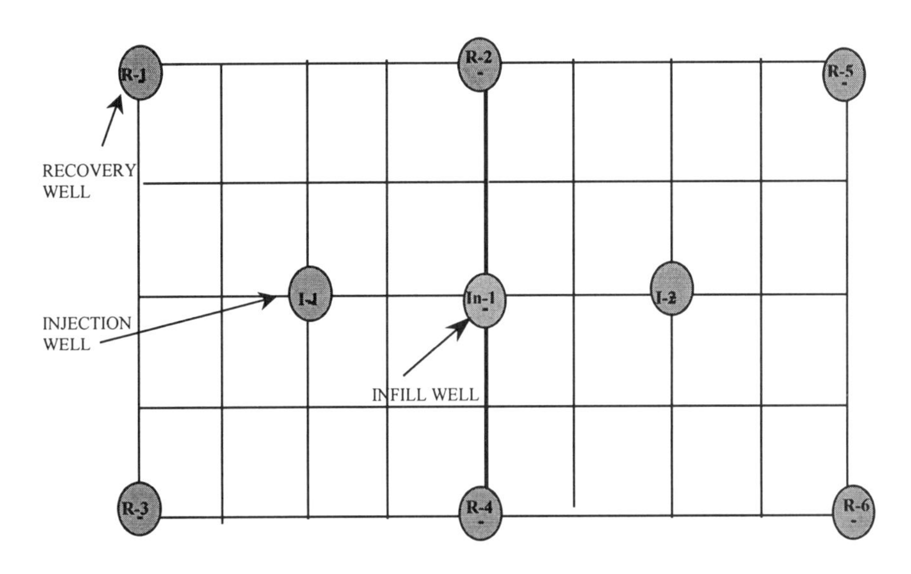

Figure 1. LAYOUT OF RECOVERY AND INJECTION WELLS IN TWO CONTIGUOUS SQUARE PLOTS AT TEST CASE SITE

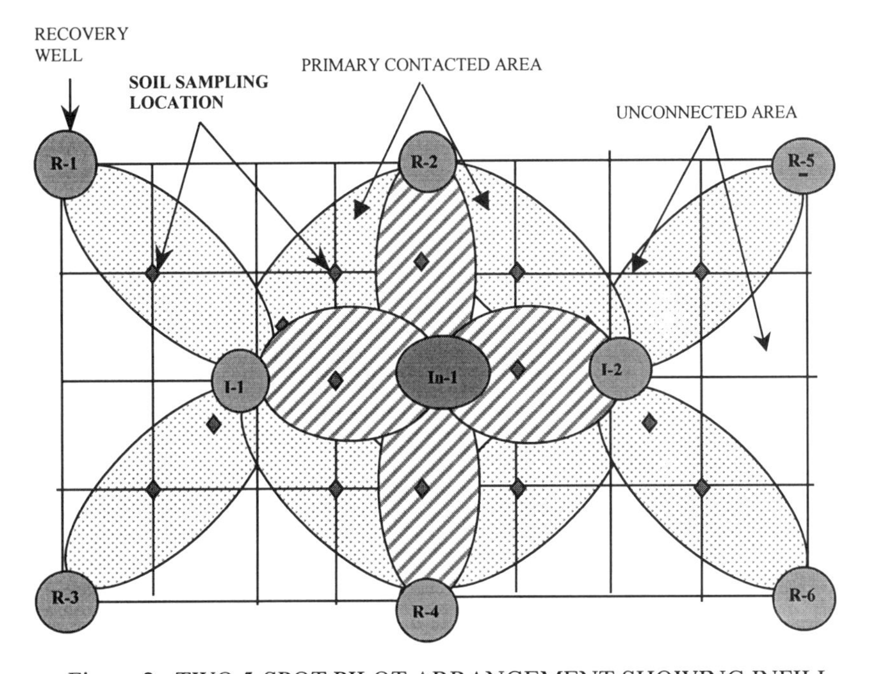

Figure 2. TWO 5-SPOT PILOT ARRANGEMENT SHOWING INFILL WELL, PRIMARY AND SECONDARY CONTACTED AREAS AND 16 SOIL SAMPLING LOCATIONS

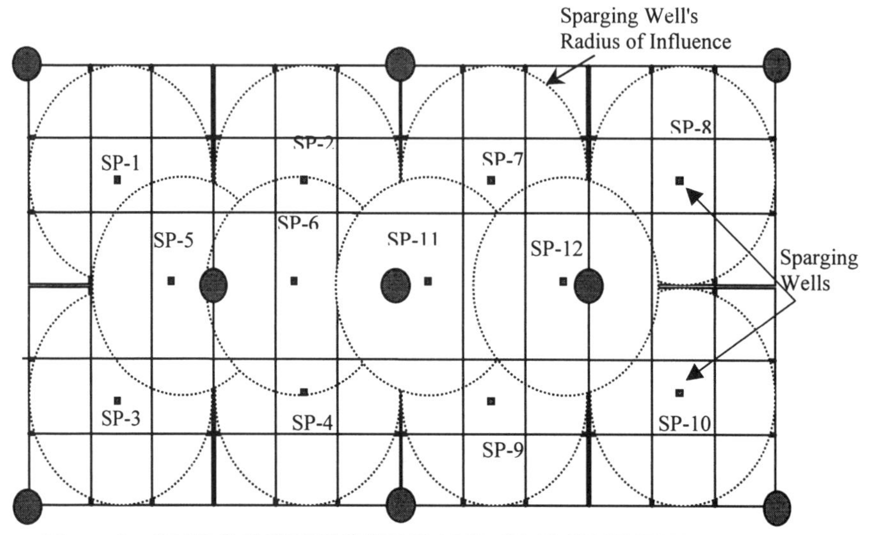

Figure 3. SPARGING/RECOVERY AND SPARGING WELL LOCATIONS AND RADII OF INFLUENCE

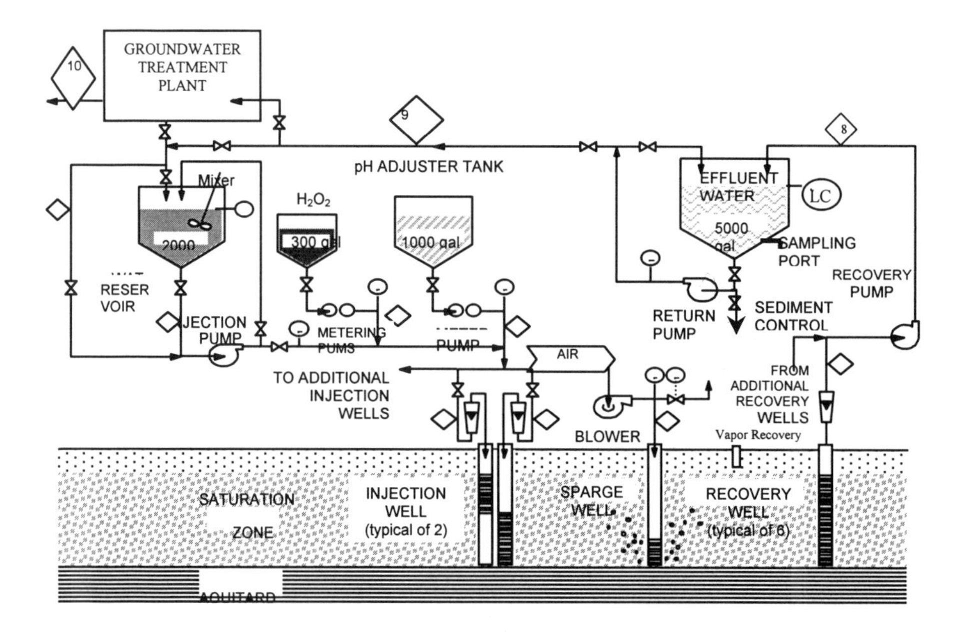

Figure 4. SURFACE FACILITY FLOW DIAGRAM

The MGP-REM Process for the Bloomington site will consist of biological treatment followed by chemical treatment. The biological treatment will consist of:

1) adjusting soil pH to approximately 7.0 (by injecting chemicals)
2) injecting liquid nutrients
3) delivering oxygen by air sparging within the contaminated zone

For the purposes of the pilot test, treatment fluids will be injected at 4,000 gallon/day per well (or 8,000 gal/day total) evenly divided between the 2 injection wells (I-1 and I-2) while simultaneously withdrawing fluid from the 6 recovery wells (at an average rate of 1,660 gal/well.) Initially, a total of 28,000 gallons (approximately 2 pore volumes) of nutrients dissolved in water will be circulated within the pilot plots over 7 days.

Following the pH adjustment and the addition of nutrients, air sparging will begin. Ambient air at low pressure will be sparged into the subsurface to supply up to 10 mg/l of oxygen. Alternating the pulsing of air into 6 wells for a period of 2 days, and then switching to the other 6 wells for the next 2 days will provide efficient air sparging. Initially, an approximate air flow rate of 7,500 cubic feet/day/well will be maintained (total pressure not to exceed 5 to 7 psig). Based on the pre-pilot tests conducted in the field, the airflow rate will be adjusted to maintain a dissolved oxygen (DO) level of 10 mg/liter.

Nutrient addition and pH adjustment will continue throughout the biological treatment phase. Monitoring of liquid and soil samples will assess bacterial growth and activity. Analysis of PAH concentrations, colony forming units (CFU) measurements and other bioactivity tests will be used to determine the level of microbial activity. Dissolved oxygen, nitrogen, phosphorus and potassium (NPK) and pH measurements will be examined to maintain proper levels of nutrients and oxygen for bacterial growth. The degree of PAH reduction and/or bacterial growth will be used as the criteria for proceeding to the next treatment phase.

The chemical treatment program will be based upon results of the pre-pilot injectivity testing. For the purposes of the preliminary design, chemical treatment will consist of injecting 2 to 3 pore volumes of hydrogen peroxide (H_2O_2) and chelated iron into the subsurface through the injection wells. The initial injection rate will be 4,000 gal/day. Hydrogen peroxide will be injected in slugs preceded and proceeded by the chelated iron injections. The overall concentration of H_2O_2 will be maintained at approximately 2% of the soil weight. During chemical treatment, soil and water pH and H_2O_2 concentrations in retrieved water will be closely monitored. If sufficient removal of leachable contaminants is not achieved through the initial chemical treatment, an additional 2 to 3 pore volumes of H_2O_2 may be injected.

Following the chemical treatment, a second "polishing" biological treatment may executed to eliminate any remaining or recharged leachable contaminants from the soil or water. Procedures for the polishing biological treatment are identical to the initial biological treatment program.

The success of the pilot will depend on the degree of contact between contaminants in the subsurface and biological and chemical agents used as part of the MGP-REM Process. A rigorous sampling plan has been designed to monitor overall program effectiveness and guide the pilot operation. Results of these analyses will determine the effectiveness of the MGP-REM Process at the Bloomington site. Biological and chemical monitoring is essential for determining process performance. Continuous review of hydrogeological and operational parameters is necessary for efficient delivery of the treatment regime to the field pilot.

Based upon given assumptions and knowledge of the site, full remediation of the site using the MGP-REM Process could potentially be performed for $30 to $50 per cubic yard.

CONCLUSIONS

The three-phase treatability study protocol used to evaluate Bloomington site soils offers a potentially cost effective, expedited method for assessing the applicability of particular technology at a specific site. It can also predict the effectiveness of a specific technology, such as MGP-REM, for removing contaminants from soils under certain conditions. The innovative phased approach also rapidly and reliability characterized the soil matrix and contaminants of interest, identified achievable treatment endpoints and provided key engineering criteria for the implementation of MGP-REM technology for cleanup of Bloomington site soils.

The combination of innovative technology with phased technology testing is powerful because key parameters necessary for application of MGP-REM to the remediation program are elucidated rapidly and cost-effectively. Performance of the technology is first defined in the laboratory and then further developed for onsite installation. Although the procedures for technology testing were originally developed for remediation of former manufacturing gas plant (MGP) sites contaminated with PAHs, it may also be used for testing other technologies which remove organic contaminants such as PCBs, TPH, BTEX, chlorinated solvents and other compounds.

BIODEGRADATION OF 6 PAHS IN KUWAITI, JAPANESE, AND THAI SOILS

Toshio Omori, Masae Horinouchi, Hideaki Nojiri, Hisakazu Yamane
(The University of Tokyo, Tokyo, Japan)
Kanchana Juntongjin (Chulalongkorn University, Bangkok, Thailand)

ABSTRACT: Most of polycyclic aromatic hydrocarbons (PAHs), such as naphthalene (NAH), phenanthrene (PHE), acenaphthene (ACE), anthracene (ANT), fluoranthene (FLU), and pyrene (PYE), were degraded in Japanese or Thai soils, while only 0 to 40 % of PAHs except NAH were degraded in Kuwaiti soil. Remaining amount of ACE decreased from 80 % to 50 % by coexistence of other PAHs in Kuwaiti soil. Degradation of ANT and PYE was not influenced by coexistence, while degradation of PHE and FLU was retarded in Kuwaiti soil. In Japanese soil, ACE degradation was enhanced while PYE degradation was retarded. The number of bacteria showing *meta*-cleavage activity in the soil samples supplemented with 6 PAHs increased to about 10^6 cells/g soil after 20 days incubation in both soils. The accumulated number was kept in Japanese soil, while that began to decrease from 20 days and became undetectable after 50 days in the Kuwaiti soil. By the addition of the Japanese soil, the degradation activity of PHE, ACE, and ANT in Kuwaiti soil was enhanced, while the enhancement of the PYE degradation activity was almost negligible.

INTRODUCTION

PAHs are a class of ubiquitous environmental pollutants which have been detected in numerous aquatic and terrestrial ecosystems. PAHs has been released into the environment in many ways, including combustion or accidental spilling of petroleum, fallout from urban pollution, and coal liquefaction and gasification processes (Cerniglia, 1992). Crude oil contains not only low molecular weight PAHs like NAH or PHE, but also high molecular weight PAHs. Because the high molecular weight PAHs are more recalcitrant in soil, the contamination of these compounds is considered to be hazardous. Because PAHs possess common structures, it is known that some bacteria can degrade high molecular weight PAHs which are not utilized as a carbon source when other PAHs are furnished for growth (Keck *et al.*, 1989). As low molecular weight PAHs such as NAH and PHE are known to be degraded relatively easier than other PAHs, we expected the degraders of these low molecular weight PAHs might be able to cooxidize high molecular weight PAHs such as PYE and FLU. Based on this consideration, in this study, we added a mixture of low and high molecular weight PAHs to the oil-contaminated Kuwaiti soil, the Japanese farm soil, or Thai field soil, and monitored PAHs degradation and population of PAH-degraders. On the other hand, because of low PAH-degrading activity of the oil-contaminated Kuwaiti soil, enhancement of the PAH-degrading activity is quite important for its remediation. For investigation of the method to remediate PAH-contaminated soil with low PAH-degrading activity, we also estimated the effects of the addition of Japanese soil which has higher PAH-degrading activity to Kuwaiti soil on the enhancement of PAH-degrading activity.

MATERIALS AND METHODS

Soil. The Kuwaiti soil used in this study was sandy and light-contaminated with crude oil. The total petroleum hydrocarbon of this soil was about 1.87 %. The Japanese soil used in this study was collected from the farm of The University of Tokyo. Japanese soil contained a lot of humid and was not contaminated with PAHs. Thai soil used in this study was collected near the Chulalongkorn university in Bangkok and with slightly contaminated with oil.

Supplement with PAHs. The crystals of PAHs, such as NAH, ACE, PHE, ANT, FLU and PYE, were ground down into powder by mortar and pestle, and 100 mg (for each PAH, total amount was 600 mg) of the resultant powder was supplemented as a model substrate to 100 g of each soil. The resultant soil mixtures were homogenized by the mill (SCM-40A model; Shibata scientific technology LTD., Tokyo, Japan). Ten ml of sterile distilled water, supplemented with NH_4NO_3 (1% [wt/wt] in water) and Na_2HPO_4 (1% [wt/wt] in water) was added to the soil mixture. The resultant soil mixtures were mixed well again.

Incubation of soil samples. The homogenized soil mixture was put into a sealed plastic box (8 cm i.d.× 5 cm) and incubated at 30°C.

Estimation of PAHs remaining. In appropriate intervals, 1 g of soil sample was extracted with 3 ml of methanol by mixing over 1 min using vortex mixer. After centrifugation (1,500 rpm, 15 min), the supernatant was directly subjected to HPLC on a reverse-phase column (DOCOSIL [C-22], 6 mm i.d. x 25 cm; Senshu Scientific Co., Ltd., Tokyo, Japan) that was eluted with 80 % aqueous methanol at 40℃ at a flow rate of 1.0 ml min^{-1}. The PAHs in the elutes were detected by monitoring the absorption at 270 nm. The quantification of PAHs remaining in the soils was conducted by comparing the peak areas with those of authentic PAHs.

Monitoring bacteria with *meta*-cleavage activity in soil samples. Three ml of sterile water was added to 1 g of the soil and mixed well by vortex mixer over 1 min. The resultant mixture was left for 10 min and the diluted supernatant were spread onto nutrient broth plates or carbon-free mineral medium (CFMM) plates with NAH given in vapor. CFMM is consisting of the following components (g/L): Na_2HPO_4 (2.2), KH_2PO_4 (0.8), NH_4NO_3 (3.0), $MgSO_4 \cdot 7H_2O$ (0. 1), $FeCl_2 \cdot 6H_2O$ (0.05), and $CaCl_2 \cdot 2H_2O$ (0.1). NAH was given in vapor by placing crystals in the lid of an inverted plate. The number of growing colonies on nutrient broth plate were defined as a total population of bacteria. Ethanol solution of catechol derivatives were sprayed on the colonies grown on CFMM plate with NAH, and then yellow colonies were counted. As catechol derivatives, 3-methylcatechol and 2,3-dihydroxybiphenyl (5 g each/L ethanol) were used.

RESULTS AND DISCUSSION

Biodegradation of PAHs which were added simultaneously. Because many kinds of aromatic compounds exist as a complex mixture in the oil-contaminated soil, we examined a degradation profiles of 6 PAHs (NAH, PHE, ACE, ANT, FLU, and PYE) which were simultaneously added into Kuwaiti, Japanese, or Thai soils, and compared them to those of the corresponding PAH which was solely added.

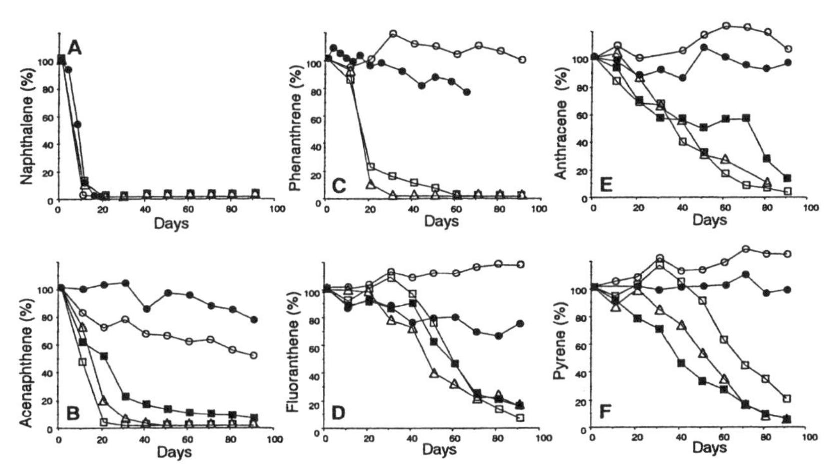

FIGURE 1. Effects of the simultaneous addition of PAHs on the degradation rate of NAH (A), ACE (B), PHE (C), FLU (D), ANT (E), and PYE (F) in Kuwaiti, Japanese, and Thai soil at 30°C. Each substrate was solely added to Kuwaiti (●) or Japanese (■) soils, or simultaneously added to Kuwaiti (○), Japanese (□), or Thai (△) soils with other 5 PAHs.

As shown in Fig. 1A, NAH disappeared in all the soils within 20 days incubation. In the Kuwaiti soil, about 40 % of ACE which was added with other PAHs simultaneously was degraded in 90 days incubation (Fig. 1B), while the amount of degraded ACE added solely in Kuwaiti soil was less than 20 % (Fig. 1B). These results suggest that the degradation was enhanced by coexistence of other PAHs. In Japanese soil, degradation of ACE was also enhanced when 6 PAHs were added together (Fig. 1B). These enhancements were considered to be caused by cooxidation of ACE by other PAH-degrading bacteria. The rate of degradation of ACE added simultaneously with other 5 PAHs in Kuwaiti soil was lower than in Japanese or Thai soil in which most of ACE was degraded after 30 days incubation (Fig. 1B).

The amounts of PHE and FLU did not decrease in 90 days incubation in the case of the simultaneous supplementation, while they decreased gradually in the case of the addition of the respective substrate alone in Kuwaiti soil (Figs. 1CD). From these results, it is considered that PHE and FLU were not degraded by other PAH-degraders, and the changes of microflora originated from the simultaneous addition with other PAHs had inhibitory effect on growth of degraders of these compounds. In Japanese soil, degradation profile of FLU was considered to be almost the same between the cases in which FLU was added simultaneously and solely (Fig. 1D). Degradation profiles of both PHE and FLU were vary similar between in Japanese soil and in Thai soil, though it was a little bit faster in Thai soil than in Japanese soil.

Almost all ANT and PYE were remaining in the Kuwaiti soil regardless of simultaneously added with other PAHs or solely added (Figs. 1EF). These results indicate that simultaneous addition of 6 PAHs have no effect on degradation of the respective PAHs in Kuwaiti soil. In Japanese soil, degradation rate of PYE was almost the same between in soils supplemented with 6 PAHs and with PYE alone, though the start of degradation in the soil supplemented with 6 PAHs was delayed (Fig. 1F). As the reasons for this delay of degradation, the following possibilities could be considered: (i) NAH or other PAHs were rather preferably utilized by PYE-degraders before they utilized PYE. (ii) Because other supplemented PAHs were toxic to PYE-degraders as NAH is reported to be sometimes toxic for high molecular weight PAH-degrader (Bouchez *et al.*, 1995).

Population of PAH-degrading bacteria. For effective bioremediation of PAH-contaminated soils, it is important to know bacterial population of PAH-degraders in soils. As the first step to investigate the bacterial population, we examined the ratio of the bacteria having *meta*-cleavage activity in total bacteria, because many bacterial strains which has the ability to degrade aromatic compounds including PAHs have been reported to have *meta*-cleavage activity of catechol derivatives (Guerin and Jones, 1988; Williams and Sayers,1994; Daly *et al.*, 1997). As a result of *meta*-cleavage of catechol derivatives, the yellow compounds are produced. By spraying an ethanol solution of a mixture of 3-methylcatechol and 2, 3-dihydroxybiphenyl, the color of bacteria which have *meta*-cleavage enzyme turn to yellow to brown. Because the degradation profile of PAH in Japanese soil was quite similar to that in Thai soil, we compared the population of PAH-degrading bacteria between Kuwaiti and Japanese soils. Six PAHs were added together to Kuwaiti or Japanese soil as described above. After appropriate intervals, we counted total population of bacteria and bacteria with *meta*-cleavage activity.

The total population of bacteria increased to 10^7 cells/g soil in the Kuwaiti soil and to 10^9 cells/g soil in Japanese soil after 20 days incubation, then they kept their populations (Fig. 2). At the start of incubation, the numbers of bacteria showing *meta*-cleavage activity were nearly negligible both in Japanese soil and Kuwaiti soil (Fig. 2). After 20 days incubation, bacteria showing *meta*-cleavage activity increased to about 10^6 cells/g of both soils, and then kept the population in the Japanese soil. On the other hand, in Kuwaiti soil, increased bacteria showing *meta*-cleavage activity began to decrease after 20 days incubation, and became undetectable 50 days after the start of incubation. In the Kuwaiti soil, it is considered that NAH-degraders firstly grew utilizing NAH as a carbon source and then began to die after NAH was run out, because only NAH was degraded effectively in the degradation experiment as described previously (Fig. 1). On the other hand, all PAHs used in this study were degraded in Japanese soil (Fig. 1). Therefore, the reason why the number of bacteria with *meta*-cleavage activity in the Japanese soil did not decrease is likely to be that, after the growth of NAH-degrader, degraders of other PAHs with the ability to utilize NAH increased.

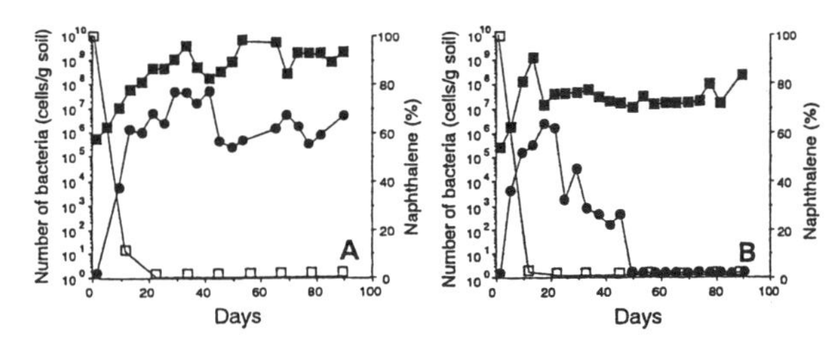

FIGURE 2. Changes of the bacterial population in Japanese soil (A) or Kuwaiti soil (B) supplemented with 6 PAHs. (■) indicates the number of total bacteria, (●) indicates the number of bacteria with *meta*-cleavage activity, and (□) indicates remaining amount of NAH.

Effect of addition of Japanese soil on PAH-degrading activity in Kuwaiti soil. Kuwaiti soil has poor ability to degrade PAHs (Fig. 1). To remediate PAH-contaminated Kuwaiti soil, we tried out the addition of Japanese soil with high activity to degrade PAHs.

The more amount of Japanese soil were added to Kuwaiti soil, the more amounts of PHE, ACE, and ANT were degraded (Fig. 3). Though the degradation of FLU was almost negligible when the ratio of Japanese soil was under 30 %, about 30 % of FLU was degraded in soil whose ratio of Japanese soil was 50 % (Fig. 3). On the other hand, PYE did not degraded in all the mixed soils even though it was degraded in the Japanese soil (Fig. 3). These results indicate that PYE-degraders living in Japanese soil were not able to grow in the sandy and oil-contaminated Kuwaiti soil. Generally, there exist a large number of bacteria in soil that can degrade low molecular weight PAHs, while only a small number of specific bacteria can degrade high molecular weight PAHs (Cerniglia, 1992). Among the various bacteria, a restricted number of bacteria seem to adapt to foreign soils. Therefore, it is unlikely that the small number of specific high molecular weight PAH-degraders can adapt to foreign soils and show their respective degrading activity. This is considered to be the reason why FLU or PYE was hardly degraded in the Kuwaiti soil by the addition of the Japanese soil. Under these considerations, it is important to isolate the high molecular weight PAH-degrading microorganisms which can adapt to the environmental condition in oil-contaminated Kuwaiti soil. If those microorganisms are added to the Kuwaiti soil, the enhancement of PAH-degrading activity of the soil will be possible.

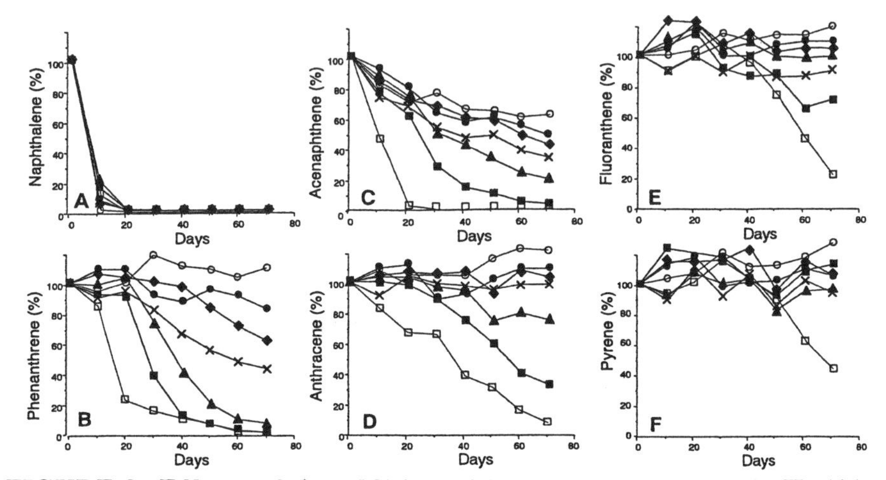

FIGURE 3. Effects of the addition of Japanese soil on the NAH- (A), PHE- (B), ACE- (C), ANT- (D), FLU- (E), and PYE-degrading activities of Kuwaiti soil. The ratio of Japanese soil are 0 (○), 1 (●), 5 (◆), 10 (×), 30 (▲), 50 (■), and 100 % (□) (wt/wt).

REFERENCES

Bouchez, M., D. Blanchet, and J. P. Vandecasteele. 1995. "Degradation of polycyclic aromatic hydrocarbons by pure strains and by defined strain associations: Inhibition phenomena and cometabolism." *Appl. Microbiol. Biotechnol. 43*(1): 156-164.

Cerniglia, C. E. 1992. "Biodegradation of polycyclic aromatic hydrocarbons." *Biodegradation 3*: 351-368.

Daly, K., A. C. Dixon, R. P. J. Swannell, J. E. Lepo, and I. M. Head. 1997. "Diversity among aromatic hydrocarbon-degrading bacteria and their *meta*-cleavage genes." *J. Appl. Microbiol. 83*(4):421-429.

Guerin, W. F., and G. E. Jones. 1988. "Mineralization of phenanthrene by a *Mycobacterium* sp." *Appl. Environ Microbiol. 54*(4):937-944.

Keck, J., R. C. Sims, M. Coover, K. Park, and B. Symons. 1989. "Evidence for cooxidation of polynuclear aromatic hydrocarbons in soil." *Wat. Res. 23*(12):1467-1476.

Williams, P. A., and J. R. Sayers. 1994. "The evolution of pathways for aromatic hydrocarbon oxidation in *Pseudomonas*." *Biodegradation 5*(3-4):195-217.

EFFECT OF HUMIC ACIDS ON THE BIOAVAILABILITY OF PAHS FROM WEATHERED SOILS.

S. Lesage, W.C. Li, K. Millar and D. Liu.
National Water Research Institute, Burlington, Ont., Canada. L7R 4A6

ABSTRACT: This is the last report on a comprehensive multi-year study on the effect of humic acids on the dissolution and biodegradation of aromatic hydrocarbons of petroleum origin. Early phases of the work showed that the addition of humics acids enhanced the dissolution of hydrocarbons from light fuels such as diesel, and thus indirectly helped the bioremediation, because dissolution is often rate limiting. The second phase of the work using radiolabelled phenanthrene, pyrene and benzo-a-pyrene showed that humic acid retarded the degradation of the pure PAHs spiked onto soil, but that this effect was reversed when the petroleum product was also added. This indicated that biodegradation was dependent on the relative sorption of PAHS onto soil, or humic acids. Sandy soils were predominantly used in these early studies. In the current work, humic acids were added to highly weathered clay soils. A subset also including freshly spiked PAHs was included as a control. The hypotheses to be tested were whether humic acids could displace the PAHs on the clay surface and enhance bioremediation and whether weathered PAHs were more difficult to degrade than freshly introduced contaminants. The results showed that humic acids had no noticeable effect on the biodegradation of either weathered or spiked soil. The most important factor was the presence of a microbial population capable of effecting the degradation.

INTRODUCTION

In our previous studies, the addition of humic acids was found to increase the rate of dissolution of PAHs from diesel fuel in groundwater (Xu et al., 1994). The dissolved plume was also found to undergo biodegradation (Lesage et al.,1995). However, this initial experiment did not address the effect of different concentrations of humic acids on the biodegradation process itself and on higher molecular weight PAHs present in heavier fuels or crude oil. A separate study was undertaken in the laboratory, using microcosms, to look at the effect of humic acids on phenanthrene, pyrene and benzo-a-pyrene as representatives of the PAHs. In the first phase of the study (Lesage et al.,1996; 1997a), where the pure compounds were spiked to a sandy soil, the presence of high concentrations of humic acids was found to decrease the amount of ^{14}C labelled phenanthrene that was mineralized. However, when crude oil was added to the mixture, this tendency was reversed. It was therefore postulated that the lower amount of mineralization observed was due to sorption of either the parent compound or its metabolites onto the humic molecule.

In this last phase of the project, the effect of adding humic acid to a weathered clay soil was examined. Humic acids have been shown to bind strongly with clays (Filip and Alberts, 1994; Spark et al., 1997). Therefore, this experiment was designed to verify whether this binding would interfere with that of the PAHs, potentially displacing the aromatic molecules from the clay surface and increasing their bioavailability.

The sample came from a composite of soils from a decommissioned refinery. The levels of PAHs were relatively low, and most of the easily degraded hydrocarbons were absent from this fine clay soil. It was therefore decided to run two parallel sets of microcosms, one with the soil as received, and the other with an additional amount of PAHs spiked in. This would also show whether there was a significant difference in the bioavailability of spiked vs. weathered residues of PAHs. In addition to a mineral medium, a mixed bacterial consortium that has been found to degrade phenanthrene was added to ensure that at least some degradation would be observed. A non-inoculated control was also used. The degradation was followed by doing a solvent extraction and a GC/MS analysis of the parent compounds.

METHODS

The contaminated soil was given to us by Imperial Oil of Canada Ltd. and came from a decommissioned refinery undergoing remediation. The soil was a highly weathered clay from various areas on the refinery grounds, which had received various petroleum contaminants over a few decades. The soil was air-dried, ground and sieved through a 2-mm mesh sieve, then homogenised. A 1 kg portion of the soil was also spiked with acetone solutions containing phenanthrene (500 mg), pyrene (50 mg) and benzo-a-pyrene (50 mg).

The microcosms were made up in 500 mL Mason jars. Each contained 100 g dried soil which was mixed with 25 mL of mineral medium and sodium humate (Aldrich Chemicals, Milwaukee) at 0, 50, 500 or 2000 mg/L. A 10 mL bacterial inoculum (0.08 absorbance at 600 nm) of a consortium that has been selected for its ability for growing on phenanthrene was also added. The final soil moisture content was approximately 80% of field capacity. Abiotic controls were made by adding 0.02% sodium azide and by omitting bacterial inoculation. The samples labelled "control" were not inoculated and did not receive sodium azide.

The microcosms were opened and mixed on a weekly basis to enhance the oxygen availability. Periodic sampling consisted of removing 20 g of soil (wet wt.) and stopping the biodegradation by adding 5 mLs of a 2000 ppm solution of sodium azide. The samples were air-dried and ground before extraction. The samples (10g dry wt.) were extracted by Soxhlet for 8 hours using a 1:1 mixture of hexane/acetone. The extracts were concentrated to 1 mL using a Snyder column and then made up to 20 mL with hexane. A 2 mL aliquot was cleaned up on a column of activated silica gel (11 x 300 mm). The aliphatic hydrocarbons were eluted with hexane and the PAHs with hexane containing 20% dichloromethane. Both fractions were analysed by GC/MS, using a 25 m x 0.2mm i.d. x 0.33μm film Ultra-2 (Hewlett-Packard) capillary column. The oven temperature was: 50°C to 180°C at 30°C/min, then to 300°C at 3°C/min - held for 20 min. d10-Anthracene was used as the internal standard.

RESULTS

The results of the degradation of phenanthrene in the spiked and unspiked soil are shown in Figure 1 a and b respectively. In the spiked sample, there was no loss of phenanthrene in the abiotic control or in the abiotic control to which humic acid had been added. The degradation was rapid, > 90% in two weeks, in all the other samples, regardless of the presence of humic acid. The degradation of the non-inoculated control lagged behind the other samples.

Similar results were obtained with the unspiked samples, although the difference between the inoculated samples and the control was not as dramatic, because of the much lower initial concentration of phenanthrene. The degradation was faster with all the inoculated samples compared to the non-inoculated control, except for the highest concentration of HA, which was similar to the control at two weeks. Overall, the most significant factor in the degradation of phenanthrene seems to be the addition of the acclimated bacteria. There was no significant difference in the half-life of phenanthrene between the spiked and weathered samples, although the residual concentration after 20 weeks was higher in the spiked samples (average 8 vs. 1 μg/g).

The results for pyrene are combined in one graph (Figure 2). The degradation of pyrene in the weathered soil was negligible in all samples within the 20 week period of the study, but there was a 50 percent reduction in concentration in the spiked samples within the same period. There was however no difference between humic acid treatments. Similar to what was observed with phenanthrene, the slowest degradation was in the non-inoculated control. The addition of a bacterial consortium, even if it wasn't of bacteria specifically acclimated to pyrene did have a positive effect, although it was much less important than with phenanthrene.

The degradation of benzo-a-pyrene (results not shown) was not significant in either the weathered or the spiked samples. This is not very surprising, because as was shown in previous work (Lesage et al., 1997b), benzo-a-pyrene has been found to be recalcitrant to bacterial degradation because of its combined low solubility and the difficulty to enter the cell. Fungal degradation has obtained much more success.

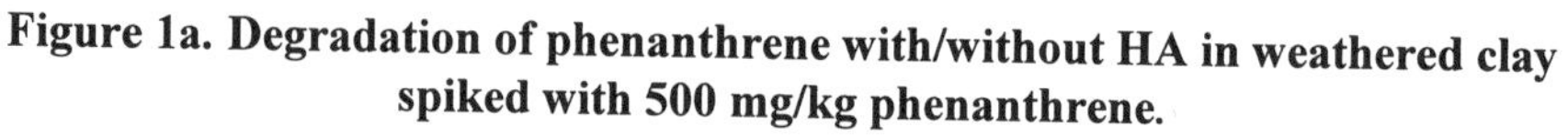

Figure 1a. Degradation of phenanthrene with/without HA in weathered clay spiked with 500 mg/kg phenanthrene.

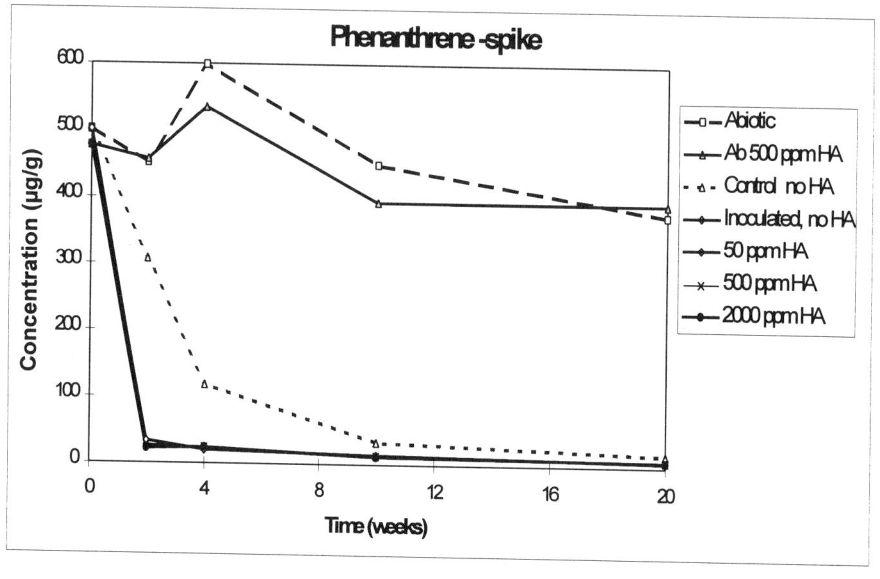

Figure 1b Degradation of phenanthrene with/without HA in weathered clay.

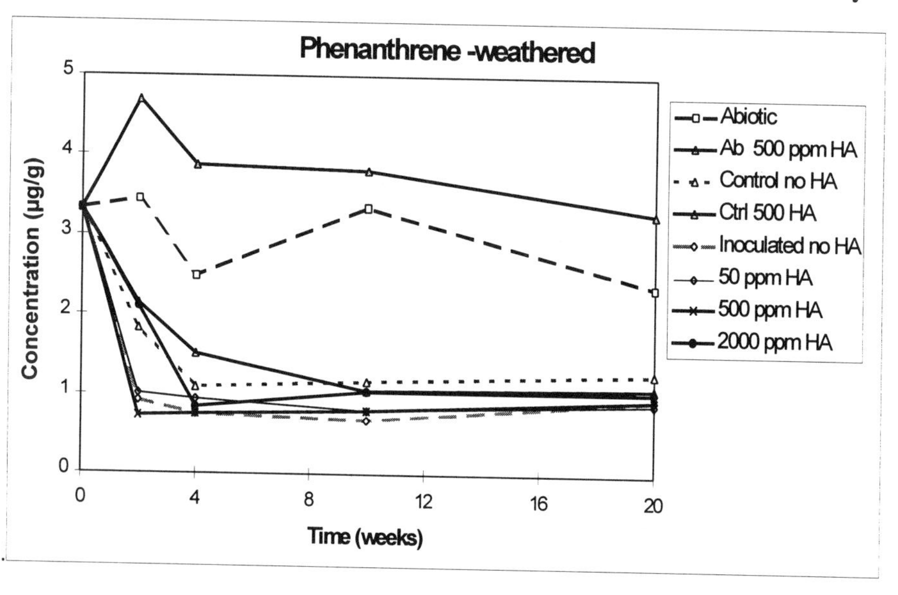

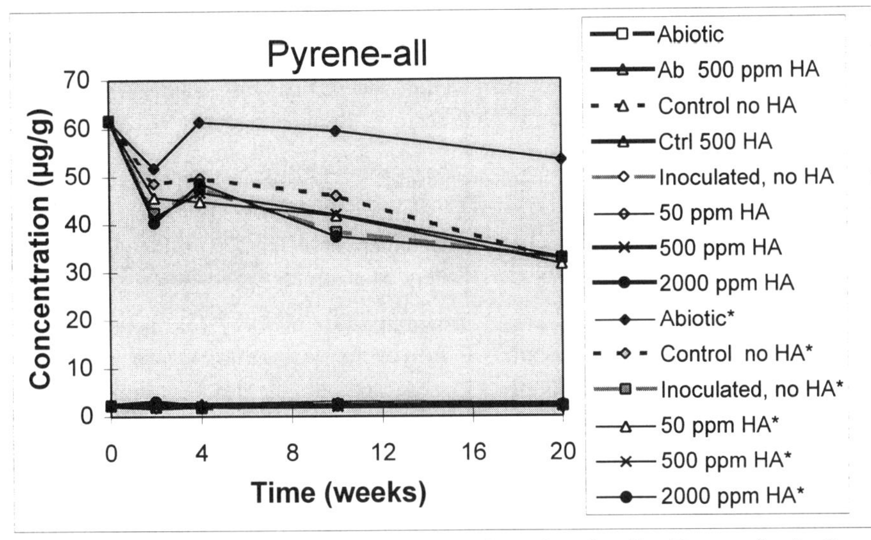

Figure 2. Degradation of pyrene in weathered and spiked* samples in the presence of increasing concentrations of humic acid (HA).

CONCLUSION

In summary, the addition of humic acid to a clay soil did not improve the bioavailability of the PAHs. The rate of degradation of individual compounds seems to be governed mostly by the presence of a bacterial population capable of degrading them. In the case of phenanthrene, the rate of degradation was not very different between the spiked and weathered sample. The degradation of pyrene was only significant in the spiked sample, and no significant degradation of benzo-a-pyrene was observed in the 20-week period. Apart from a slight decrease in the degradation of phenanthrene in the presence of 2000 ppm of humic acid in the weathered sample, humic acids had no discernible effect on the degradation of PAHs.

REFERENCES

Lesage, S., H. Xu, K. S. Novakowski, S. Brown and L. Durham. 1995. "Use of humic acids to enhance the removal of aromatic hydrocarbons from contaminated aquifers Part II: Pilot Scale." *5th Annual Symposium on Groundwater and Soil Remediation.* (Toronto, Ontario, October 2-6). Proceedings on CD/ROM, 10 p.

Lesage, S. , H. Xu , K. S. Novakowski and S. Brown. 1996 "Humic Acids as an Alternative to Surfactants for the Remediation of Fuel Spills." Emerging Technologies for Hazardous Waste Management VIII, , Birmingham AB, Sept 9-11. American Chemical Society. Extended abstract, p. 743-746.

Lesage, S., W. C. Li, K. Millar, S. Brown and D. Liu 1997. "Influence of Humic Acids on the Bioremediation of Polycyclic Aromatic Hydrocarbons from Contaminated Soil and Aquifers." Proceedings of the 6th Symposium and Exhibition on Groundwater and Soil Remediation in Montreal, March 18-21, 1997. p. 325-338.

Lesage, S., H. Hofmann, W.C. Li, K. Millar, D. Liu and H. Seidel. 1997. "Effect of Humic Acids on the Biodegradation of PAHs by Bacteria and Fungi." In *In Situ* and On-Site Bioremediation, B.Alleman and A. Leeson, chairs. Proceedings of the Fourth International Symposium on In Situ and On Site Bioremediation, New Orleans, Louisiana, April 28-May 1, 1997. Vol 4(2), 185-191.

Filip, Z. and Alberts, J. J. 1994. Adsorption and transformation of salt marsh related humic acids by quartz and clay minerals. The Science of the Total Environment 153, 141-150.

Spark, K. M., Wells, J. D. and Johnson, B. B. 1997. Characteristics of the sorption of humic acid by soil minerals. Aust. J. Soil Res. 35, 103-112.

Xu, H., S. Lesage and L. Durham, 1994. "The use of humic acids to enhance removal of aromatic hydrocarbons from contaminated aquifers. Part I: Laboratory studies." *4th Annual Symposium on Groundwater and Soil Remediation.* (Calgary studies." *4th Annual Symposium on Groundwater and Soil Remediation.* (Calgary, Alberta, September 21-23). pp 635-666.

ACKNOWLEDGEMENTS

Funding for this project was received from the Panel on Energy Research and Development (PERD) of Natural Resources Canada and from Environment Canada. The authors are indebted to Imperial Oil of Canada Limited in Montreal, Canada, for providing the contaminated soil.

IN SITU BIOREMEDIATION OF DEGRADABLE CONTAMINANTS FROM A FORMER WOOD TREATMENT FACILITY

Barry C. O'Melia, Randy Siegel, Gary Dupuy, Geoffrey C. Compeau, (URS Greiner Woodward-Clyde Seattle), RueAnn Thomas (International Paper Memphis)
Tim Syverson (GeoSyntec Consultants)

ABSTRACT. A cleanup action using biosparging and bioventing at a former wood treatment site was implemented in October 1998. The objective is to remediate mobile waste components of creosote and petroleum hydrocarbons including monoaromatic and polycyclic aromatic hydrocarbons (PAHs). Two site source areas subject to treatment are separated by a nearly contiguous silt layer. The groundwater in the lower system is anoxic under existing conditions. Sparged air is being injected in the lower saturated groundwater and upper sand. Baseline field data suggest that populations of hydrocarbon-degrading microbes exist in most site areas. Additional operational data, including respiration data and nutrient analysis, are being used to continually optimize and evaluate system effectiveness.

INTRODUCTION

Cleanup of soil and groundwater impacted during past wood treatment operations requires an approach addressing a unique and complex mixture of recalcitrant and degradable organic contaminants. Carcinogenic polynuclear aromatic hydrocarbon (cPAH) compounds typically encountered at these sites are comprised of 4–7 condensed aromatic rings and are difficult to biodegrade. However, these contaminants are relatively insoluble with limited mobility in the environment. When exposure pathways to human and environmental receptors are mitigated or eliminated, impacted soil and groundwater can often be left in place. The key to this strategy is to effectively treat the more mobile PAHs (2–4 condensed aromatic rings) at wood treatment sites, which are amenable to in situ biodegradation.

Background. The site is a former wood treatment facility with creosote, pentachlorophenol and petroleum hydrocarbon impacted soil and groundwater. Investigations showed that groundwater migration from this site is extremely slow and containment of the soils and groundwater with a slurry wall surrounding an area covered with an engineered low permeability cover effectively eliminates pathways. In situ enhanced biodegradation using bioventing/biosparging is also being implemented to further mitigate and reduce concentrations of contaminants of concern (COCs). Site COCs have varying physical properties affecting mobility (Table 1).

TABLE 1. Contaminants of Concern and Select Physical and Chemical properties.

Compound	Aqueous Solubility (1) (mg/L)	Vapor Pressure(1) (mm hg)	Highest Groundwater Concentration(2) (μg/L)
Acenaphthylene	3.93E+01	2.90E-01	791
Benzo(a)anthracene	1.20E-01	2.20E-07	187
Naphthalene	3.00E+02	7.96E-01	6,700
Pentachlorophenol	1.40E+02	1.10E-03	13,500

Notes: (1) EPA 1986
(2) September 1998 System Baseline Data

Objectives. The specific objectives of the bioventing/biosparging system are to reduce the mobility of the contaminant mass through enhanced natural degradation of carrier hydrocarbons and mobile aromatic constituents and to continue to where possible to promote reduction of recalcitrant toxic constituents and potentially reduce contaminant concentrations. The COCs include single-ring aromatic hydrocarbons and multiple-ring PAHs. Low biodegradation rates (2.1 to 6.3 mg/kg/day) for single-ring aromatic and double-ring PAHs have been observed during in situ bioventing (Gentry and Simpkin 1995) at wood-preserving sites. Biosparging has been shown to be effective in laboratory tests of total PAHs from wood preserving sites (Mueller et al. 1995). However, high molecular weight and particularly cPAH compounds undergo limited degradation under bioventing and biosparging remediation approaches (Gentry and Simpkin 1995). Therefore, significant remediation of cPAHs would not be anticipated under this cleanup action. A pilot test at a wood treatment site indicated that oxygen utilization rates were slightly higher one year after system start-up. Increased aerobic metabolism was attributed as typical for sites with relatively low initial oxygen levels and stressed populations of aerobic microorganisms (Alleman et al. 1998). Evaluation of indigenous metabolism will be monitored by measuring aerobic respiration rates at the subject site in the fall of 1999, after one year of full operation.

System Design. The bioventing/biosparging system includes a series of eleven passive bioventing wells screened in the upper sand unit and eight biosparging/venting wells screened in the upper portion of Aquifer A in the remedial action areas (Figure 1). Bioventing wells are not screened across the

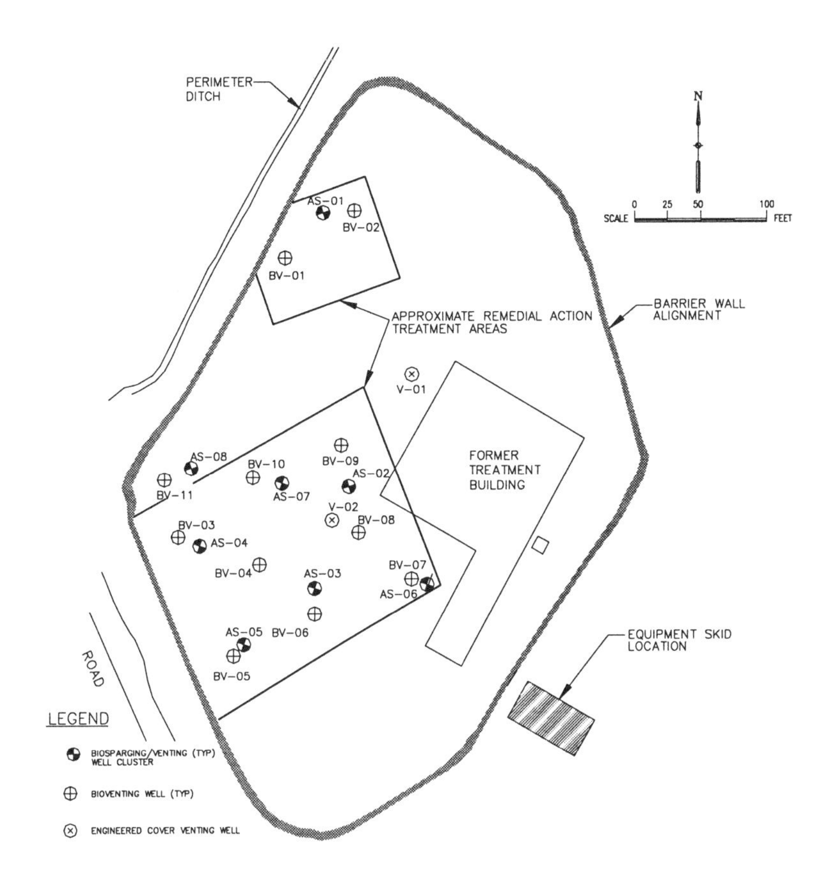

FIGURE 1. Conceptualized system cross-section showing biosparge and biovent well relationships.

upper silt layer to limit potential cross contamination. The venting wells for the biosparging system, which are located in the upper portion of Aquifer A, may be in a saturated zone much of the year due to seasonal groundwater occurrence and flucuation. However, these wells will still provide an escape pathway for air injected into the aquifer via the sparging wells (Figure 2). Valves on these vents will be adjusted to limit the discharge of O_2 and/or COCs. Two additional vents were installed as part of the engineered cover. The placement of biosparging wells was modified based on visual contamination encountered during construction of the barrier wall. Biosparging and bioventing wells near the west side of the barrier wall were moved closer to the contamination observed.

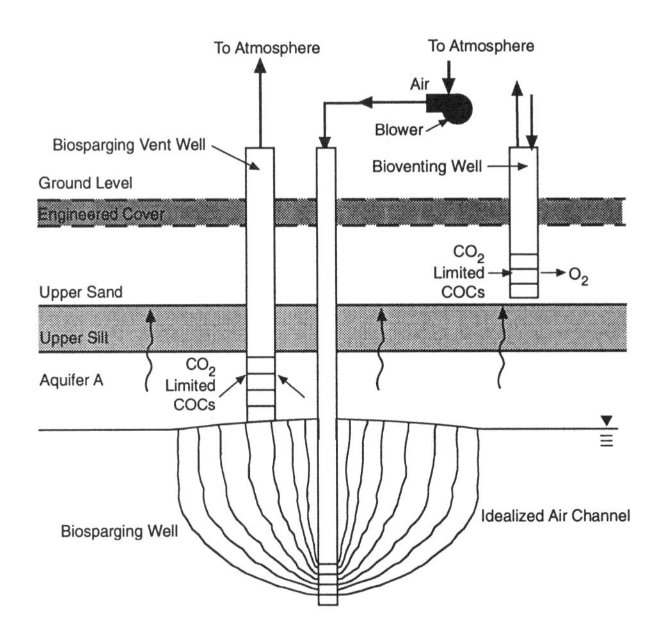

FIGURE 2. Idealized section view of biotreatment system.

A significant hydraulic gradient across the site is not anticipated, due to the barrier wall. As such, the most effective electron acceptor for bioremediation (oxygen) is supplied in Aquifer A, utilizing air sparging wells in the source areas. Atmospheric air is sparged approximately 15 feet below the upper silt layer. Oxygen will dissolve into the aqueous phase within the radius of influence of the sparging well. When oxygen levels in the groundwater are continuously near saturation, the air sparging may be cycled off until oxygen levels are reduced through microbial metabolism to below 5 mg/L. The cycling of wells will likely have the benfit of minimizing well fouling. The Aquifer A venting system will discharge carbon dioxide generated during aerobic biological activity as well as low levels of COCs, which may volatize into the sparged air.

The sparging wells have been placed in the source areas, based on an assumed radius of influence of less than 30 feet. Radius of influence is defined as the distance from an air sparging well that defines where air flow can be detected or the effects of air contact, groundwater mixing or groundwater oxygenation are detectable and consistent (API 1995). Radius of influence will be determined predominately by dissolved oxygen (DO) concentrations. Biosparging can directly impact the DO concentrations of water in contact with the sparged air or through transfer within the saturated zone, due to the mixing associated with the sparging system startup and shutdown (API 1995).

No nutrient addition is initially proposed, based on data from other wood-preserving sites (Mueller et al. 1995). However, baseline and operation nutrient analysis was performed during system start-up. Nutrient analysis included nitrate, nitrate as nitrogen, total phosphorus and sulfate as SO_4.

Flexibility was designed into the system to accommodate a range of soil permeabilities, and a piping system was installed for the bioventing wells. In addition, the biosparging system and the bioventing system can accommodate the addition of gas phase nutrients. The future use of an active bioventing system will be evaluated after one year of operation.

RESULTS

Microbal Population. Hydrocarbon-degrading bacteria were enumerated using diesel fuel hydrocarbons as a general indicator of metabolic potential. Hydrocarbon-degrading bacteria ranged from less than $1x10^3$ cfu/mL in AS-03 to $110x10^3$ in AS-02 and AS-07. These measurements are relative and will become meaningful in comparison to future measurements. The highest hydrocarbon-degrading bacteria levels in groundwater were found in the east corner of the site near the highest concentrations and nearest to former treatment operations. Total heterotrophic microbal population ranged from $1.2x10^3$ CFU/ml to $3,000x10^3$ CFU/ml. The highest total heterotrophic population was observed in the well with one of the lowest hydrocarbon-degrading bacteria populations (AS-04).

Dissolved Oxygen and Reduction/Oxidation Potential. Groundwater parameters indicated the system was anoxic (DO less than 0.6 mg/l) prior to system startup. After one week of injecting air into four of the eight biosparging wells, significant increases in the in-well DO concentrations and/or reduction/oxidation potential were measured. The DO levels in the wells range from approximately 0.6 to over 8 mg/l after three months of operation, from initial levels less than 0.6 mg/l. The reduction/oxidation potential has increased from <-100 mV for most wells to between <-100 and +10 mV after approximately two months of operation. The DO in AS-01, AS-02, and AS-08 decreased with significantly higher water elevations a month after startup prior to flow adjustments.

Contaminate Concentrations. Groundwater from four of the biosparging wells was sampled and analyzed for PAHs and total petroleum hydrocarbons (TPH). The TPH concentrations as diesel ranged from 0.63 mg/l in AS-08 to 30 mg/l in AS-03. Higher TPH concentrations in groundwater were typically found in wells with the highest PAH concentrations. The highest groundwater contaminant concentrations tend to be in the center of the site in sparging wells AS-02 through AS-06 (Table 1). Light nonaqueous phase liquid was observed in AS-06 and will be removed prior to initiating sparging activities in this well.

Off Gas. Off gas is measured for O_2, CO_2, and CH4 concentrations and flow. Initial off gas data from the bioventing wells has shown that air flows (between 0 and 3 standard cubic feet per minute) at the bioventing wells have changed from slightly into the wells prior to startup to measurable flows out of the bioventing wells. This may be indicative of potential air flow pathways from Aquifer A through the upper silt and/or atmospheric influences. Three of the air sparging vent wells, which are located beneath the upper silt, have noted a significant increase in oxygen in the off gas. These vents will be adjusted to limit the loss of usable O_2 in Aquifer A.

The combination of physical and biological techniques for treating waste components at the study site represents an approach for sites containing mobile, recalcitrant constituents. The construction of a barrier wall and low permeability cover effectively isolates these constituents within a localized area and eliminates pathways to receptors. Biotreatment methods applied within the isolated area is designed to accelerate degradation of these compounds. The effectiveness of the biotreatment methods is being evaluated by comparison of temporal variations of microbal populations, dissolved oxygen content, redox potential, and other parameters. Ultimately, the effectiveness of the system will be determined based on concentrations of COCs.

REFERENCES

Alleman, Bruce C., Paul T. McCauley and Richard C. Brennan. 1998. *Bioventing PAH Contamination at the Reilly Tar Site*. Battelle. Unpublished.

American Petroleum Institute (API). 1995. *In Situ Air Sparging: Evaluation of Petroleum Industry Sites and Considerations for Applicability, Design and Operation*. Publication No. 4609.

EPA – see United States Environmental Protection Agency.

Gentry, Jeff L. and Thomas J. Simpkin. 1995. "Experience with bioventing at wood-preserving sites." In *In Situ Aeration: Air Sparging, Bioventing, and Related Remediation Processes*. Edited by Robert E. Hinchee, Ross N. Miller, and Paul C. Johnson. Battelle Press.

Mueller, James G., Michael D. Tischuk, Mitchell D. Brourman, and Garet E. Van De Steeg. 1995. "In situ bioremediation strategies for organic wood preservatives." In *In Situ Aeration: Air Sparging, Bioventing, and Related Remediation Processes*. Edited by Robert E. Hinchee, Ross N. Miller, and Paul C. Johnson. Battelle Press.

United States Air Force (USAF). 1992. Test Plan and Technical Protocol for a Field Treatability Test for Bioventing. AFCEE, Brooks AFB, Texas.

United States Environmental Protection Agency (EPA). 1986. *Superfund Public Health Evaluation Manual*. EPA/540/1-86/060.

Zimmermen, et al. 1997. "Principles of Passive Aeration for Biodegredation of JP-5 Fuel." *In-Situ and On-Site Bioremediation: Volume 1*. Battelle Press.

MEASUREMENT AND COMPARISON OF MONOD BIODEGRADATION PARAMETERS OF PAHS

Christopher D. Knightes and Catherine A. Peters
(Princeton University, Princeton, NJ)

ABSTRACT: The goal of this study was to determine the biodegradation rate parameters for a range of polycyclic aromatic hydrocarbons (PAHs). For each of ten PAHs, a series of sole-substrate biodegradation experiments were performed measuring the PAH concentration and biomass concentration over time. The parameters determined were q, the rate coefficient [mg PAH/mg protein/hour], K_S, the half-saturation coefficient [mg/L] and Y, the yield coefficient [mg protein/mg PAH]. The results demonstrate that although the q and K_S values ranged over almost three orders of magnitude, the ratio of q/K_S ranged from 0.12 to 1.22 L/mg/hour (except for acenaphthene). The similarities of q/K_S demonstrate that these PAHs have similar biodegradation rates despite differences in size and structure. There was a mismatch in the timing of substrate depletion and biomass growth, and the values for the yield coefficient, Y, were determined to be lower than typically found for aerobic heterotrophs. This may indicate the rise of dead-end products and metabolites. Dead-end products would deprive the microorganisms of substrate, and the transformation of metabolites might be the rate-limiting step in biodegradation.

INTRODUCTION

PAHs are an important class of organic compounds because of their suspected carcinogenicity. Research has demonstrated the biodegradability of PAHs, which encourages the feasibility of using naturally occurring microorganisms for the remediation of PAH-contaminated sites. Individual PAHs are important for different reasons, e.g. some tend to dissolve in groundwater and some are highly toxic (Peters *et al.*, 1996). Thus, it is important to understand their individual biodegradabilities. Most research on PAH biodegradation has focused on pathways (Cerniglia, 1992) and the kinetics of biodegradation in complex multiphase microcosms (Park *et al.*, 1990). This research serves to isolate the process of biodegradation from other processes that may occur in the field such as slow desorption or other local processes, so that the parameters governing the biodegradation rate kinetics can be determined without being confounded.

Ten PAHs were selected (Table 1) to represent a range of PAHs which have different properties and structures. The compounds range from the simpler two-ring structures (e.g. NPH) up to the four-ring structures (e.g. PYR). These compounds have aqueous solubilities that span four orders of magnitude, and include compounds with a cyclopentane ring and with alkyl functional groups. By using a reactor consisting solely of a mixed aqueous phase with parallel abiotic experiments, the process of biodegradation can be independently studied. By performing identical experiments, the determined biodegradation parameters can be compared across the range of PAHs. The results allow inference on how the structure of the molecule may affect the biodegradation parameters.

Table 1. Structure and chemical characteristics of compounds in this study.

Compound	Abbrev	Structure	Molecular Weight	Aqueous Solubility [a] [mg/L]	Detection Limit[b] [mg/L]
Naphthalene	NPH	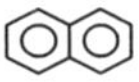	128	31	0.005
1-Methylnaphthalene	1MN	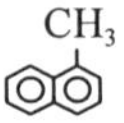 CH_3	142	28	0.005
2-Methylnaphthalene	2MN	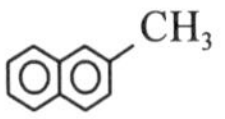 CH_3	142	24.6	0.001
Acenaphthene	ACE		154	3.8	0.001
2-Ethylnaphthalene	2EN	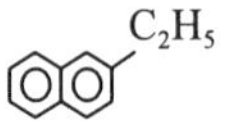 C_2H_5	156	8	0.001
Fluorene	FLR	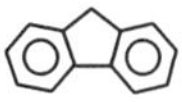	166	1.9	0.005
Phenanthrene	PHN	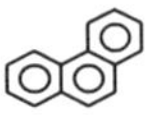	178	1.1	0.001
Anthracene	ANTH	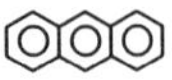	178	0.05	0.0005
Fluoranthene	FLN	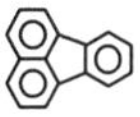	202	0.26	0.001
Pyrene	PYR		202	0.13	0.005

[a]at 25° C from Mackay, et al., 1992
[b]Detection limit from HPLC-UV/FLD analysis

MATERIALS AND METHODS

There were two parallel series of experiments performed for each PAH. Fifteen 35 mL serum bottles were used as the reactor vessels for each series. Each reactor vessel was given 25 mL of a nutrient buffer stock solution with a PAH concentration at approximately aqueous solubility. The biotic series was inoculated with 1 mL of the *Pseudomonas* consortium. The abiotic series of experiments were performed under identical conditions but with no inoculation. The aqueous phase was completely mixed throughout the duration of the experiment via magnetic stir bars. Each reactor was then crimp-sealed with a teflon-lined rubber stopper. All reactors were covered with aluminum foil to prevent photo-oxidation losses.

Fifteen minutes after the initiation of the experiments, the first vessel was sacrificed to determine partition coefficients for sorption to the apparatus and to bio-related material. Samples were then taken at time intervals dictated by the apparent rate of biodegradation. For each vessel, the aqueous phase PAH concentration was measured. For the biotic experiments, the biomass concentration was additionally measured. The dissolved oxygen concentration was measured to verify that the oxygen was not limiting.

Measurement techniques. The aqueous phase was sampled by using a pre-conditioned Hamilton syringe. The biotic samples were filtered through an Anotop 10 Whatman inorganic, hydrophilic 0.1 μm filter to remove any of the biomass or bio-related material. The sample was analyzed using a Hewlett Packard Series 1050 HPLC using a HP Spherisorb ODS 2 (5 μm, 250 x 4 mm) column followed by ultraviolet and fluorescence detection. Each method of HPLC-UV-FLD analysis was optimized for each PAH to allow for order of magnitude ranges of detectable concentrations (see Table 1). The biomass concentration was measured using the Bio-Rad Protein Assay, based on the Bradford method (Bio-Rad Laboratories), with bovine gamma globulin as the protein standard.

Monod Formulation and Non-linear Parameter Estimation. Bacterial growth kinetics were modeled using the Monod equation. The system of equations used in this analysis are

$$\left(1 + K_{app} + K_{bio}X\right)\frac{dC}{dt} = -q\frac{C}{K_S + C}X - k_{abio}C \qquad \text{(EQN 1)}$$

$$\frac{dX}{dt} = -qY\frac{C}{K_S + C}X - bX \qquad \text{(EQN 2)}$$

where C is the PAH concentration [mg/L], X is the biomass concentration [mg/L], t is time [hours], q is the rate coefficient [mg PAH/mg protein/hour], K_S is the half-saturation coefficient [mg/L], Y is the yield coefficient [mg protein/mg PAH], K_{app} is the distribution coefficient for partitioning to the apparatus [dimensionless], K_{bio} is the distribution coefficient for partitioning to bio-related material [L/mg protein], k_{abio} is the first order rate coefficient for abiotic losses [per hour], and b is the endogenous decay rate [per hour].

K_{app} and K_{bio} are parameters determined by using the total PAH concentration in the system and the aqueous PAH and biomass concentrations of the first sample (t = 0.25 hours). The parameter b is determined from an independent batch experiment with no substrate present. The remaining parameters of q, Y and K_S are determined by using the Maximum Likelihood Method for non-linear parameter estimation where the error in X is assumed to be distributed normally and the error in C is assumed to be distributed log-normally.

The data from the abiotic experiments were used to determine if there was any statistically significant abiotic loss over time. After transforming the concentration data into a log-normal space, a linear regression was performed to determine the coefficient for any possible abiotic losses. The results of the linear regression were then statistically compared to the error in the abiotic data. NPH was the only compound for which the k_{bio} was seen to be significantly different from zero.

RESULTS AND DISCUSSION

The data are plotted in Figure 1 as PAH concentration (log scale) and as Biomass concentration (linear scale) versus time for each PAH studied. The estimated Monod parameters are listed in Table 2. The results of this work demonstrate that although the q and K_S values range over almost three orders of magnitude, the q/K_S values are in a much more limited range of one order of magnitude. This seems to suggest that despite the range of the size of the molecules and the presence of functional groups each PAH has the same biodegradation rate when in the first-order region of Monod kinetics (when $C << K_S$). This research shows that these PAHs have comparable biodegradation rates, while studies in soils have found that PAH biodegradation rates vary over orders of magnitudes from two to four ring structures (Park *et al.*, 1990). The differential rates that are determined in soil samples and in the field are likely due to the physical and chemical processes that limit bioavailability. All the PAHs were degraded except acenaphthene.

Table 2. Parameter results for compounds.

Compound	q [mg/mg/hour]	K_s [mg/L]	Y [mg/mg]	q/K_s [L/mg/hour]
Naphthalene	4.13	3.40	0.0835	1.22
1-Methylnaphthalene	4.11	4.33	0.0809	0.950
2-Methylnaphthalene	n/a[a]	n/a	0.147	0.519
Acenaphthene		no detectable degradation		
2-Ethylnaphthalene	n/a	n/a	0.473	0.205
Fluorene	0.170	1.52	0.129	0.112
Phenanthrene	2.87	4.22	-0.313[b]	0.679
Anthracene	0.0426	0.0731	0.582	0.755
Fluoranthene	0.541	1.52	0.126	0.356
Pyrene	1.21	1.51	0.121	0.805

[a] n/a: these could not be determined as unique values.
[b] the yield for PHN was not seen to be statistically different from zero,the modeling result is present here.

The concentration of acenaphthene did not decrease over the 120 hours of the experiment. It seems that this *Ps.* consortium was unable to initiate biodegradation along the pathway that has been proposed by Schocken and Gibson (1984). Both fluorene and fluoranthene have a cyclopentane ring in

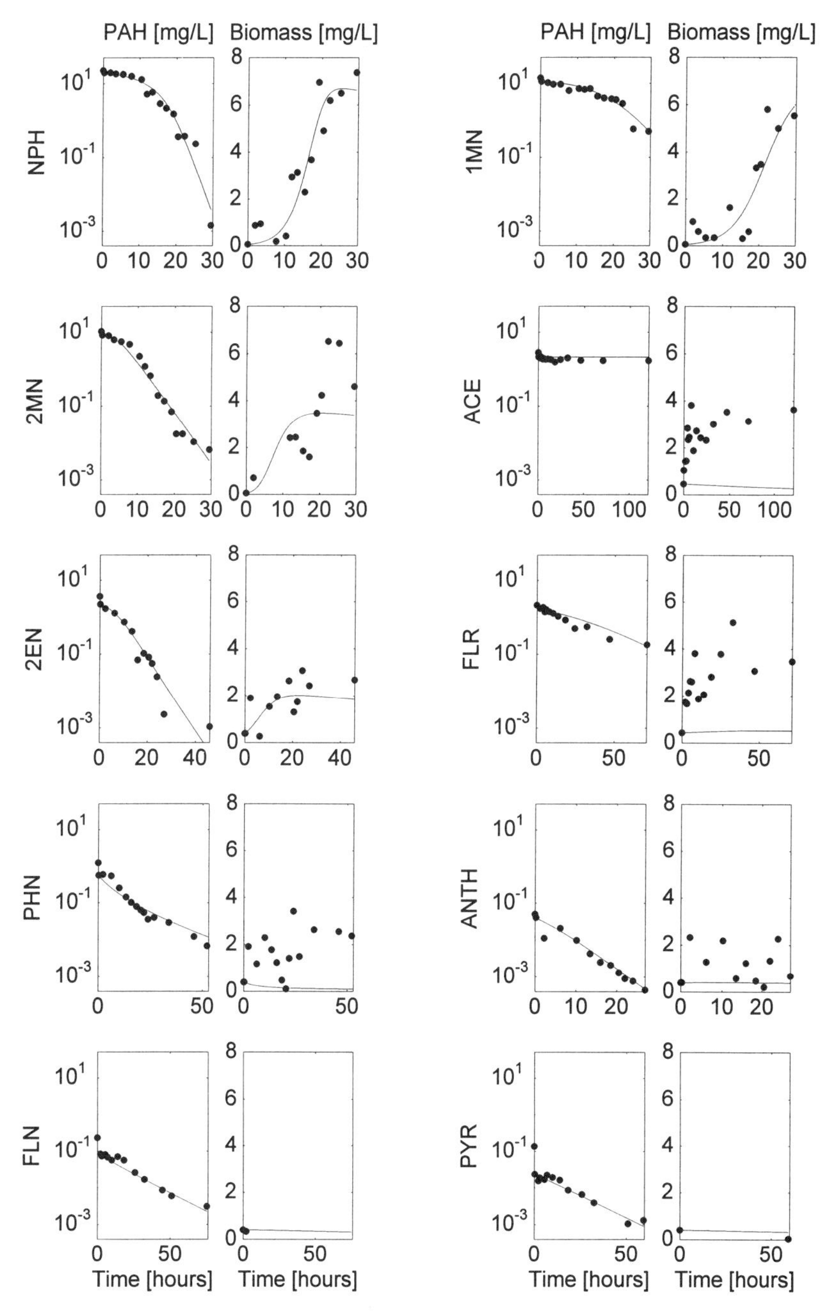

Figure 1. PAH (log axis) and Biomass (linear axis) Concentrations versus time: experimental measurements and modeling fits.

their structure, and each was biodegradable. It seems, therefore, that this consortium was unable to use acenaphthene as a sole carbon source because of the location of the cyclopentane ring.

Aerobic heterotrophs typically have yields in the range of 0.74 - 1.1 mg protein/mg carbon. The compounds here are seen to generally have yields well below these values. Because PAHs are large molecules, the biodegradation pathways have many steps. Some PAHs have pathways that may lead to end products other than mineralization. Furthermore, some intermediates can undergo non-enzymatic rearrangement that results in the formation of a quinone, which is unusable by the consortium. These possibilities result in a lower yield than that which is typically possible, which may be the cause for the observed low yields.

In addition, there seems to be a mismatch in the timing of substrate depletion and biomass growth. This is particularly evident in the data for 2MN. The biomass concentration undergoes a lag in growth, and then it continues to increase after the parent compound is depleted. This suggests that metabolites form and their transformation is the rate limiting step in biodegradation. This would also cause an apparent low value for yield. The HPLC chromatograms of the PAH samples have shown a rise in different elution peaks as the parent compound is depleted. This further supports the theory of the rise of metabolites and/or dead-end products. The investigation of these possibilities is a topic of ongoing research in our laboratory.

REFERENCES

Cerniglia, C. E. 1992. "Biodegradation of Polycyclic Aromatic Hydrocarbons." *Biodegradation. 3*: 351-368.

Mackay, D., W. Y. Shiu, and K. C. Ma. 1992. *Illustrated Handbook of Physical-Chemical Properties and Environmental Fate for Organic Chemicals*, Vol 2. Lewis Publishers: Boca Raton, FL.

Park, K. S., R. C. Sims, R. R. Dupont, W. J. Doucette, J. E. Matthews. 1990. "Fate of PAH Compounds in Two Soil Types: Influence of Volatilization, Abiotic Loss and Biological Activity." *Environmental Toxicology and Chemistry*, 9: 187-195.

Peters, C. A., P. A. Labieniec, C. D. Knightes. 1996. "Multicomponent NAPL Composition Dynamics and Risk," *Proceedings of the ASCE Annual Convention, Conference: NAPLs in the Subsurface Environment: Assessment and Remediation*; ASCE: Washington, DC, Nov 1996: 681-692.

Schocken, M. J., D. T. Gibson. 1984. "Bacterial Oxidation of the Polycyclic Aromatic Hydrocarbons Acenaphthene and Acenaphthylene," *Applied and Environmental Microbiology*, *48 (1)*: 10-16.

BIODEGRADATION OF FLUORANTHENE AND BENZO[a]PYRENE RELATED TO MICRO-AGGREGATION OF SOIL

Herm Zweerts (Katholieke Universiteit Leuven, Belgium)
Bert Torbeyns, Erik Smolders and Roel Merckx (Laboratory of Soil Biology and Soil Fertility, K.U. Leuven, Belgium)

ABSTRACT: A batch experiment was carried out with three different soils to determine the influence of micro-aggregation on the degradation of polycyclic aromatic hydrocarbons (PAHs). Two PAHs (fluoranthene (225 mg/kg); benzo[a]pyrene (40 mg/kg)) were added to these soils and subject to aerobic degradation for 112 days. Over this period fluoranthene decomposed on average to 3 % of the initial concentration, whereas benzo[a]pyrene only decomposed to 40 %. Differences between soils were observed for fluoranthene, corresponding to the organic carbon content of the different soils, whereas for benzo[a]pyrene this was not the case. On 14 DAA (days after addition) and 112 DAA the soils were wet-sieved into four size-fractions (< 2 µm, 2-53 µm, 53-250 µm and > 250 µm) after a mild dispersion to preserve micro-aggregation. Fractions with high organic carbon contents generally retained more fluoranthene and benzo[a]pyrene than fractions with low organic carbon content. To separate this organic carbon effect from a true micro-aggregation effect, PAH concentrations were normalized and their distribution amongst size separates re-investigated. Differences in biodegradation rate of the PAH between the different size separates vanished as a consequence. We conclude that differential degradation kinetics can be entirely ascribed to an association of the PAHs and soil organic matter contents of different aggregate classes.

INTRODUCTION

Most PAHs can be degraded by indigenous soil micro-organisms (bacteria, fungi). Despite the ubiquitous presence of these micro-organisms, a part of the PAHs seems to resist biodegradation (Scribner et al., 1992). Key processes governing this reduced bioavailability are often referred to as ageing (Hatzinger and Alexander, 1995) and may be ascribed to soil pore diffusion (Pignatello, 1989).

Protective mechanisms preventing decay in soil are well documented in soil science. Differences in organic matter content between soils are often related to differences in protection capacity. Organic matter in soil is known to be protected by aggregate formation. Soil aggregates can be divided in macro-aggregates (> 250 µm) and micro-aggregates (< 250 µm). Where micro-aggregates are formed by the binding of the individual soil particles by organic matter and other cementing agents like iron- and aluminum-oxides (Tisdall and Oades, 1982), organic matter may aggregate them to macro-aggregates.

Objective. The objective of this study was to determine the influence of micro-aggregates on the bioavailability of PAHs, using a similar line of thinking as in "conventional" soil science. Therefore three different soils were spiked with PAHs and incubated over 112 days. The PAH contents of the total soil and several micro-aggregate fractions were determined at two specified time intervals.

MATERIALS AND METHODS

Soil Sampling and Analysis. The top layer (0-10 cm) of three arable soils of different texture and organic matter content (loam, sand, peat) was sampled in two replicates. The samples were sieved (< 2 mm), dried to 7 % moisture and stored at 5 °C until usage. Organic carbon concentrations were determined by dry combustion. During incubation subsamples of 10 g were taken at 0, 7, 14, 28, 56 and 112 DAA for PAH concentration analysis by U.S. EPA SW-846 Method 3541/8100.

Incubation. The aerobic degradation experiment was carried out in 2 L glass jars that contained 1.2 kg of soil. After an initial incubation time of 60 hours at 25 °C, the soil was remoistened to 70 % of field capacity with aqua dest.. Subsequently the soil was spiked with a solution of PAHs in dichloromethane. This resulted in contamination levels of 225 mg/kg fluoranthene (FLU) and 40 mg/kg benzo[a]pyrene (B[a]P). As a control, soils without added PAHs and autoclaved soils with PAHs were incubated.

The soil was kept at 25 °C during the whole experiment. To avoid CO_2 concentrations higher than 2% the glass jars were opened at specified times and the soil was homogenized. CO_2 was monitored regularly in a 10 ml 1 M NaOH trap present in the glass jar (Santruckova and Simek, 1997).

Soil Fractionation. Subsamples of the soils on 0, 14 and 112 DAA were fractionated according to Monrozier et al. (1991). A 250 ml centrifuge tube is filled with 30 g soil, 200 ml aqua dest. and 5 quartz stones of 1.4 g. Macro-aggregates are reduced to their constituting micro-aggregates by shaking the tubes end-over-end for 16 hours. The contents of two tubes are combined and sieved over 250 and 53 μm. Destilled water is applied on the sieve until the outflow is clear. The fraction < 53 μm is subdivided in a fraction 53-2 μm and < 2μm by centrifuging and calculating the sedimentation time with Stokes' law.

TABLE 1. Selected Soil Characteristics.

	Peat	Sand	Loam
Organic Carbon (%)	9.1	3.1	0.8
pH ($CaCl_2$ 10^{-2} M)	4.4	4.7	9.7
CEC ($cmol_c$/kg)	20.6	9.7	10.2
Texture (% of mineral soil)			
2000 - 53 μm	65.3	79.9	15.0
53 - 2 μm	10.7	7.9	66.0
2 - 0 μm	4.7	2.0	19.0

RESULTS AND DISCUSSION

Total Soil. Degradation of fluoranthene and benzo[a]pyrene for the whole soil is shown in Figure 1. No decline in concentrations is observed at 14 DAA. After this lag period a steady decline of fluoranthene is observed. The decline in loam is somewhat faster than in sand or peat. Benzo[a]pyrene concentrations show a slow but steady decline in all three soils. Nevertheless, it is still present in significant amounts after 112 days. The values of fluoranthene after 112 days can be ranked according to the organic carbon content of the corresponding soil. Peat retains the highest concentrations while the lowest concentrations of fluoranthene are measured in the loam soil.

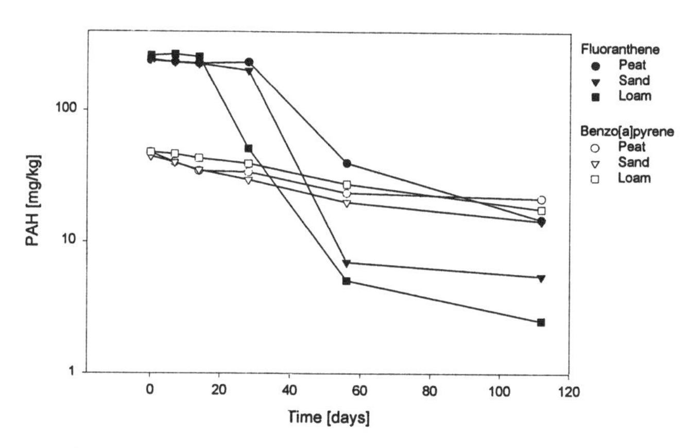

FIGURE 1. Degradation of Fluoranthene and Benzo[a]pyrene in three different soils.

Ageing. The gradual decrease of benzo[a]pyrene in al three soils is probably mainly due to microbial degradation and not too ageing processes. This was confirmed in two ways. First, a more rigid extraction (acetone, 48 hours) at day 112 resulted in only 6 mg/kg more benzo[a]pyrene. Second, extraction efficiencies of the sterile, spiked control soils remained constant in time (data not shown). However, we are aware that 112 days is perhaps too short a time period to observe ageing processes.

PAH Distribution in Soil Separates. Organic carbon percentages of the soil separates and the results of the PAH determinations at 14 and 112 DAA are shown in Table 2. The smallest fractions have the highest percentages of organic carbon. Loam is somewhat exceptional because all separates have a low carbon content and the fraction > 250 μm is negligible. In all < 2 μm separates differences between replicates are quite large. This is due to the small weight of this separate which causes more experimental variability.

The loss of PAHs in 112 days is clearly shown in all three soils and in all their separates. Differences in degradation rates between size separates become apparent and are inversely correlated to their organic carbon content. All fluoranthene concentrations are linearly correlated to organic carbon content after 112 days of incubation: FLU = 1.90 * OC, $r^2 = 0.85$. For benzo[a]pyrene such linearity is less clear but still present: B[a]P = 4.96 * OC, $r^2 = 0.55$. In these equations OC stands for the percentage organic carbon, FLU and B[a]P for the respective PAH concentrations.

TABEL 2. Distribution of Organic Carbon percentage and FLU and B[a]P over different size separates of three different soils at 14 and 112 DAA.

Fraction [µm]	OC % [-]	Fluoranthene [mg/kg dry fraction]				Benzo[a]pyrene [mg/kg dry fraction]			
Peat		*day 14*		*day 112*		*day 14*		*day 112*	
Total soil	9.12	226	226	15	15	36	32	17	26
> 250	7.78	124	103	10	10	20	23	16	16
250 – 53	6.69	179	176	12	12	22	27	20	21
53 – 2	16.66	437	487	38	34	65	94	84	72
< 2	17.24	615	290	24	50	130	85	81	155
Sand									
Total soil	3.11	220	216	5.5	5.5	41	28	16	2.0
> 250	2.23	89	84	2.9	3.7	16	8.5	13	7.5
250 – 53	1.36	69	70	1.9	1.4	12	12	4.5	9.5
53 – 2	9.70	567	548	18	17	121	93	85	91
< 2	10.57	358	824	9.9	17	66	161	9.5	12
Loam									
Total soil	0.85	233	233	2.8	2.5	41	44	20	15
250 – 53	1.07	38	38	1.4	1.3	17	7.8	4.7	7.3
53 – 2	0.59	163	179	2.4	1.6	28	19	8.4	8.7
< 2	2.00	539	751	3.1	8.3	16	56	11	53

Organic Carbon Normalized Distribution. To eliminate the influence of organic matter, PAH concentrations are divided by their organic carbon content. This allows to isolate the influence of micro-aggregation. After normalization the distribution of the PAHs over the micro-aggregates is given in Figure 2. Major differences in PAH distribution between 14 and 112 DAA vanished after normalization. The micro-aggregate distribution itself did not change during the experiment.

"Ab initio" all soils show a decline of 5 to 10 percent of their biggest fraction size (peat, sand: < 250 µm; loam: 250-53 µm) and a relative enrichment of the 53-2 µm fraction. For peat and sand this enrichment seems to increase with time.

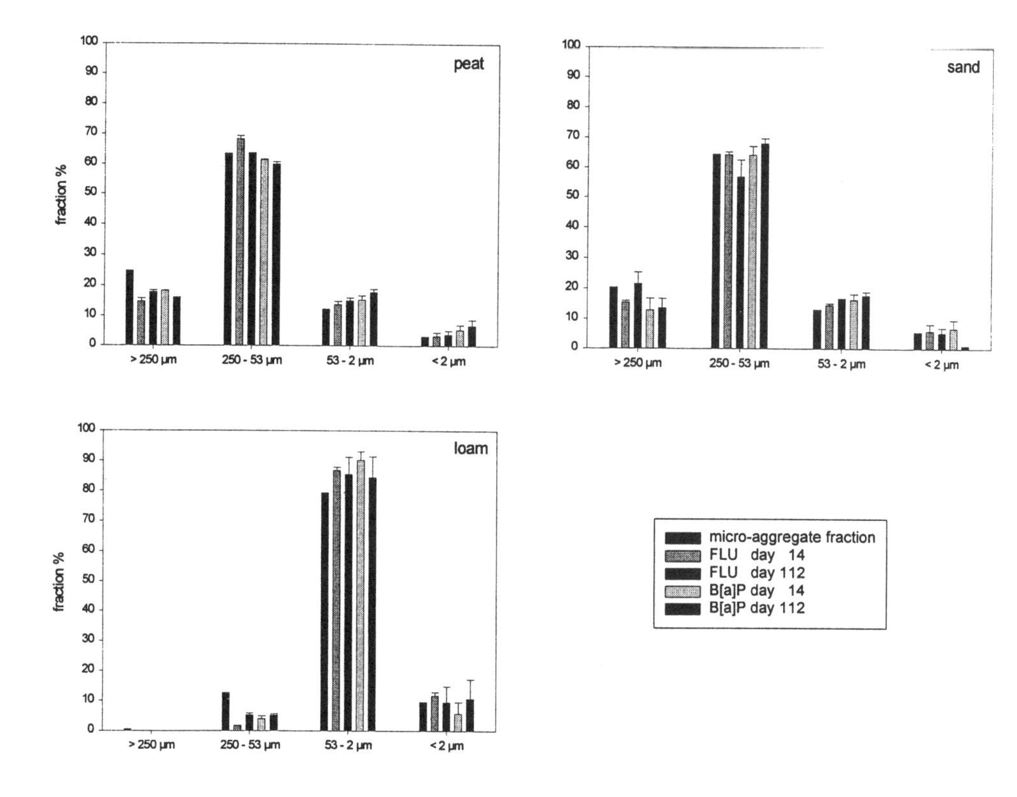

FIGURE 2. The relative distribution of micro-aggregates compared with the relative average distribution of Fluoranthene and Benzo[a]pyrene.
(n=2; error bar = maximum)

Conclusions. Degradation of fluoranthene in a 112 days aerobic incubation was inversely related to the organic carbon content of the soil (Figure 1). The fluoranthene concentrations in the different size separates show a similar relationship with organic carbon content. Benzo[a]pyrene concentrations are much less predictable by the organic carbon content (Table 2). However, on a whole, fractions with high organic carbon contents retained more PAHs than fractions with low organic carbon content.

To eliminate this organic carbon effect, all PAH concentrations were normalized (Figure 2) and their distribution amongst the size separates re-investigated. As a consequence, differences in degradation rate between size separates disappeared. So, during the time period of this experiment there is no reason to assume a micro-aggregation effect on the distribution of PAHs, apart from its indirect effect through the different organic carbon contents.

REFERENCES

Hatzinger, P.B., and M. Alexander. 1995. "Effect of aging of chemicals in soil on their biodegradability and extractability." *Environ. Sci. Technol. 29*: 537-545.

Monrozier, L.J., J.N. Ladd, R.W. Fitzpatrick, R.C. Foster, and M. Raupach. (1991). "Components and microbial biomass content of size fractions in soils of contrasting aggregation." *Geoderma 49*: 37-62.

Pignatello, J.J. 1989. "Sorption dynamics of organic compounds in soils and sediments." In Sahwney,B.L. and K. Brown (Eds.), *Reactions and movement of organic chemicals in soils.* pp. 45-80.Soil Science Soc. of America, Madison, WI.

Santruckova, H., and M. Simek. 1997. "Effect of soil CO_2 concentration on microbial biomass." *Biol. Fertil. Soils 25*: 269-273.

Scribner, S.L., T.R. Benzing, S. Sun, and D.A. Boyd. 1992. "Desorption and bioavailability of aged simazine residues in soil from a continuous corn field." *J. Environ. Qual. 21*: 115-120.

Tisdall, J.M., and J.M. Oades. 1982. "Organic matter and water-stable aggregates in soils." *Journal of Soil Science 33*: 141-163.

RELATIONSHIP BETWEEN DETOXIFICATION AND DECONTAMINATION DURING THE TREATMENT OF A PAH CONTAMINATED SOIL

Katia Santini, Jacques Bureau, ***Louise Deschênes***
NSERC Industrial Chair in site Bioremediation, École Polytechnique de Montréal, Montréal (Québec), Canada

ABSTRACT. The relationship between detoxification and decontamination has been studied through the bioaugmentation and the biostimulation of a PAH-contaminated soil. Residual toxicity was measured through earthworm mortality (*Eisenia foetida*) and growth of watercress (*Lepidium sativum*). The chemical analysis allowed quantification of the 16 USEPA priority PAHs. Results showed that for the different treatments studied, total PAH concentration decreased by at least 50 % over a 45 week period. Among the different amendments studied (*P. chrysosporium*, sawdust and mineral salt), nutrient addition showed the greatest influence on the reduction of the toxicity. Bioaugmentation with *P. chrysosporium*, did not increase the soil toxicity nor the rate of PAH transformation.

INTRODUCTION

Polycyclic aromatic hydrocarbons (PAHs) are found in high concentrations on many industrial sites associated with petroleum or wood preserving industries. Because of the carcinogenicity and mutagenecity of some PAHs, their removal from contaminated soils have received increasing attention. Bioremediation, using microorganisms to breakdown hazardous organic pollutants into harmless compounds, is a cost effective technique that preserves the soil as a living system, and was shown to offer a good alternative for the removal of most PAHs (Atlas and Cerniglia, 1995). Bioremediation can be done through the biostimulation of indigenous microorganisms or by adding specific microorganisms in the soil (bioaugmentation). However, some limitations, mainly associated with the bioavailability of the contaminants in soil, may prevent the complete biodegradation of high molecular weight PAHs (Deschênes et al., 1996; Andersson and Henrysson, 1996) and the formation of metabolites, which may be more toxic than the parent compounds, may prevent the detoxification of the treated soil.

So far, the efficiency of biological treatment has been based on achievement of chemical criterion (Hund and Traunspurger, 1994). That requires an exhaustive chemical characterization. This chemical characterization presents however several weaknesses. First of all, it does not take into account site-specific factors such as contaminant sorption/desorption and bioavailability, nor does it take into account the potential complexity of contaminant types and sources (Hamilton et al., 1994). Furthermore, because only the disappearance of initial compound is measured, chemical analysis may not indicate the fraction of the initial contaminants that has been mineralized relative to the proportion that has been transformed into metabolites (Baud-Grasset et al., 1994). To complement the lack of chemical analysis, biological assays should be performed. These assays allow the detection of the toxicity induced by the presence of compounds non-detected by chemical analysis such as biodegradation products and metabolites, and the toxicity induced by the consequences of contaminant mixtures (antagonism, synergism, additivity) (Baud-Grasset et al., 1994; Renoux, 1995). The purpose of this work was to evaluate the relationship between the detoxification and the decontamination of a PAH-contaminated soil during its treatment by both white rot fungi addition (bioaugmentation) and biostimulation.

MATERIALS AND METHODS

Soil. The weathered PAH-contaminated soil used in all experiments was collected near Montréal, Québec (Canada). A bulk sample of surface soil (200 L) was collected at a 5-20 cm depth. The wet soil was homogenized, sieved to 5 mm and kept at 4 °C in the dark until use. Soil characterization showed high concentrations of high molecular weight PAHs (table 1). Concentration of many PAHs was above the C Quebec criteria, indicating that remediation must be performed. The water content was 21 % (soil wet basis) and the pH was 7.5.

TABLE 1. Soil PAH and C_{10}-C_{50} concentrations. (n=20).

Compound	Concentration (mg/kg)	Quebec Criteria C (mg/kg)
Acenaphtylene	88	100
Acenaphtene	8	100
Phenanthrene	3	50
Anthracene	16	100
Fluoranthene	41	100
Pyrene	**163**	**100**
Benzo[a]anthracene	**29**	**10**
Chrysene	**50**	**10**
Benzo[b+j+k]fluoranthene	**46**	**10**
Benzo[a]pyrene	**41**	**10**
Dibenzo[ah]anthracene	5	10
Indeno[1,2,3-cd]pyrene	**12**	**10**
benzo[ghi]peryene	**17**	**10**
dibenzo[a,i]pyrene	4	10
dibenzo[a,h]pyrene	1	10
dibenzo[a,l]pyrene	4	10
C_{10}-C_{50}	**33 000**	**3500**

Experimental design. To study the relationship between detoxification and decontamination during the treatment of the weathered PAH contaminated soil, specially designed 20 L aerated (0.1 m^3 air/m^3 soil/hour) reactors were used. Six different experimental conditions were studied in duplicate: (1) soil inoculated with a white rot fungi (*Phanerochaete chrysosporium*) grown on a mixture of sawdust (10 % v/v) and mineral nutrients (MSM) (Greer et al., 1990), (2) soil sterilized prior to the inoculation, (3) soil biostimulated with MSM and sawdust (10 % v/v), (4) soil amended only with MSM, (5) soil receiving no treatment (control), and (6) an abiotic control (sodium azide added at 0.2 % (w/w)). To each reactor, 12-kg of soil with respective amendments were added. The soil water content was adjusted to 25 % (w/w) with sterile MSM solution in the first four treatments. Every week, reactors content was homogenized to prevent the formation of dead zones and the environmental conditions (water content, air flow rate, and temperature) were monitored. The 45 week incubation was performed at 20 °C in the dark. Each reactor was sampled after 0, 5, 14, 20 and 45 weeks. The soil was mixed before each sampling period. For each sampling, 65 samples were collected from different locations in accordance to a statistically designed plan and mixed together. Each sample was then split into two fractions, one to be used for the biological assays and the other for the chemical analysis. Prior to their analysis, all samples were stored at 4 °C.

Chemical analyses. The method used to extract PAHs from the soil was based on EPA SW-846 Method 8270D modified for GC/MS quantification in SIM mode (Selective Ion Monitoring) (EPA, 1998a). The modification consisted of using an ultrasonic extraction with a solution of 20 % acetone and 80 % dichloromethane (EPA, 1998b). The extraction was followed by a gel-permeation clean-up (EPA, 1998c) and the extract was then analyzed by GC-MS.

Biological assays. The residual toxicity of treated soil was measured by earthworm mortality (14 days) (*Eisenia foetida*), and growth (17 days) of watercress (*Lepidium sativum*). To avoid the alteration of soil characteristics such as the sorption-phenomenon by the dilution with an artificial soil, the analysis was made at a unique concentration of soil (100 %). The method used to perform the earthworm mortality assay was based on EPA/600/3-88/029 (EPA, 1989). The soil sample was divided in 15 replicates of 40 g each (2 organisms in each sample to avoid serial mortality) and the water content was adjusted at 75 % of the water holding capacity. The method used to perform the watercress assay was based on ISO 11269-2 (ISO, 1995). The soil sample was divided in 4 replicates of 130 g each (10 seeds in each sample) and the water content was adjusted at 55 % of the water holding capacity.

RESULTS AND DISCUSSION

PAH biodegradation. Among all studied PAHs, the biodegradation was more important for the four-ring PAHs (Figure 1A). After 45 weeks, the four-ring PAH concentrations in the soil were reduced by at least 80 % in the bioaugmented and biostimulated treatments and by 70 % in the control (soil without amendment).

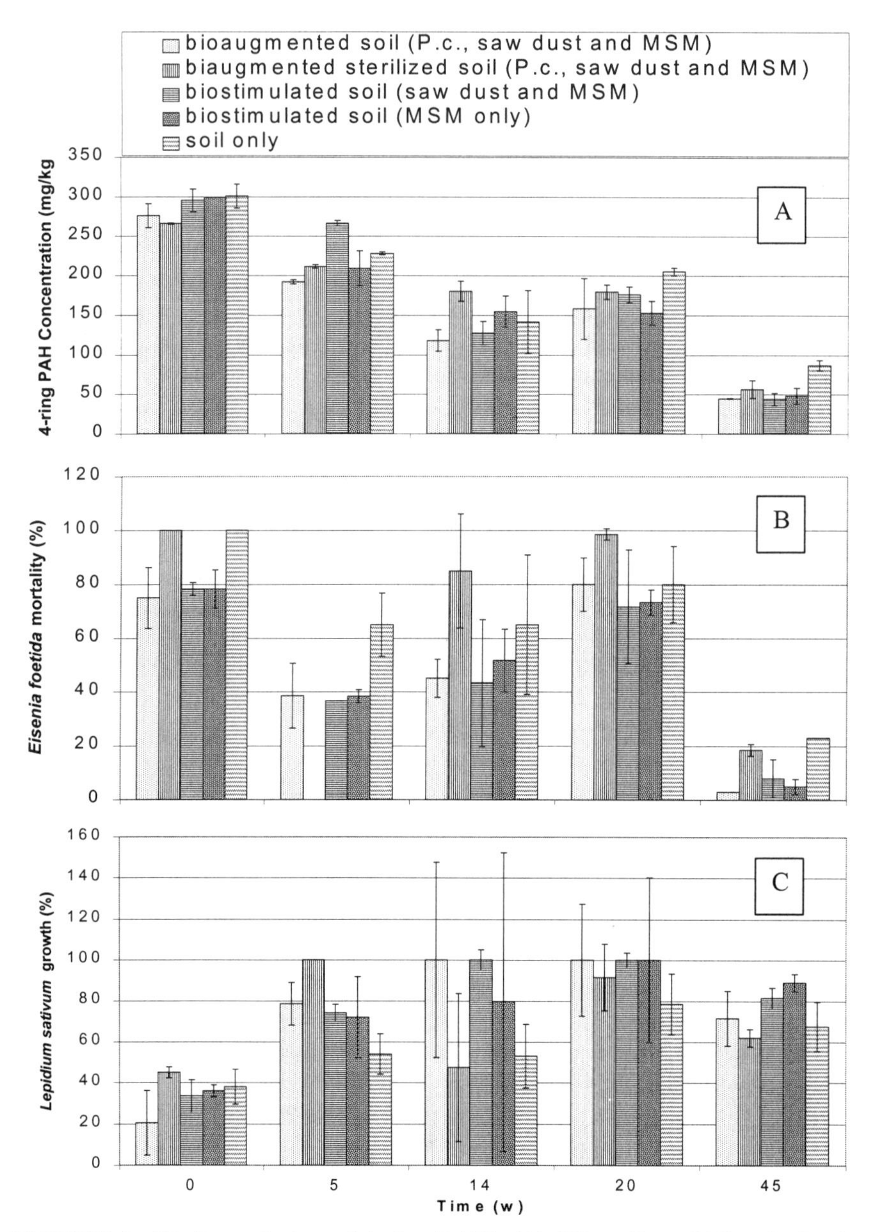

FIGURE 1. Biodegradation of 4-ring PAHs (A), Mortality of *Eisenia foetida* after 14-days of incubation (B), and Growth of *Lepidium sativum* after 17-days of incubation (C) in the presence of the different amendments *P. chrysosporium*, saw dust and MSM). *Errors bars represent* one standard deviation about the mean of duplicate.

The concentration of these PAHs was then below the Quebec criteria C. In comparison and after the same period, the higher-molar weight PAHs (5-6 rings) were reduced by a maximum of 20 % , and the concentration of 3-ring PAHs by 40 %. The concentration of most of the 3-ring PAHs was below the detection threshold except for the acenaphtylene, which is the most volatile of the 3-ring PAHs. The total PAH removal was about 50 % for the control treatment and nearly 60 % for all the other treatments. The concentration of the studied PAHs in the abiotic control has remained constant (near its initial value) throughout the 45 weeks. Among the different amendments studied, MSM addition showed the greatest influence on the reduction of PAH concentration. On the other hand, bioaugmentation with *P. chrysosporium*, did not increase the rate of PAH transformation. As the PAH concentrations decreased in all samples as the same way observed in the control, this could mean that the control of environmental conditions, such as aeration and water content as well as a good homogeneisation of the soil were sufficient to trigger PAH biodegradation.

Biological assessment. For all studied treatments, the biological assays have shown a reduction in soil toxicity. In fact, the toxicity decreases at a similar rate in all treatments (earthworm assay), but after the 45th week, it appears that the MSM addition reduced the toxicity to a lower level (5 %) compared with the soil receiving no treatment (23 %) (Figure 1B). Bioaugmentation with *P. chrysosporium*, did not increase the soil toxicity either in the earthworm and the watercress assays (Figure 1B and 1C).

In conclusion, in spite of the fact that the concentration of a few PAH had not reach the selected chemical criterion, an important reduction of the toxicity was observed. It is then advisable to use the relationship between detoxification and decontamination to better assess a bioremediation process.

ACKNOWLEDGMENTS

The authors acknowledge the financial support from the industrial Chair partners: Alcan, Bodycote/Analex, Bell Canada, Browning-Ferris Industries, Cambior, Centre québécois de valorisation de la biomasse et des biotechnologies (CQVB), Hydro-Québec, Natural Science and Engineering Research Council (NSERC), Petro-Canada and SNC-Lavalin and Prof. Réjean Samson for valuable discussion.

REFERENCES

Anderson, B. E., and C. E. Cerniglia. 1996. "Accumulation and Degradation of Dead-End Metabolites during Treatment of Soil Contaminated with Polycyclic Aromatic Hydrocarbons with Five Strains of White-rot Fungi." *Appl. Microbiol. Biotechnol. 46*: 647-652.

Atlas, R. M., and C. E. Cerniglia. 1995. "Bioremediation of Petroleum Pollutants : Diversity and Environmental Aspects of Hydrocarbon Biodegradation." *BioScience. 45*: 332-338.

Baud-Grasset, F., S. I. Safferman, S. Baud-Grasset, and R. T. Lamar. 1994. "Demonstration of Soil Bioremediation and Toxicity Reduction by Fungal Treatment." In R. E. Hinchee, A. Leeson, I. Semprini, and S. K. Ong (Eds.), *Bioremediation of Chlorinated and PAH Compounds,* pp. 496-500. Lewis, Boca Raton, FLA.

Deschênes, L., P. Lafrance, J.-P. Villeneuve, and R. Samson. 1996. "Adding sodium dodecyl sulfate and Pseudomonas aeruginosa UG2 biosurfactants inhibits polycyclic aromatic hydrocarbon biodegradation in a weathered creosote-contaminated soil." *Appl. Microbiol. Biotechnol. 46*: 638-646.

EPA. 1989. Protocols for Short Term Toxicity of Hazardous Wastes Sites : Eartworm Survival (*Eisenia foetida*). *EPA Technical Report : EPA/600/3-88/029 (modified).*

EPA. 1998a. EPA Series SW-846 Methods : 3550B *Ultrasonic Extraction.*

EPA. 1998b. EPA Series SW-846 Methods : 8270D *Semivolatil Organic Compounds by Gas Chromatography/Mass Spectrophotometry (GC/MS).*

EPA. 1998c. EPA Series SW-846 Methods : 3640A *Gel-Permeation Cleanup.*
Greer, C. W., J. Hawari and R. Samson. 1990. " Influence of environmental factors on 2,4-dichlorophenoxyacetic acid degradation by Pseudomonas cepacia isolated from peat. *Arch Microbiol.* 154. 317-322.

Hund, K., and W. Traunspurger. 1994. "Ecotox - Evaluation strategy for soil bioremediation exemplified for a PAH - Contaminated site." *Chemosphere. 29*: 37-55.

ISO. 1995. 11269-2 Soil quality - Determination of the effects of pollutants on soil flora - Part 2 : Effects of chemicals on the emergence and growth of higher plants.

Renoux, A. Y., R. D. Tyagi, Y. Roy, and R. Samson. 1995. "Ecotoxicological Assessment of Bioremediation of a Petroleum Hydrocarbon-Contaminated Soil." In R. E. Hinchee, F. J. Brockman and C. M. Vogel (Eds.), *Microbial Processes for Bioremediation,* pp. 259-263. Lewis, Boca Raton, FLA.

NAPHTHALENE DEGRADATION AND MINERALIZATION BY NITRATE-REDUCING AND DENITRIFYING PURE CULTURES

Karl J. Rockne (Chemical Engineering, Rutgers University, Piscataway, NJ)
Joanne C. Chee-Sanford (Animal Science, University of Illinois, Urbana, IL)
Robert A. Sanford (Civil Engineering, University of Illinois, Urbana, IL)
Brian Hedlund (Microbiology, University of Washington, Seattle, WA)
James T. Staley (Microbiology, University of Washington, Seattle, WA)
Stuart E. Strand (Forest Resources, University of Washington, Seattle, WA)

ABSTRACT: Although most PAHs with fewer than five rings are known to be biodegraded under aerobic conditions, most contaminated sediments and soils are anaerobic. In these environments, aerobic bioremediation may be difficult to implement because of the problems associated with oxygen delivery to the subsurface. With recent results clearly demonstrating some bicyclics and PAHs can be degraded without oxygen, further progress in understanding this process will be achieved by the identification of pure cultures of anaerobic PAH degrading bacteria. We attempted to isolate pure bacterial cultures from a denitrifying enrichment that anaerobically degrades phenanthrene, naphthalene, and biphenyl with stoichiometric nitrate reduction. After enrichment and screening, four pure cultures were obtained. The isolates were assayed for the ability to degrade naphthalene or phenanthrene as the sole source of carbon and energy in the presence of nitrate. Two pure cultures demonstrated unambiguous naphthalene biodegradation ability, designated NAP-3-1 and NAP-4. Both NAP-4 and NAP-3-1 transformed naphthalene, and the transformation was nitrate dependent. No significant removal of naphthalene occurred in nitrate-limited incubations or in cell-free controls. Both cultures partially mineralized naphthalene, representing 12-15% of the initial added ^{14}C-labeled naphthalene. Isolate NAP-3-1 was a denitrifier, as shown by gas production in a denitrification assay. In contrast, NAP-4 did not produce gas, but did produce significant amounts of nitrite. The complete 16s rDNAs of NAP-3-1 and NAP-4 were sequenced and compared to other PAH- and non-PAH degrading bacteria. NAP-4 was phylogenetically closely related to *Vibrio* spp. and NAP-3-1 was phylogenetically closely related to *Pseudomonas* spp., results which suggest that anaerobic bicyclic degradation ability may be more widely spread within the proteobacteria.

INTRODUCTION

Over the last decade our understanding of anaerobic aromatic biodegradation has increased greatly. With the isolation of denitrifying and sulfate reducing pure cultures able to degrade toluene, several biodegradation pathways and phylogenetic relationships between strains have been determined. These studies led to the finding that this metabolic ability is relatively widespread. In contrast, until recently anaerobic biodegradation of unsubstituted aromatics like benzene and polycyclic aromatic hydrocarbons (PAHs) has not been widely reported (with a few exceptions, e.g. Mihelcic and Luthy, 1988). In fact, anaerobic biodegradation of these compounds was thought unlikely to be biochemically possible due to the lack of ring constituents such as the methyl group on toluene (Evans and Fuchs, 1988). In the last three years however, anaerobic biodegradation of bicyclics and PAHs such as biphenyl, naphthalene, phenanthrene, fluorene, and fluoranthene have been conclusively demonstrated in marine sulfate-reducing and denitrifying enrichment cultures derived from Puget Sound, WA (Rockne and Strand, 1998), Arthur Kill, NJ (Zhang and Young, 1997), and San Diego Harbor, CA (Coates et al., 1997). At present it is unknown why this activity is seemingly more prevalent in marine environments.

Although a pathway for initial PAH ring oxidation has been found using a mixed sulfate-reducing enrichment culture (Zhang and Young, 1997), detailed phylogenetic analysis and further biochemical pathway elucidation would be greatly enhanced by studies with pure cultures. In the work reported here, we give an overview of the isolation of pure cultures of anaerobic PAH-degrading bacteria derived from a denitrifying enrichment culture previously demonstrated to degrade PAHs anaerobically (Rockne and Strand, 1998).

MATERIALS AND METHODS

The source of inoculum for the isolation procedures was a highly enriched denitrifying culture previously shown to degrade naphthalene, biphenyl, and phenanthrene with near stoichiometric amounts of nitrate reduction under strictly anaerobic conditions (Rockne and Strand, 1998). The culture was initially enriched on a mixture of the PAHs in a fluidized bed reactor (FBR), but was subcultured on individual PAHs in batch culture studies demonstrating stoichiometric biodegradation and mineralization of the PAHs. A schematic of the entire isolation procedure used in this work is shown in Figure 1. A complete description of the isolation procedure is found in Rockne et al. (1999). Briefly, cells were first transferred to agar plates containing nitrate-amended R2A denitrifying artificial seawater (ASW) media. R2A has a variety of electron donors for selection of heterotrophic bacteria. Cells from individual colonies on the R2A plates were transferred to liquid media with individual PAHs. Liquid cultures demonstrating growth (by turbidity) were transferred to agar shake (AS) tubes fed PAH by diffusion through the agar from a hexadecane phase as described previously (Rockne et al., 1999). Individual colonies from the AS tubes were transferred to liquid culture containing nitrate and PAHs. Cultures showing growth with PAH and nitrate transformation were plated twice on R2A agar and the resulting colonies were assayed for PAH degradation, denitrification, and mineralization as described previously (Rockne et al., 1999).

Colonies of selected strains grown anaerobically on R2A plates were harvested and genomic DNA was isolated as described previously (Rockne et al., 1999). 16s rDNA was amplified from the genomic DNA preparations by the PCR using standard techniques (Sambrook et al., 1989), sequenced, and compared with several sequences accessed from the Ribosomal Database Project (RDP, Larsen et al., 1993).

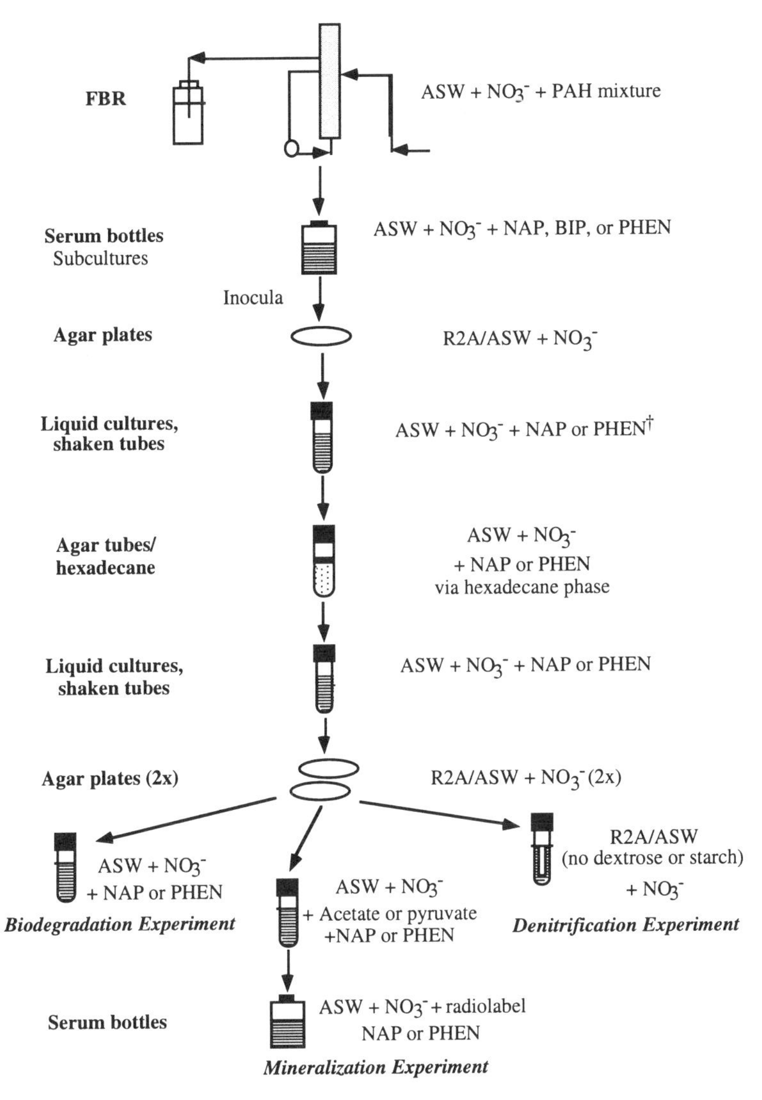

FIG. 1. Schematic of biodegradation experiments and procedures used to isolate nitrate-reducing PAH-degrading bacteria. †- No growth was observed in biphenyl-fed liquid subcultures.

RESULTS AND DISCUSSION

Screening of AS tube Colonies for PAH and Nitrate Transformation. No colonies grew in liquid cultures inoculated from the initial R2A plating of the biphenyl-fed culture. A total of 13 colonies were detected in the agar shake (AS) tubes. These colonies were incubated in liquid culture with either naphthalene or phenanthrene and nitrate and screened for the ability to transform PAH and nitrate.

All of the 13 cultures could transform nitrate (Table 1). However, only three naphthalene-fed (NAP-3-1, NAP-3-2, and NAP-4) and one phenanthrene-fed culture (PHEN-2-3) could transform PAH. This was not entirely unexpected because the R2A agar is a rich media with several electron donors which would likely have selected for a variety of non-PAH-degrading nitrate-reducers and/or denitrifiers.

Table 1. Summary of initial PAH biodegradation and nitrate transformation screening assay results for putative isolate cultures.

Culture	Transform PAH	Transform Nitrate
	Naphthalene-Fed Cultures	
NAP-2-1	-	+
NAP-2-2	-	+
NAP-3-1	+	+
NAP-3-2	+	+
NAP-4	+	+
	Phenanthrene-Fed Cultures	
PHEN-1-1	-	+
PHEN-1-2	-	+
PHEN-1-3	-	+
PHEN-2-1	-	+
PHEN-2-2	-	+
PHEN-2-3	+	+
PHEN-3-1	-	+
PHEN-3-2	-	+

PAH Biodegradation Assays. The four cultures able to transform PAH and nitrate were then plated twice on R2A to ensure purity. These cultures were assayed for PAH biodegradation, mineralization, and denitrification ability.

All of the naphthalene-fed cultures could degrade and mineralize naphthalene (Table 2). In addition, there was no significant naphthalene transformation in nitrate-free incubations of NAP-3-1 and NAP-4, demonstrating nitrate-dependence (there was no nitrate-free control with NAP-3-2). The total amount of mineralization varied from 7% for NAP-3-2 to 15% for NAP-3-1. NAP-3-1 appeared to be a denitrifier, as shown by the gas production assay (Table 2). No significant gas production was observed with the other naphthalene-fed cultures. Interestingly, NAP-3-2 did not significantly transform nitrate with naphthalene as the sole carbon and energy source.

The results with the phenanthrene-fed culture were largely inconclusive. Although the culture exhibited phenanthrene degradation and nitrate transformation, significant transformation of phenanthrene ($40 \pm 15\%$ removal) was observed in nitrate-free incubations (Table 2). In addition, only one of the two duplicate incubations showed significant mineralization ability. Subsequent studies with this culture suggested it was not a pure culture, but probably a co-culture of two or more organisms which grew in single colonies (data not shown).

Table 2. Summary of biodegradation, denitrification, and mineralization assay results for pure cultures.

Culture	Degrade PAH/ NO_3^- dependent[1]	Transform Nitrate[2]	Mineralize PAH (%)[3]	Gas Production
		Naphthalene-Fed Cultures		
NAP-3-1	+/+	+	+ (15)	+
NAP-3-2	+/NT[4]	-	+ (7)	-
NAP-4	+/+	+	+ (12)	-
		Phenanthrene-Fed Culture		
PHEN-2-3	+/-	+	+/-[5] (0.2)	+

[1]-No significant PAH transformation in nitrate-free control.
[2]-With PAH as sole carbon source.
[3]-% of initial ^{14}C-labeled PAH carbon mineralized to CO_2 after 57 days (average of duplicates).
[4]-Not tested with nitrate-free control.
[5]-Only one of two duplicates had significant activity.

Phylogenetic Characterization. NAP-3-1 and NAP-4 were selected for phylogenetic characterization because they both demonstrated unambiguous nitrate-dependent naphthalene transformation and mineralization ability. Comparison of the 16s rDNA from both NAP-3-1 and NAP-4 with sequences in the RDP demonstrated that they were both phylogenetically related to bacteria in the γ proteobacteria (Table 3). NAP-4 was phylogenetically closely related to *Vibrio* spp. *Vibrio* are characterized by a fermentative metabolism and do not produce gas (Holt, 1994), consistent with the results from the denitrification assay. This suggests that NAP-4 may possess a fermentative or nitrate-reductive metabolism. NAP-3-1 is phylogenetically very closely related to several *Pseudomonas stutzeri* spp. Many Pseudomonads are denitrifiers, consistent with the denitrification assay results (Dagher et al., 1997).

Table 3. Percent homology of 16s rDNA sequences of NAP-3-1 and NAP-4 with bacteria from the RDP closest in similarity.

Culture	NAP-3-1	NAP-4
Vibrio parahaemolyticus	-	0.992
Vibrio alginolyticus	-	0.992
Pseudomonas stutzeri AN10	0.997	-
Pseudomonas stutzeri DSM 50227	0.997	-

CONCLUSIONS

This is the first report we know of demonstrating anaerobic degradation of an unsubstituted aromatic compound by pure cultures. The biodegradation assays demonstrated that anaerobic PAH biodegradation metabolism is diverse, with bacteria from both denitrifying and fermentative or nitrate-reductive metabolisms capable of naphthalene mineralization. These isolates will allow further understanding of the metabolic pathways, nutritional requirements, and genetics of anaerobic aromatic degradation. This knowledge will be critical for the assessment of anaerobic PAH biodegradation as a potential active bioremediation or natural attenuation remediation option.

ACKNOWLEDGMENT

Part of this work was supported by grant ONR N00014-92-j-1578 from the Office of Naval Research.

REFERENCES

Coates, J. D., J. Woodward, et al. 1997. "Anaerobic degradation of polycyclic aromatic hydrocarbons and alkanes in petroleum-contaminated marine harbor sediments." *Appl. Environ. Microbiol.* 63:3589-3593.

Dagher, F., E. D'eziel, et al. 1997. "Comparative study of five polycyclic aromatic hydrocarbon degrading bacterial strains isolated from contaminated soils." *Can. J. Microbiol.* 43: 368-77.

Evans, W. C. and G. Fuchs. 1988. "Anaerobic degradation of aromatic compounds." *Ann. Rev. Microbiol.* 42: 289-317.

Holt, J. G. 1994. *Bergey's manual of determinative bacteriology*. Williams & Wilkins, Baltimore, MD.

Larsen, N., G. J. Olsen, et al. 1993. "The ribosomal database project." *Nucleic Acids Res.* 21: 3021-3023.

Mihelcic, J. R. and R. G. Luthy. 1988. "Degradation of polycyclic aromatic hydrocarbons compounds under various redox conditions in soil-water systems." *Appl. Environ. Microbiol.* 54: 1182-1187.

Rockne, K. J. and S. E. Strand. 1998. "Biodegradation of bicyclic and polycyclic aromatic hydrocarbons in Anaerobic Enrichments." *Environ. Sci. Technol.* 32: 3962-3967.

Rockne, K. J., J. C. Chee-Sanford, R. A. Sanford, B. Hedlund, J. T. Staley, and S. E. Strand. 1999. "Anaerobic Naphthalene Degradation by Microbial Pure Cultures under Nitrate-reducing Conditions." *Appl. Environ. Microbiol* In review.

Sambrook, J., E. F. Frisch, and T. Manniatis. 1989. *Molecular cloning: a laboratory manual.* Cold Spring Harbor Laboratory Press, Plainview, NY.

Zhang, X. and L. Y. Young. 1997. "Carboxylation as an initial reaction in the anaerobic metabolism of naphthalene and phenanthrene by sulfidogenic consortia." *Appl. Environ. Microbiol.* 63: 4759-4764.

BIOTRANSFORMATION OF TEN PAH'S BY MEANS OF AN AEROBIC MIXED CULTURE

Sixto Pérez (Instituto Mexicano de Tecnología del Agua. Jiutepec, Morelos. MEXICO) and ***Luis G. Torres*** (Instituto Mexicano del Petróleo. México, D.F. MEXICO).

ABSTRACT Even when there is not enough safe information regarding the presence of polycyclic aromatic hydrocarbons (PAHs) on waterbodies or underground waters in Mexico, some recent reports of works developed in different reservoirs show how sizable this problem has become. Our group has worked on biodegradation of some chlorophenols and pesticides, using low-cost biofilters. Our aim is to select aerobic microorganisms able of degrading PAHs with complex chemical structures (i.e., with three or more aromatic rings), and to study the abilities of these microrganisms. As the next step, this aerobic consortia can be immobilized in a low-cost packaging column and be used in the degradation of streams containing one or more of the selected PAHs.

INTRODUCTION

Polycyclic aromatic hydrocarbons (PAHs) are widely distributed environmental contaminants that have detrimental biological effects, including acute and chronic toxicity, mutagenicity, and carcinogenecity (Shuttleworth and Cerniglia, 1997). Because of the ubiquitous occurrence, recalcitrance, bioaccumulation, potential, suspected carcinogenic activity of PAHs many researchers have developed procedures for the bioremediation of PAH-contaminated sites. Even when there is not enough safe information regarding the presence of polynuclear aromatic hydrocarbons (PAHs) on waterbodies or underground waters in Mexico, some recent reports of works developed in different reservoirs show the problem's size

Most of the published papers dealing with PAHs biodegradation have been developed for contaminated soils, and very frequently involves the use of white-rot fungi (Andersson and Henrysson, 1996). Bacteria genera as *Pseudomonas* have demonstrated high capabilities in the degradation of PAHs (Johnsen et al., 1996). Some papers have reported the problem of aquifer contamination by PAHs and its potential solutions (Nielsen and Christensen, 1994).

Objective. The aim of this work is to propose the use of an aerobic mixed culture for degradation of PAHs.

MATERIALS AND METHODS

The employed aerobic mixed culture was Polytox, a commercially available liophilyzed bacterial culture used for rapid biological toxicity screen tests (Polybac Co., Bethlehem, PA USA). The powder was rehydrated in the saline solution defined for the BOD test in Standard Methods (1.6 g of dried powder in 250 ml of BOD saline solution). PAHs concentrations were measured using a gas chromatography/mass spectrometry (GC/MS) system. Bacterial growing was evaluated indirectly by observing optical density (O.D.) changes at 610 nm. All the tests were carried on 250 ml flasks with cotton plugs. 5 ml of the Politox inoculum were added to 45 ml of the BOD test saline medium. Blanks of PAHs only (no cells) and of cells only (no hydrocarbons) were run all the time. Samples were taken every 24 hours for as long as 144 hours. The flasks were agitated during that time at 32+/-2° C. The PAHs were chosen by the following criteria: (a) their significant presence in Mexican waterbodies and (b) their molecular and toxicological properties. It was decided to work with 3- to 6-ring compounds to try to correlate biotransformation, chemical structure, and toxic level. The selected PAHs were: benzo(g,h,i)perylene, benzo(k)fluoranthene, benzo(a)pyrene, fluoranthene, dibenzo(a,h)anthracene, anthracene, pyrene, benzanthracene, chrysene, and indenopyrene (all at concentrations of 50 mg/L). The benzo(a)pyrene and dibenzo(a,h)anthracene were chosen for 5, 20, and 50 mg/L tests to investigate the posible effects of inicial concentrations. Some significative physical-chemical properties of the hydrocarbons are displayed on Table 1.

RESULTS AND DISCUSSION

Biomass growth in presence of PAHs. On Table 2, the maximum growth values in presence of the PAHs are summarized. Reported values are an average of a duplicate. The value for the blank (no hydrocarbon present) also is included. In all the assessments, microbial growth occurred. Cell growth was more notorious in the case of anthracene, benzanthracene, benzo(a)pyrene, and chrysene. Optical densities values up to 0.198 were achieved, in comparison with the 0.016 blank value (a twelve-fold increase). In contrast, low cell growth was observed in the presence of indenopyrene, fluoranthene, and benzo (g,h,i) perylene (O.D. = 0.027-0.043).

PAH biodegradation. Figure 1 shows a typical degradation profile found in the assessments, in this case for the benzo (a) pyrene (initial concentration of 50 mg/l). This assessment was stopped at 72 hours, since no changes where observed after this point. Bars indicate the test variability and full points indicate the average concentration value.

TABLE 1. Some Physical-Chemical Properties of the selected PAHs

Compound	CAS number	Molecular weight	Melting point (°C)	Boiling point (°C)	Vapor pressure Torr @ T(°C)	Solubility in water mg/l @ T(°C)
Anthracene	120-12--7	178.23	216.3	340		1.29@25
Benzanthracene	56-55-3	228.29	162	435	5E-09@20	0.01
Benzo(a)pyrene	50-32-8	252.32	179.3	488-495	5E-09@20	0.0038@25
enzo(g,h,i)perylene	50-32-8	276.34	222	>500	1E-10@20	2.6E-04@25
nzo(k)fluoranthene	207-08-9	252.32	217	480	9.59E-11@20	
enzo(a,h)anthracene	53-70-3	278.35	266-267	524		0.0005@25
Chrysene	218-01-9	228.29	225	448	6.3E-09@25	0.006@25
Fluoranthene	206-44-0	202.26	107-110	375		0.265@25
Indenopyrene	193-39-35	276.34	160	536		
Pyrene	129-00-0	202.26	156	393	2.5@200	0.016@26

TABLE 2.Maximum Growth Values in Presence of the PAHs. (Average of a Duplicate).

Compound	Biomass maximum value (O.D)	Compound	Biomass maximum value (O.D.)
Blank (no compound)	0.016	Dibenzo(a,h)anthracene	0.099
Anthracene	0.182	Chrysene	0.141
Benzanthracene	0.198	Fluoranthene	0.030
Benzo(a)pyrene	0.193	Indenopyrene	0.027
Benzo(g,h,i)perylene	0.043	Pyrene	0.153
Benzo(k)fluoranthene	0.104	-	-

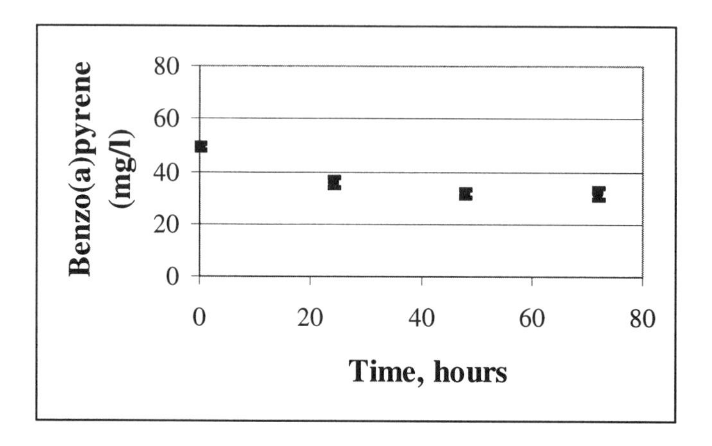

FIGURE 1. Typical Degradation Profile (Benzo(a)pyrene, 50 mg/l)

On Figure 2, the removals rates are compound for the different PAHs assessed. The best biodegraded compounds were fluoranthene (75.5%), dibenzo (a,h) anthracene (36.96%) , benzo (a) pyrene (35.35%), and benzo(g,h,i) perylene (31.06%). Total PAHs were in the range of 3.23-75.5%. The lowest degradation value corresponds to chrysene. It is important to note that the initial compound concentration was approximately 50 mg/L, a high concentration if compared with values used in works reported on literature. In all the cases growth and biodegradation occured.

Effect of the initial PAHs concentration. To investigate the effect of the initial hydrocarbon concentration, experiments with concentrations of 5 and 20 mg/L were carried out with benzo (a) pyrene and dibenzo (a,h) anthracene. Table 3 shows the results for these assessments. As shown, the trends are very similar for both compounds. Regarding the maximum biomass growth, the higher the PAH concentration, the higher the cell load achieved. All O.D. values were higher than the no compound assessment. At the other hand, PAH removals were also dependent on the initial PAH concentrations. For the benzo(a)pyrene, the higher the PAH concentration, the higher the compound biodegradation. For the dibenzo (a,h) anthracene, the maximum PAH removal was achieved at the initial concentration of 20 mg/L.

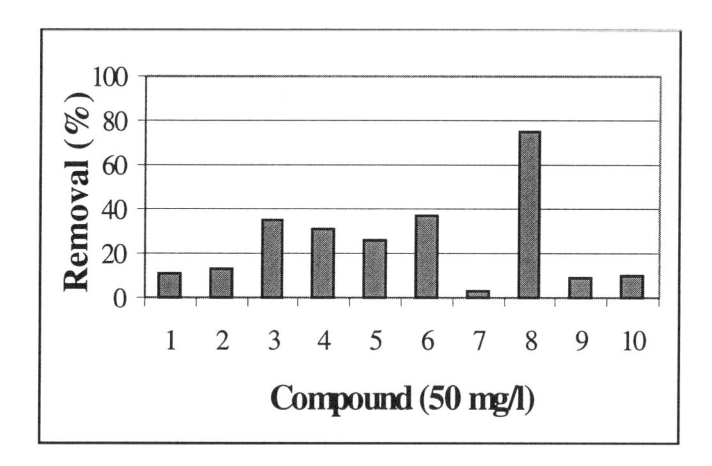

FIGURE 2. Removal Rates for the Different PAHs Assessed: 1. anthracene, 2. benzanthracene, 3. benzo(a)pyrene, 4. benzo(g,h,i)perylene, 5. benzo (k) fluoranthene, 6. dibenzo (a,h) anthracene, 7. chrysene, 8. fluoranthene, 9. indenopyrene and 10. pyrene.

TABLE 3. Effect of Initial PAH concentration Over Growing and Compound Removal.

Compound	Initial concentration (mg/l)	Maximum biomass growth O.D.	PAH removal (%)
Blank (no compound)	-	0.016	-
Benzo(a)pyrene	5	0.023	5.33
Benzo(a)pyrene	20	0.045	13.50
Benzo(a)pyrene	50	0.193	35.35
Dibenzo(a,h)anthracene	5	0.018	2.22
Dibenzo(a,h)anthracene	20	0.033	50.25
Dibenzo(a,h)anthracene	50	0.099	36.96

The results showed that significant bacterial growth occurred in the presence of every PAH assessed. The best growth rate was found with benzanthracene, benzo(g,h,i)perylene, anthracene, and pyrene. PAH removal rates between 3% and 75% were achieved for the 50 mg/L tests. The more biotransformed compounds were dibenzo(a,h)anthracene, fluoranthene, benzo(a)pyrene, and benzo(g,h,i)perylene. The different concentrations' tests confirmed a clear concentration effect over the PAH removal. In the case of the benzo(a)pyrene, the higher the compound concentration (for the 5-50 mg/L range), the higher the hydrocarbon removal. For the dibenzo(a,h)anthracene, the maximum PAH removal was found at an initial compound concentration of 20 mg/L. It is worth noting that regarding the bacterial growth, the higher the hydrocarbon concentration, the higher the O.D. achieved.

Our next goals are: (a) to identify the bacteria present in the aerobic culture, (b) the identification of some PAHs degradation intermediates, and (c) immobilize of the bacterial system in packed columns for the improvement of waterbodies or underground water treatment.

ACKNOWLEDGEMENTS

This work was carried out at IMTA. Thanks to Manuel de la Torre for the English style revision.

REFERENCES

Andersson B.E. and T. Henrysson. 1996. Accumlation and degradation of dead-end metabolites during treatment of soil contaminated with polycyclic aromatic hydrocarbons with five strains of white-rot fungi. *Applied Microbiology and Biotechnology*, *46*: 647-652.

Johnsen K., S. Andersen, and C.S. Jacobsen. 1996. Phenotypic and genotypic characterization of phenantrene-degrading fluorescent *Pseudomonas* biovars. *Applied and Environmental Microbiology*, *62*(10): 3818-3825.

Nielsen P.H., and T.H. Christensen. 1994. Variability of biological degradation of aromatic hydrocarbons in an aerobic aquifer determined by laboratory batch experiments. *Journal of contaminant Hydrology, 15*: 305-320.

Shuttleworth K.L. and C.E. Cerniglia. 1997. Practical methods for the isolation of polycyclic aromatic hydrocarbons (PAH)-Degrading microorganisms and the determination of PAH mineralization and biodegradation intermediates. In: *Manual of Environmental Microbiology* (C.J. Hurst, edit). ASM Press, Washington, D.C.

BIODEGRADATION OF ANTHRACENE OIL IN SOIL

Jolanta Turek-Szytow, Korneliusz Miksch, (Environmental Biotechnology Department, Silesian Technical University, Gliwice, Poland)

ABSTRACT: In natural agricultural soil and in soil modified with wastewater sludge, the biodegradation of anthracene oil at a concentration of 1% and 3% was carried out. Experiments lasted 61 days.
The examinations carried out shown that:

- the mass decrement of anthracene oil introduced after 61 days is 53% up to 75% depending on soil modification ;
- the quantity of huminic acid increases in all modified systems;
- the sum of bases in sorptive complex as well as hydrolytic acidity decreases over the experiment duration;
- C:N ratio increases;
- in case of hydrocarbon mixture, the compound mixtures of alcohols and carboxylic acids are obtained, whereas primary alcohols dominate.

INTRODUCTION

When introduced into an eco-system, crude oil and its derivatives create a series of disadvantageous phenomena. The degradation effect of petroleum derivatives consists mainly in creating an acute nitrogen and phosphorous demand, a water condition disturbance, and a toxic and mutagenic effect to environment. Removal of hydrocarbons from soil is possible using biological methods, when concentration of these contaminations does not exceed 5-10%. At lower concentrations (i.e. within the range 0.5 %-1 %), decomposition rate does not depend on hydrocarbon concentration (Ola• czuk-Neyman, 1994).

MATERIALS AND METHODS

Objective. The objective of the examinations carried out was to track anthracene oil change in natural soil and in soil modified with sludge, contaminated with oil at a concentration of 1% and 3%. The experiment was carried out over 61 days. The following soil systems were analysed:

G - natural soil
G+1% - soil with the addition of anthracene oil at a concentration of 1%
G+3% - soil with the addition of anthracene oil at a concentration of 3%
G+O - soil with sludge (modification),
G+O+1% - soil with the addition of sludge and anthracene oil at a concentration of 1%,
G+O+3% - soil with the addition of sludge and anthracene oil at a concentration of 3%.

For the examination, the soil was used from the agricultural terrain of Gliwice located away from industrial emission sources (pH=6.7). Soil was taken from surface layer at a depth of not exceeding 30 cm. To obtain uniform soil, the soil was screened and then activated at room temperature (about 25°C), keeping humidity at a level of 25%.

Physical and chemical analyses of soil included the following findings: mass decrement of oil, C:N ratio, environment reaction (pH), content of humus, huminic acids, sum of bases, and hydrolytic acidity.

Analysis of anthracene oil using high-performance liquid chromatography (HPLC) showed that it has, among other substances, 40% of polycyclic aromatic hydrocarbons (Table 1).

Table 1: Components of anthracene oil

Compound	Concentration[mg/L]	%
Naphthalene	0.089	4
Phenanthrene	0.240	10
Anthracene	0.028	2
Fluoranthene	0.246	11
Pyrene	0.317	14

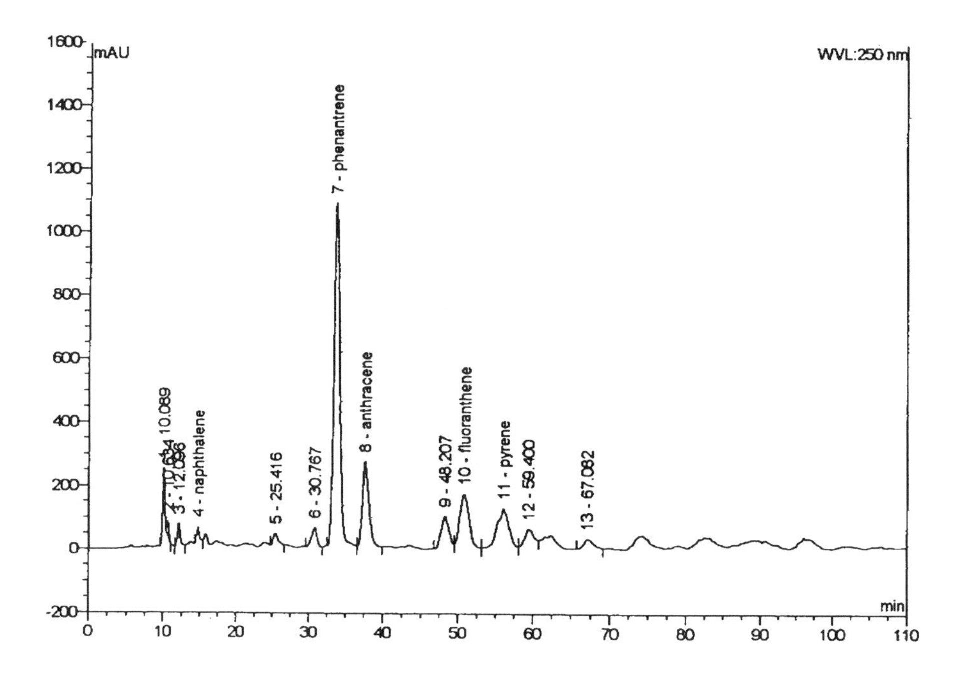

To separate the decomposition products of the examined comound from the soil, 70 g samples were taken and extracted with cyclohexane in a Soxhlet apparatus. Analyses of created products were made using spectrophotometry in infrared. For this purpose, solutions of dry residual mass were prepared in carbon tetrachloride. Interpretation of spectra was made by means of analysis of the whole infrared area (4000-700 cm^{-1}), dividing it into smaller areas.

RESULTS AND DISCUSSION

Oil may be removed from the environment by means of biotic and abiotic processes, (i.e. evaporation, sorption in soil material, and decomposition) as a result of chemical and photochemical oxidation as well as biodegradation which occurs under anaerobic and aerobic conditions.

It was found, taking into consideration all soils, that the mass decrement of anthracene oil introduced after 61 days of process is of 53% up to 75%. In soil contaminated with 1% of anthracene oil in the last week of examination, the decrement was 64%. In a case of soil contaminated in amount of 3% it was maintained at the constant level 53% and it did not decay. The same process was proceeded other way in soil modified with sludge. In both cases removal followed, whereas in a sample contaminated with the dose of 1% it reached 61%. However, at dose of 3% of oil, the process was run faster and in the final stage the removal at level of 75% was achieved.

Infrared spectrum. The detail analysis of the the infra red spectrum allows as to find the biochemism of transition of introduced oil.

Infrared analysis of crude anthracene oil.
3070 ÷ 3010 cm^{-1} oscillation of C-H bonds
2940 ÷ 2916 cm^{-1} oscillation of CH2 group in alkanes (R-CH2-R)
2930 ÷ 2920 cm^{-1} oscillation of aromatic nucleus
about 1560 cm^{-1} CH3 oscillation coupled with CH2
1540 ÷ 1000 cm^{-1} oscillation of C-O, NO2, P-O-R

Analysis of IR of exctracted decomposition products of anthracene oil after 61 days of biodegradation.

G+1%
- decreasing bands intensity within the range of oscill of C-H group, i.e. 3070 ÷ 3010 cm^{-1} , 2940 ÷ 2916 cm^{-1}, 2930 ÷ 2920 cm^{-1}
- decay of oscill of C-H bonds at 1856 ÷1800cm^{-1} , decreasing of osillations intensity 1740 ÷ 1715 cm^{-1} , which are characteristic for esters and 1640 ÷ 1540 cm^{-1} ketones;

G+3%
- disappearance of bands within the range of 3060 ÷ 2920 cm^{-1}
- increase of intensity of bands, which are characteristic for C=O bonds in ketones (1640 ÷ 1540 cm^{-1}) and within the range 1260 ÷ 1200 cm^{-1} C-C as well as 1020 ÷ 960 cm^{-1});
- distinct peaks within the range of 1100 ÷ 1000 cm^{-1} in alcohols;

G+O+1% distinct distinguishing and increase of intensity of bands, which are characteristic for esters (at 1740 ÷ 1715 cm^{-1}) of C=O group and 1310 ÷ 1160 cm^{-1} of C-O group as well as ketones (within the range 1725 ÷ 1700 cm^{-1} of C=O group).

G+O+3% alike in G+3%, increase of band within the range of 1725 ÷ 1705 cm^{-1} and (1640 ÷ 1540 cm^{-1}) of C=O group, which is characteristic for ketones.

In the case of a hydrocarbon mixture, we obtained compound mixtures of alcohols and carboxylic acids, whereas primary alcohols dominate [Klug and Markovetz, 1987]. That can be explained by the fact that the increase of bands intensity was observed during process duration within the range 1210 ÷ 1000 cm^1, which is characteristic for alcohol. Oxidation of a hydrocarbon chain, by means of indirectly created alcohols and carboxyl acids, may lead to the creation of citric acid, which may be mineralised to carbon dioxide and water. Because carboxyl acids subject to metabolism more easily than alcohols, significant amounts of them were not found in samples.

In the first stage of the examination of the process it was found that in all types of soil, the decrease of the intensity of bands characteristic of CH_3 as well as occurence of bands which show evidence of the creation of oxygen compounds. The occurrence of tension oscillations within the range being charcteristic of hydrogen bonds and alcohols, show an intensive degradation process of introduced oil. The express peak within the range of C=O bonds suggests that ketones, aldehydes and esters or acids were created [Steward and Kallio, 1959]. In the experiment, creation of epoxy compounds was not observed, this being the result of oxidation of unssaturated and aromatic hydrocarbons.

The essential criteria of soil degradation estimation is the ratio of carbon content to nitrogen in the humic layer of soil. A lack of C:N balance in natural soil was found. Chemical analysis results show that this was soil poor in organic substances, and the C:N relationship was 13 [Siuta, 1995]. In oil contaminated soil, the ratio between organic carbon and nitrogen as changes radically, introduced in the form of sludge. The acute lack of these macro-elements caused by a surplus of carbon means that microorganisms may not fully utilise the energy included in oil. Chemical analyses of control soils and contaminated ones in consecutive stages of examination showed a decrement of nitrogen. This could be the result of microorganism activity as well as abiotic processes. Additition of sludge increased the nitrogen content in modified samples of soil. During period of examination, a decrease in content of this element took place. The percentage nitrogen decrement for separate soil systems is presented below (Table 2).

Table 2: Nitrogen decrement (%)

G	G+1%	G+3%	G+O	G+O+1%	G+O+3%
23	18	16	33	21	20

The environment reaction (pH) is a very important feature of soil; it determines the possibility of plant growth, assimilation of nutrient components and the rate, and trend of biological physical and chemical processes. During the experiments duration, the reaction decresed from pH=6.7, and in the last determination, it reached pH=5.5 in all samples.

Determination of huminic acids permits the designation the humus content. At the begining of examinations in soil not subjected to the effect of oil, the huminic acids composed 0.9% and their decrement was observed to the value of 0.68%. In soil with 1% of oil an increase of 33% was found, whereas in soil contaminated with 3 % of anthracene oil the huminic acids were maintained at the same levels of 0.91% and 0.95%. In soil with addition of sludge, in the last stage of examination, the content of huminic acids decreased by 10%. Whereas in soil with sludge and 1% and 3% anthracene oil, it increased by approximately 60%.

The value of the sum of bases is a sorptive capability coefficient of soil in relation to base cations and depends mainly on grain size, organic substance content, and the acidity of soil. The sum of bases in the test sample of soil was included within the range 20-17 [0.1 NaOH/100g] with a declining tendency. Similar tendencies were observed for contaminated soil both with 1% and 3% of anthracene oil . Addition of sludge increased by 35% the sum of bases. However the continuous decrease of bases sum was observed for soil with sludge addition and oil from initrogen value of 27 [0.1 NaOH/100g] up to 22 [0.1 NaOH/100g].

Hydrolytic acidity determines the saturation rate of a sorptive complex with hydrogen and aluminium ions. On this basis the doses of calcium are determined, which correspond to a decrease in acidity by one pH unit. A grain of soil depends on this paramameter, its acidity and the method of use. At the first stage of examination, the hydrolytic acidity in separarate tests increased proportionally with the added amount of anthracene oil. During examaninations this coefficient decreased in every case, but the greatest decrease was found after 25 days in the case of soil contaminated with 3 % of oil by 0.89 [0.1 NaOH/100g], whereas for soil with the addition of sludge contaminated with the same concentration of oil the was leved only 0.55 [0.1 NaOH/100g].

CONCLUSIONS

1. In unmodified soil contaminated with anthracene oil of concentration 1%, decrement of anthracene oil occurred (approximately 64%), increment of huminic acids was found, consumption of nitrogen followed, and the value of bases and hydrolytic acidity decreased.
2. In natural soil, with a 3% concentration of oil, oil decrement at the level of 54% was found, organic carbon and huminic acid were maintained at a constant level, a decrement of nitrogen followed, and the content of bases sum and sorbtive complex decreased.
3. In soil modified with sludge, both for 1% and 3% of contamination, an increment of huminic acids, decrement of nitrogen, and decrement of content of bases sum and acidity took place. Anthracene oil introduced was subjected to biodegradation, whereas the greatest mass decrement occurred in soil of 3% concentration (about 75%).
4. The progressive biodegradation process of oil is evidenced by bands of oscillations which are characteristic for hydrogen bonds and alcohols. The distinct oxygen peak within the range which corresponds to C+O bond suggests the creation of ketones, aldehydes or acids.
5. In soils, a mixture of partly-oxided hydrocarbon compounds was created, and a partial decrease in the intensity of bands within oxygen bonds suggests that at first oxidation of already-created oxygen compounds follows, and only later is the consecutive hydrocarbon chain metabolised.

REFERENCES

Ola• czuk-Neyman, K. 1994. "Chemical and bacteriological assessment of soil contamination of petroleum-derivative fuel reloading station." *Biotechnologia*: 50-60.

Klug M.J., A.J.Markovetz. 1987. "Degradation of hydrocarbons by members of the Genus Candida, oxidation of n-alkanes and 1-alkanes by Condida Lipolityca." *Journal of Bacteriology*: 1847-51.

Siuta J. 1995. "Soil - diagnosis of contidion and threats." *Publication of Environment Protection Institute.* Warsaw, Poland.

Steward J.E., R.E. Kallio. 1959. "Bacterial hydrocarbon oxidation, ester formation from alkanes." *Journal of Bacteriology: 726-30.*

EVALUATION OF MICROBIAL SPECIES DIVERSITY AT A FORMER TOWN GAS SITE

John A. Glaser, John Haines, Carl Potter, Ronald Herrman, and Kim McClellan (USEPA, National Risk Management Research Laboratory, Cincinnati, OH); Kathleen O'Neill (O'Neill Associates, Clinton, NJ); Steve Hinton (Exxon Research and Engineering Co., Clinton NJ).

ABSTRACT: A general approach to structural microbial diversity detection in contaminated soil was derived from phospholipid fatty acid and 16S rRNA analyses. Lipid phosphate analysis was used to screen environmental samples for viable biomass before being selected for 16S rRNA analysis. Fatty acid profiles for each sample provide an early assay of structural microbial diversity and 16S rRNA analysis offers a more thorough picture of the diversity. New analytical methodologies are required to accomplish 16S rRNA analysis for the number of samples produced from field studies. A series of labor saving techniques were investigated to increase the throughput of this very complicated analytical protocol. We are conducting a series of treatability studies on the bioremediation organic contaminants associated with town gas waste at the Bedford Indiana Town Gas Site. Samples selected from the lipid analysis are submitted for 16S rRNA analysis to provide a more thorough picture of microbial diversity at the Bedford site. Introduction to the newly developed spectrum of analytical tools for 16S rRNA analysis and critical evaluation of results derived from the field project will be presented.

INTRODUCTION

Microbial interactions in any environment are complex and often interdependent. Therefore, defining the makeup and dynamics of a microbial community is paramount to understanding any microbial activity or process in the environment. Prior to the advent of molecular analysis, the study of environmental microbial communities was severely limited by the ability to culture less than 1% of the viable microorganisms in an environment. Recent breakthroughs in DNA extraction and sequencing technology and molecular analysis, specifically of the 16S rRNA genes, have provided methods to characterize complex microbial populations without the need to cultivate bacteria. In addition the extensive database of 16S rRNA sequences generated by culture-based and environmental molecular studies has identified phylogenetically related groups of bacteria which have similar biochemical and metabolic functions. These advances make possible, for the first time, the study of the ecology of natural microbial communities. The unveiling of the vast diversity and interdependence of microorganisms holds opportunities for the understanding of biochemical novelties and natural response to environmental change.

Molecular analysis does not require cultivation or isolation of bacteria but uses ribosomal RNA genes which can be isolated directly from the environment, i.e., sediment, soil or water. The DNA extracted from the environment is a complex mixture dependent on the types and numbers of bacteria in the microbial community. The diversity of microbes in the population can be examined by sequencing DNA and analyzing the rRNA genes obtained from the extracted DNA. Phylogenetic analysis, using public databases of rRNA sequences, is used to identify relationships between the environmental sequences and known bacteria. Statistical methods are then used to determine the most similar rRNA sequence in the database to identify a closest relative. Generally, close relatives or members of a group have physiological properties in common, which is the basis for inferring properties to the rRNA-identified microbe. An inventory of the identified microbes and each assigned metabolic potential serves as a measure of microbial diversity.

The methodology for molecular analysis of environmental samples is well documented. However, the costly and time-consuming nature of these techniques, as well as numerous technical hurdles have limited the application of this molecular approach for thorough environmental assessment. Cost-effective high throughput sequencing protocols have also been developed providing high quality sequences with a minimum of time and expense. A system has been developed allowing for rapid screening and evaluation of clones using existing databases and establishing new databases for further screening. High throughput sequencing generates a large volume of sequence data that must be precisely managed in order to effectively use these sequences to evaluate an environment. In addition, phylogenetic analysis of sequences isolated from environmental samples can be difficult due to the absence of identical microorganisms in the existing databases. We have set up protocols for the systematic analysis of 16S rRNA sequences to optimize the phylogenetic analysis and quickly identify relationships within the environmental microbial communities.

MATERIALS AND METHODS

Samples. Two reference and 2 contaminated soil samples were provided for 16S rRNA analysis. A description of the sampling site and the results of FAME analysis of these samples follows: The two references were sample 1 (INA) taken outside the fence to the west at 1-3 in depth. Biomass estimate of 48.08 ± 9.62 nmol lipid phosphate/gram dry weight, moisture 20%, and organic matter 11% were characteristic of sample 1. FAME analysis (Findlay & Dobbs, 1993) shows equal levels of Gram positive bacteria, Gram-negative bacteria and fungi. Sample 2 (IND) was taken north a railroad embankment and south of a culvert (near wetland). Biomass estimate of 33.45 ± 4.53 nmol lipid phosphate/Gram dry weight, moisture 26%, and organic matter 14.4% were characteristic of sample 2. FAME analysis showed Gram positive bacteria, Gram-negative bacteria, possible sulfate reducing bacteria (SRB's), algae, and protoza, whereas fungal markers were not very strong.

The two contaminated samples designated 345-3 (CDE) and 349-3 (CDI) were soil samples contaminated with 9,000 ppm and 22,000 ppm total PAH constituents, respectively, taken from the town gas process holder area.

DNA extraction and purification. A modified direct lysis method was used to extract DNA from the soil samples (Ausubel et al., 1989; Holben et al., 1988). Extractions were performed in duplicate for each sample and the DNA extracts pooled.

PCR amplification and cloning. Ribosomal RNA genes (rDNA) were amplified by the polymerase chain reaction (PCR), using the universal 16S ribosomal RNA (rRNA) primer 1392R (5'-ACGGGCGGTGTGTRC -3'), and the bacterial primer 27F (5'-AGAGTTTGATCMTGGCTCAG-3') (numbers of the primers correspond to *Escherichia coli* positions and R = purine, M= A,C). Target rDNA was amplified using a PCR protocol of denaturing at 92°C for 30s which was annealed at 55°C for 30 s, extension at 72°C for 1 min for 30 cycles, and a final extension at 72°C for 15 min. For each sample, 10 PCR reactions were performed and the products pooled. Amplification of the target rDNA was confirmed by agarose gel electrophoresis. The PCR fragment was cloned using the Stratagene pCR script cloning kit (Stratagene, LaJolla, CA). Plasmid DNA was isolated by the alkaline lysis method modified to fit a 96 well format (Sambrook et al., 1989). Plasmids were screened for correct size insert by agarose gel electrophoresis. PCR cycle sequencing using quarter volume reactions in conjunction with the ABI 377 automated sequencer (Perkin Elmer, Foster City, CA), was used to obtain rDNA sequences from the cloned PCR products.

Screening for duplicate clones. Clones from each environmental library were screened by DNA sequencing of approximately 500 base pairs. Similarity analysis with a 95% similarity cutoff was used to determine and group identical clones, one of which was chosen for full-length sequencing. This method of screening clones not only identifies identical clones to avoid full length sequencing of duplicates but also provides sequence data which can be used to evaluate the microbial biodiversity of the environment. While 500 bases is not considered optimal for accurate phylogenetic analysis, particularly of sequences which are not well represented in the database, it can provide preliminary information on the microbial ecology of the environment and provides sufficient information for comparative analysis between different environments.

Phylogenetic analysis. Sequences were analyzed for chimera formation, after which phylogenetic analysis was performed using maximum likelihood, similarity matrix, and maximum parsimony analyses (Felsenstein, J. 1993, Larsen, et al. 1993). Bootstrap analysis was performed to provide an estimate of the accuracy of the phylogenetic trees.

RESULTS AND DISCUSSION

Currently, 135 clones from sample 1 (INA) and 137 clones from sample 2 (IND) have been partially sequenced. Similarity analysis (using a >95% similarity cutoff) of these partial sequences was used to group the clones. Sixty-nine clones from sample 1 and 60 clones from sample 2 were chosen as representative groups and were completely sequenced. Full length sequencing, alignment, editing, and phylogenetic analysis has been completed on all INA and IND clones.

Sequencing and analysis of the contaminated soil sample clones has also been completed. As described above partial sequencing and similarity analysis of 114 clones from sample 349-5 and 39 clones from sample 345-3 were used to group clones. Forty-four clones from soil sample 349-5 and 6 clones from 345-3 were chosen for sequencing.

Biodiversity of contaminated soil samples. Environments, restricted by physical, chemical, or nutritional conditions, have a less diverse microbial community. Pollution has also been shown to decrease the diversity of environmental microbial communities. Therefore, the diversity of a microbial community may provide a means of assessing the health of an ecosystem. The Shannon-Weaver index:

$$(2.3/N)(N \log_{10}N - \Sigma n_i \log_{10}n_i)$$

where N is the total number of individuals and n_i is the number of individuals in the ith species, was used to calculate the diversity of the index soil samples. Since the concept of a bacterial species is problematic in reference to 16S rRNA the >95% similarity used in screening clones was also used for the diversity analysis. This provides a standardized means of comparing the relative microbial diversity of the samples.

The Shannon-Weaver index indicates that the Indiana reference soil samples contain diverse microbial communities. The diversity in both samples was similar with indices of 3.90 and 3.99 for INA and IND respectively. The complexity of the microbial communities is indicative of a healthy, nutritionally complex environment. Each contaminated soil sample showed a decrease in biodiversity of the microbial communities, in contrast with reference samples, with indices of 0.47 and 2.12 for samples 345-3 and 349-5, respectively.

The first step in phylogenetic analysis is to determine whether the sequences are chimeric molecules. Chimeras can occur during PCR amplification of a mixed population when partial sequences (due to sheared DNA or incomplete extension of a PCR product) from one microorganism act as primers for the rDNA gene from a different microorganism. This forms a PCR product made up of sequences from two different microorganisms. The *Check Chimera* program from the RNA database project is used to determine whether sequences are chimeras. Three of the completed sequences from reference soil sample 1 (INA), 3 sequences from reference soil sample 2 (IND) and 2 sequences from contaminated soil sample 349-5 (CDI) were identified as chimeric molecules and removed from further analysis.

Phylogenetic analysis was performed on the non-chimeric sequences using the fastDNAml maximum likelihood program available through the RNA Database Project and Maximum Parsimony and Bootstrap programs in *Phyllip* (Felsenstein, 1993). Sequences from both reference soil samples placed in the *Proteobacteria, Verrucomicrobium, Cytophagales, Acidobacterium*, and Gram-positive divisions (Table 1). Thirty-nine sequences are only distantly related to any known bacterial divisions and are considered unclassified at this time. The contaminated soil samples were much less diverse. No sequences from either site placed within the *Acidobacterium* or *Verrucomicrobium* divisions. The more highly contaminated soil sample 345-3 was even more limited in diversity with only three bacterial divisions being represented. The sequences from the contaminated soil in these divisions were generally not closely related to those isolated from the reference soil samples. Sequences in the contaminated soil samples were identified as representing bacteria common to the reference soil samples, which are able to tolerate and perhaps utilize hydrocarbon contamination as a nutrient source. These sequences placed within the *Beta Proteobacteria, Gamma Proteobacteria, Delta Proteobacteria,* and the Gram positive low G+C bacteria (Figure 1).

Table 1. Phylogenetic distribution of sequences from Indiana soil samples.

	Indiana Soil Samples		
	Reference soil samples	Contaminated soil 349-5	Contaminated soil 345-3
Bacterial Division	*Cytophagales* *Alpha Proteobacteria* *Beta Proteobacteria* *Gamma Proteobacteria* *Delta Proteobacteria* *Gram Positive* Unclassified *Acidobacterium* *Verrucomicrobium*	*Cytophagales* *Alpha Proteobacteria* *Beta Proteobacteria* *Gamma Proteobacteria* *Delta Proteobacteria* *Gram Positive* Unclassified	*Alpha Proteobacteria* *Beta Proteobacteria* Unclassified

Conclusions. 16S rDNA analysis has shown that the Indiana reference soil samples contain a diverse group of microorganisms. The complexity of the microbial communities is indicative of a health, nutritionally complex environment. The contaminated soil samples show a marked contrast to the reference soil samples with a greatly decreased diversity in the microbial communities. Biodiversity indices, which are calculated using partial sequences, may provide a rapid measure of ecosystem health. Sequences, which are present in the non-contaminated soil but absent in the contaminated soil, such as the *Acidomicrobium or Verrucomicrobium* divisions, may serve as indicator sequences for contamination. Sequences present in the contaminated soil may represent hydrocarbon degrading bacteria potentially useful in remediation. An extensive database of sequences from the Indiana soil samples has been developed. This database will prove invaluable in further studies at this site and also as a comparison for sequences isolated from other sites.

Figure 1. Sequences from Indiana soil samples common to contaminated and non-contaminated soils A = Beta Protobacteria subdivision, B = Gamma Proteobacteria subdivision, C = Delta Proteobacteria subdivision, D = Gram Positive low G+C subdivision. Sequences in bold typeface represent sequences from Indiana soil. Clone designations: INADC = sequence found in all soil samples, INA = sequence found in one reference soil sample, CD = sequence found in contaminate soil sample. Trees infered by maximum parsimony analysis and edited from more extens analysis for simplicity.

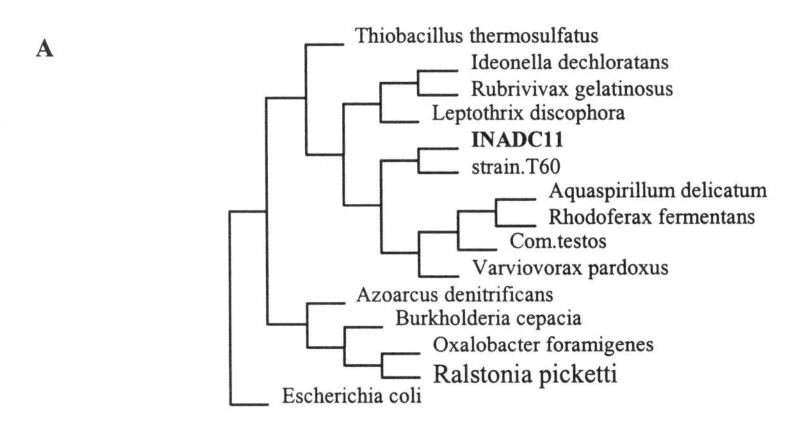

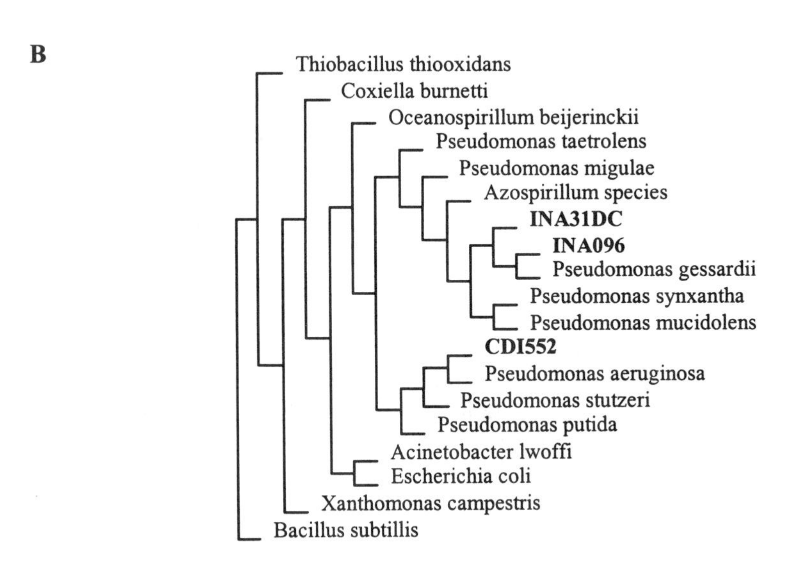

REFERENCES

Ausubel, F.M., R. Brent, R.E. Kingston, D.P. Moore, J.G. Seidman, J.A. Smith, and K. Struhl. 1989. *Short protocols in molecular biology.* John Wiley & Sons, Inc., New York.

Felsenstein, J. 1993. *PHYLIP* (Phylogeny Inference Package), version 3.5, Department of Genetics, University of Washington, Seattle, WA.

Findlay, R.H. and F.C. Dobbs. 1993. "Quantitative description of microbial communities using lipid analysis". In: *Handbook of Methods in Aquatic Microbial Ecology*. Kemp, P.F., B.F. Sherr, E.B. Sherr, and J.J. Cole. Lewis Publishers, Boca Raton, FL. Chapt. 32, pp 271-284.

Holben, W.E., J.K. Jansson, B.K. Chelm, and J.M. Tiedje. 1988. DNA probe method for the detection of specific microorganisms in the soil bacterial community. *Appl. Environ. Microbiol. 54*:703-711.

Larsen, N., G.J. Olsen, B.L. Maidak, M.J. McCaughey, R. Overbeek, T.J. Macke, T.L. Marsh, and C.R. Woese. 1993. "RNA database project". *Nuc. Acids. Res. 21*:191-198.

Sambrook, J., E.F. Fritsch, and T. Maniatis. 1989. *Molecular cloning: a laboratory manual. 2nd ed.* Cold Spring Harbor Laboratory, Cold Spring Harbor, NY.

ASSESSMENT OF ACUTE PAH TOXICITY USING PROKARYOTE BIOSENSORS.

B.J. Reid, C.J.A. MacLeod, ***K.T.Semple*** (Lancaster University, Lancaster, U.K.).
H.J. Weitz, G.I. Paton (University of Aberdeen, Aberdeen, U.K.).

ABSTRACT: The aim of this study was to assess the acute toxicity of phenanthrene, pyrene and benzo[a]pyrene using *lux*-marked bacterial biosensors. Standard solutions of phenanthrene, pyrene and benzo[a]pyrene were produced using 50 mM hydroxpropyl cyclodextrin (HPCD) solution. Four microorganisms containing the *lux* cassette were used as the test biosensors and over the incubation time period (280 min), there was no significant decrease in bioluminescence in any of the biosensors. This study has shown that the three PAHs tested are not acutely toxic to the prokaryotic biosensors, although acute toxicity has been shown in other bioassays. These results question the rationale for using prokaryote biosensors to assess the toxicity of hydrophobic compounds, such as PAHs, from more complex environmental matrices.

INTRODUCTION

Polycyclic aromatic hydrocarbons (PAHs) enter soils and waters via atmospheric deposition and industrial and agricultural activities and, as such, constitute a significant health risk (Miller and Miller, 1981). Although it is well established that PAHs are genotoxic, the literature is divided on the level of acute toxicity of PAHs. For example, Hund and Traunspurger (1994) used a battery of biological indicators to assess the toxicity of PAH-contaminated soil before and during bioremediation. The authors found that as the PAHs were degraded, there was a decrease in toxicity measured using Microtox. Pradhan *et al.* (1997) determined the effectiveness of an innovative chemical/biological treatment process for soils contaminated with PAHs. The authors found that this process reduced the toxicity of the soil by a factor of 50 as measured by Microtox. However, in all these studies, what is unclear is the effect that other potentially toxic compounds, which may also be present in the samples, were having on the acute toxicity test, namely Microtox, as mentioned by Hauser *et al.* (1997).

A problem that is often encountered with organic pollutants is their hydrophobicity and the need to used organic solvents to extract/solubilize them. This is a problem as the use of organic solvents often have negative effects on biosensors (Bundy et al., 1997) and organic solvent extractions do not necessarily reflect compound bioavailability (Kelsey and Alexander, 1997). To address this, an aqueous solvent composed of hydroxpropyl-β-cyclodextrin (HPCD) has been used to increase the solubility of the PAHs (Brusseau et al., 1997).

The aim of this study was to assess the acute toxicity of selected PAHs present in an aqueous solution. This was achieved using four different bacterial biosensors containing the genes responsible for bioluminescence. Each of these organisms was selected because of their contrasting ecophysiology and association with different environmental niches (Paton et al., 1997).

MATERIALS AND METHODS

***Lux* bioassay.** Freeze-dried cultures of the *lux*-based biosensors were resuscitated according to standard laboratory protocols (Paton et al., 1997). All acute assays with the biosensors (*Ps. fluorescens* 10586r, *Ps. putida* F1, *E. coli* HB101) involved resuscitation from freeze-dried stock in 0.1 M KCl. Microtox was resuscitated in 2% NaCl. The assays (with the exception of the Microtox test) consisted of the addition of 100 μl of the appropriate cell suspension into 900 μl of the test solution respectively, in individual microcentrifuge tubes at 15 s intervals. The luminescence of the cells was then measured by taking 40 μl of the suspension from the microcentrifuge tubes and diluting to 160 μl with 1/4 strength Ringers solution in a microtitre plate. Luminescence was determined using a Lucy Anthos 1 microtitre plate reading luminometer, after exposure times (E_t) ranging from 20 to 280 min. The assays were carried out in triplicate using freeze-dried samples and the results expressed as a percentage of the luminescence measured in a non-contaminated control sample.

Test Solutions. A standard solution of phenanthrene (10 mg l^{-1}) was produced in toluene such that 1 μL would deliver 6.25 times the aqueous solubility limit of the compound to 10 mL of 50 mM HPCD solution. For standardization, pyrene (0.875 mg l^{-1}) and benzo[a]pyrene (0.25 mg l^{-1}) were also introduced at 6.25 times the solubility limit. The standard PAH test solutions (1 ml) were then incubated with the *lux*-marked microbial biosensors (1 ml of resuscitated test solution) for 280 min. All assays were carried out in triplicate and the standard error of mean (SEM) calculated and presented as error bars.

RESULTS AND DISCUSSION

E. coli (Fig 1A-C), *Ps. fluorescens* (Fig 1D-F), and *Ps. putida* (Fig 2A-C) were found to respond in a similar way to all of the pollutants tested. The response was characterised by luminescence levels throughout the duration of the study exceeding those of the control extract. It may be proposed therefore that the presence of phenanthrene, pyrene and benzo[a]pyrene exhibited no inhibitory effect, as measured by bioluminescence. These three biosensors showed evidence of increased luminescence levels across the exposure period. For *E. coli* it was found that the greatest stimulation of luminescence was observed for benzo[a]pyrene after 160 min when the levels were 136% of the control extract (Fig. 1C). There was also stimulation by pyrene, while phenanthrene had no significant effect on bioluminescence across the duration of the experiment (Fig 1A-B). For *Ps. fluorescens*, all three pollutants were found to stimulate levels of luminescence, initially the increase was greatest due to pyrene, however, after 280 min benzo[a]pyrene was associated with elevated luminescence levels (Fig. 1D-F). For the *E. coli* and *Ps. fluorescens*, bioassays, their responses to the three compounds were found to be significantly different ($P \leq 0.005$).

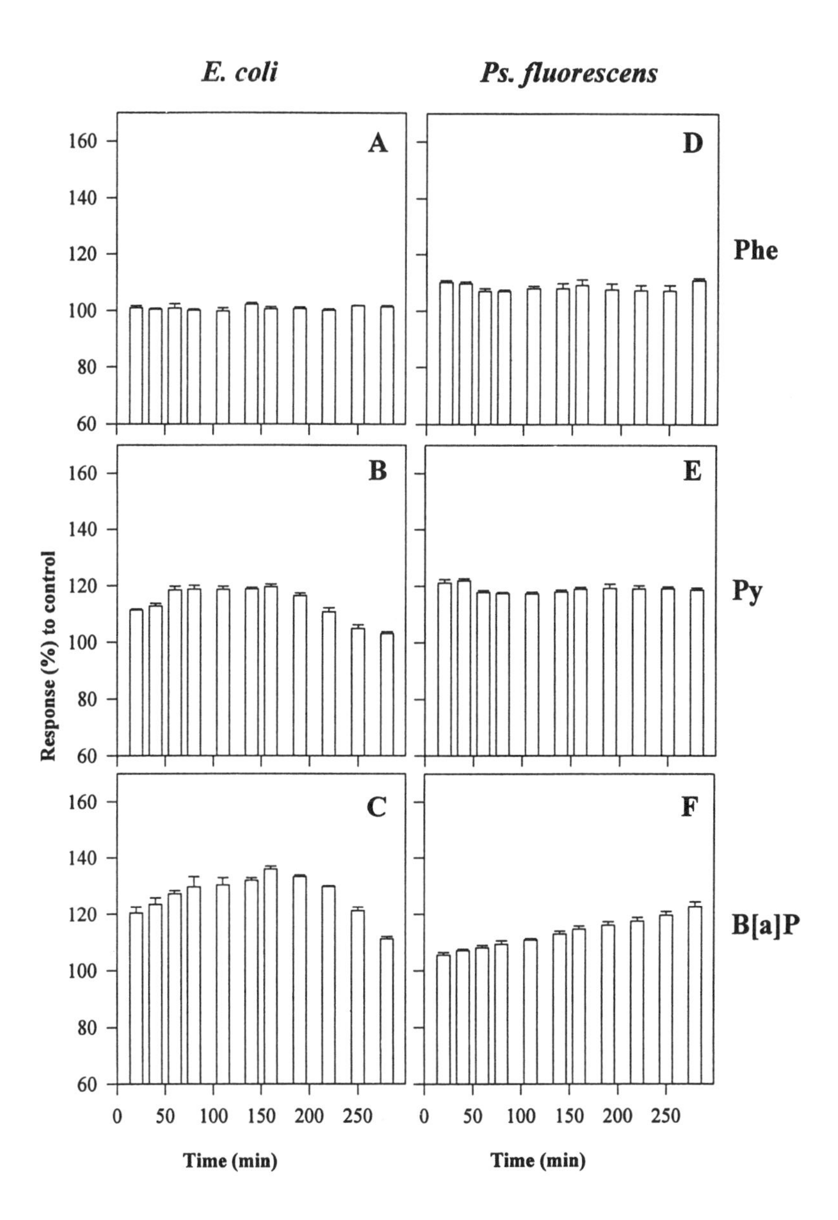

Figure 1. Changes of bioluminescence after exposure to phenanthrene (Phe), pyrene (Py) and benzo[a]pyrene (B[a]P) in 50 mM HPCD solution from *E. coli* (A-C), *Ps. fluorescens* (D-F).

In the case of *Ps. putida*, pyrene and benzo[a]pyrene caused a similar response with luminescence levels rising to between 102 and 108% of the control sample (Fig. 2A-C). There was however a much greater luminescence increase associated with phenanthrene. For the *Ps. putida* (Fig. 2A-C) bioassay, the responses between benzo[a]pyrene and phenanthrene, and phenanthrene and pyrene were found to be significantly different ($P \leq 0.005$).

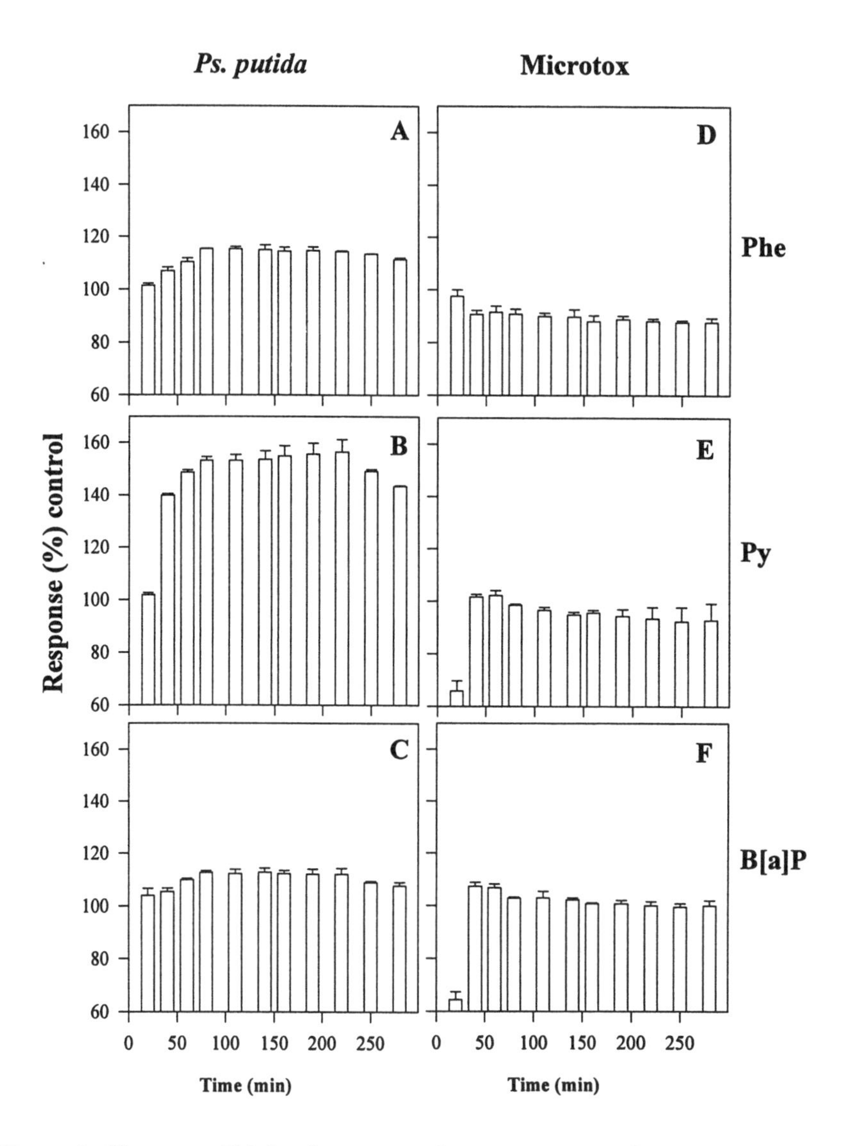

Figure 2. Changes of bioluminescence after exposure to phenanthrene (Phe), pyrene (Py) and benzo[a]pyrene (B[a]P) in 50 mM HPCD solution from *Ps. putida* (A-C) and Microtox (D-F).

Microtox activity was inhibited by benzo[a]pyrene and phenanthrene after 15 min of exposure (Fig. 2D-F), but the response to these two compounds after 30 min was similar to that observed for *E. coli*, *Ps. fluorescens*, and *Ps. putida* in that the line flattened out but the actual luminescence response was between 96 and 108% of the non-contaminant control. The response to pyrene was different in that there was initially an observed decline in luminescence of 2% which dropped and stabilised after 50 min to 92% of the control sample. For the Microtox (Fig. 2E-F) bioassay, the response between benzo[a]pyrene and pyrene was found to be significantly different ($P \leq 0.005$).

Traditionally, the determination of PAH contamination in environmental matrices and subsequent risk assessment has been achieved using exhaustive extraction methodologies giving a total concentration of pollutants. However, the relevance of 'total extractability' to the risk assessment and remediation of contaminated matrices is now in question (Kelsey and Alexander, 1997). There is now a trend towards 'less harsh' organic solvents such as butanol or ethanol-water mixtures (Kelsey et al., 1997). However, such organic solvents have been shown to be detrimental to biosensor viability (Bundy et al., 1997). These results indicate that an aqueous based solution of HPCD do not inhibit biosensor functioning. Additionally, it has been proposed that by extracting environmental samples, such as soil, bioavailable (exchangeable) hydrophobic organic molecules will transfer to the cyclodextrin cavity (Reid et al., 1998).

Further, there has been a move towards the use of more rapid bioluminescence-based microbial tests to assess environmental pollution incidents and are attractive to the end user as they are characterized by their low cost, high replication and ecological relevance. Bioluminescence, as a general reporter of toxicity, has been shown to be a sensitive indicator inorganic (Paton et al., 1995b) and organic pollutants (Boyd et al., 1996) both in simple aqueous systems and in complex environmental matrices. However, this study shows that bioluminescence-based general toxicity reporting biosensors are not sensitive to the presence of PAH suggesting, therefore, that PAHs are not toxic to the microorganisms used in this study. Additionally, the responses of the test organisms are not predictable and the presence of high levels of PAH may cause increases in the level of bioluminescence emitted making the interpretation of results very confusing.

References

Boyd, E.M., A.A. Meharg, J. Wright and K. Killham. 1996. Assessment of toxicological interactions of benzene and its primary degradation products (catechol and phenol) using a *lux* modified bacterial bioassay. *Environ. Toxicol. Chem. 16*: 849-856.

Brusseau, M.L., X. Wang and W-Z. Wang. 1997. Simultaneous elution of heavy metals and organic compounds from soil by cyclodextrin. *Environ. Sci. Technol. 31*: 1087-1092.

Bundy, J.G., J. Wardell, C.D. Campbell, K. Killham and G.I. Paton. 1997. Application of bioluminescence-based microbial biosensors to the ecotoxicity assessment of organotins. *Lett. Appl. Microbiol. 25*: 353-358.

Hauser, B., G. Schrader and M. Bahadir. 1997. Comparison of acute toxicity and genotoxic concentrations of single compounds and waste elutriates using the Microtox/Mutatox test system. *Ecotoxicol. Environ. Safety 38*: 227-231.

Hund, K. and W. Traunsburger. 1994. Ecotox - evaluation strategy for soil bioremediation exemplified for a PAH - contaminated site. *Chemosphere 29*: 371-390.

Kelsey, J.W. and M. Alexander. 1997. Declining bioavailability and inappropriate estimation of risk of persistent pollutants. *Environ. Toxicol. Chem. 16*: 582-585.

Kelsey, J.W., K.D. Kottler and M. Alexander. 1997. Selective chemical extractants to predict bioavailability of soil-aged organic chemicals. *Environ. Sci. Technol.* 31: 214-217.

Miller, E.C. and J.A. Miller. 1981. Searches for the ultimate carcinogens and their relations with cellular macromolecules. *Cancer 47*: 2327-2345.

Paton, G.I., G. Palmer, A. Kindness,C.D. Campbell, L.A. Glover, and K. Killham. 1995. The use of luminescence-marked bacteria to assess toxicity of malt whisky distillery effluent. *Chemosphere 31*: 3217-3224.

Paton, G.I., E.A.S. Rattray, C.D. Campbell, H. Meussen, M.S. Cresser, L.A. Glover, and K. Killham. 1997. Use of genetically modified microbial biosensors for soil ecotoxicity testing. in Bioindicators of Soil Health (ed. Pankhurst, C.S., Doube, B. and Gupta, V.), pp 397-418. CAB International.

Pradhan, S.P., J.R. Paterek, B.Y. Liu, J.R. Conrad. and V.J. Srivastava. 1997. Pilot-scale bioremediation of PAH-contaminated soils. *Appl. Biochem. Biotechnol. 63*: 759-773.

Reid, B.J., K.T. Semple and K.C. Jones. 1998. Prediction of bioavailability of persistent organic pollutants by a novel extraction technique. *In* ConSoil'98, Sixth International FZK/TNO Conference on Contaminated Soil, vol 2, pp 889-890. Thomas Telford Publishing, London.

BACTERIAL ASSEMBLAGE ADAPTATION IN PAH-IMPACTED ECOSYSTEMS.

Michael T. Montgomery, Thomas J. Boyd, Barry J. Spargo , Richard B. Coffin, Julie K. Steele (US Naval Research Laboratory, Washington, DC)
Dawn M. Ward (University of Delaware, Lewes, DE)
David C. Smith (University of Rhode Island, Narragansett, RI)

ABSTRACT: Natural bacterial assemblages can change in response to chronic PAH contamination. The ratios of bacterial production to PAH mineralization were compared to identify areas within field sites where the aromatic hydrocarbons may be naturally attenuating. In addition, the effect of naphthalene additions on bacterial production were used as an environmental indicator of assemblage change. These measures may be useful in identifying PAH-impacted sediment that is undergoing significant levels of natural attenuation verses those where the bacterial assemblage is metabolically inhibited.

INTRODUCTION

Determining how to accurately measure the rate of intrinsic bioremediation of a contaminant is a point of considerable interest and debate (Renner, 1998). The National Research Council (1993) has provided guidelines for gathering evidence for *in situ* contaminant biodegradation which includes using microcosms designed to mimic field conditions, using radiotracer additions, and various other measures of bacterial activity. In aquatic ecosystem studies, rates of incorporation of ^{3}H-thymidine into DNA (Furman and Azam,1982), and ^{3}H-leucine into protein (Kirchman et al., 1985), are predominantly used to measure heterotrophic growth of the natural assemblage. Despite the obvious importance of measuring bacterial metabolism to evaluate *in situ* bioremediation, little is known about productivity of contaminated sites. Kazumi and Capone (1994) found that cell specific bacterial activity was higher in pesticide-impacted sites compared to pristine areas. Others have reported on the use of ^{3}H-TdR as a measure of heterotrophic production in a wastewater (Harvey et al., 1984) and fuel-contaminated (Jensen, 1989) aquifers. The accurate measurement of bacterial productivity is important for determining whether or not the bacterial population has the metabolic capacity to account for biodegradation of the missing contaminant.

Recently, Bååth (1992) described using production as a bioassay to determine the tolerance of a bacterial community to the presence of metals. Carman et al. (1995) did not find significant changes in the bacterial community in response to PAH addition to chronically contaminated sediments. Both researchers suggest that the microbial community structure adapted to chronic contamination and that measures of heterotrophic bacterial metabolism may be useful in identifying such areas. When pollutants are released into the environment, the bacterial assemblage may change composition as sensitive strains are killed and strains that can metabolize the pollutant are given a selective advantage. The result is that the pollutant can be removed from the ecosystem, however, the time course and environmental conditions that give rise to such change are poorly understood.

Bauer and Capone (1985) found that addition of naphthalene and other organic pollutants inhibited production in aerobic sediment slurries. Over time, certain community members are given a selective advantage because they are insensitive to the toxic effects of the contaminant (Huertas et al., 1998) or they are able to metabolize the contaminant as an energy source. Some researchers have found that upon contaminant addition to a community, there is an initial decrease in overall heterotrophic activity as sensitive strains are metabolically inhibited (Bååth, 1998). This is followed by a recovery of heterotrophic activity as contaminant-insensitive strains increase their proportion of the total assemblage (Leahy and Colwell, 1990). This can be reflected as an increase in the capacity of the heterotrophic bacterial assemblage to metabolize the organic contaminant.

Objective. The objective of this field study was to identify PAH-impacted areas where the natural bacterial assemblage has adapted to the presence of elevated contaminants. The approach involved comparing rates of heterotrophic bacterial production and PAH mineralization. Inhibition of production by naphthalene additions was used to identify sites where the bacterial assemblage was sensitive to influx of PAHs. Sites that have high mineralization to production ratios and are insensitive to addition of PAHs are proposed to be candidates for intrinsic bioremediation.

Site Description. The Cooper River in the Charleston Harbor Estuarine system was the focus of this study. Sampling sites extended from three miles upriver (Station 1) of Charleston Navy Yard (CNY; Station 6) past a pulp mill that is one mile upriver of CNY (Station 4). The portion of the Cooper River adjacent to CNY is a depositional area for organic matter from the pulp mill and a petroleum storage facility (Station 5) upriver.

MATERIALS AND METHODS

Bacterial Production. Bacterial productivity in sediment samples was determined by the leucine incorporation method (Kirchman et al. 1985) as adapted by Smith and Azam (1992). The rate of incorporation of leucine into bacterial proteins is a function of the growth rate of the heterotrophic bacterial population.

PAH Mineralization. PAH mineralization was measured using a modification of Boyd et al. (1996). ^{14}C-labeled substrate (naphthalene, fluoranthene, and phenanthrene) was added to sediment samples (1 mL wet weight) in 100 X 16 mm test tubes to a final concentration of about 500 ng g^{-1}. Samples were incubated at *in situ* temperature and evolved $^{14}CO_2$ was captured on filter paper (soaked with 100 µL 2 N NaOH) suspended in the tubes. One mL 2 N H_2SO_4

was added to end the incubation or at the start of the incubation for the killed controls. Samples are then placed on a rotary shaker for 4 to 6 h (preferably overnight) to allow CO_2 in the water to equilibrate with the headspace in the tube. The filter papers with trapped $^{14}CO_2$ are removed and radioassayed.

Naphthalene Additions. The effect of PAH additions on bacterial production will be measured by adding five different concentrations (0, 5, 10, 15, 20, 25 μg mL^{-1} water or ca. 50 mg^{-1} sediment) of individual PAHs (naphthalene, phenanthrene, fluoranthene) to the bacterial production assays (1 mL final volume; 30 min incubations).

Sediment and Water Sampling. Water and sediment was sampled during surveys performed aboard the R/V Cape Hatteras. Sediments were collected with a 15 X 15 cm Eckman dredge or a Smith-McIntyre grab. Water samples were collected with an acrylic 2.2 liter Nansen bottle type sampler with conductivity, transmissivity and density (CTD) rosette mounted Niskin bottles.

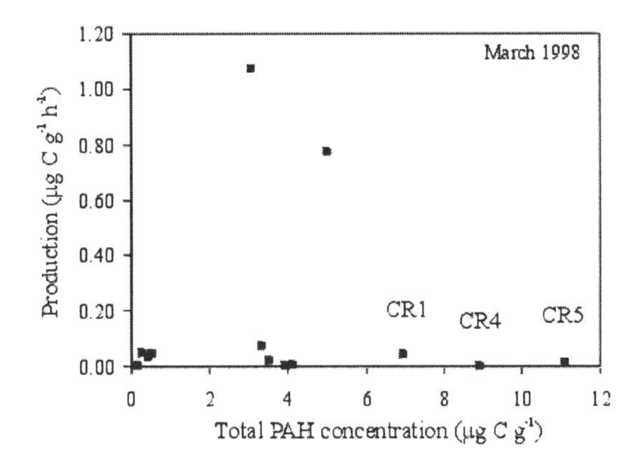

FIGURE 1. Production was generally low (e.g. Station CR4) at higher PAH concentrations (greater than 6 μg g^{-1} sediment) during the March 1998 at Charleston Navy Yard.

RESULTS AND DISCUSSION

Bacterial production was compared to both ambient PAH concentration and mineralization rates. At relatively low PAH concentrations, rates of bacterial production were highly variable. Presumably, this is because production was not affected by the ambient PAH concentration and thus other factors were important in controlling rates. However, production was generally low at higher PAH concentrations; greater than 6 μg g^{-1} sediment (Figure 1). When there are no high production values for a given PAH concentration, it is possible that the metabolic

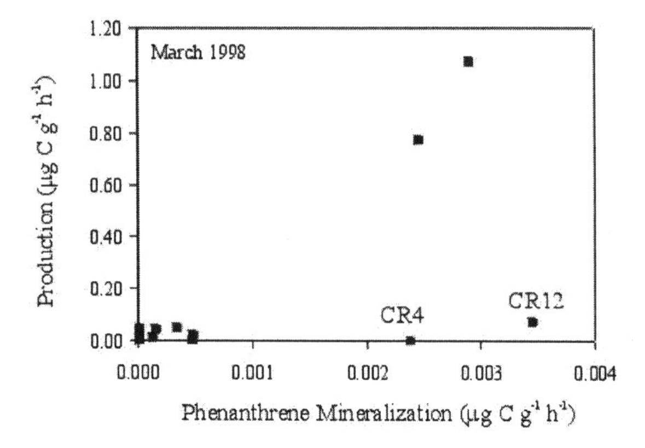

Figure 2. During the March 1998 survey, some stations with PAH concentrations that were possibly inhibiting production, also had relatively high phenanthrene mineralization rates (stations CR4 and CR12).

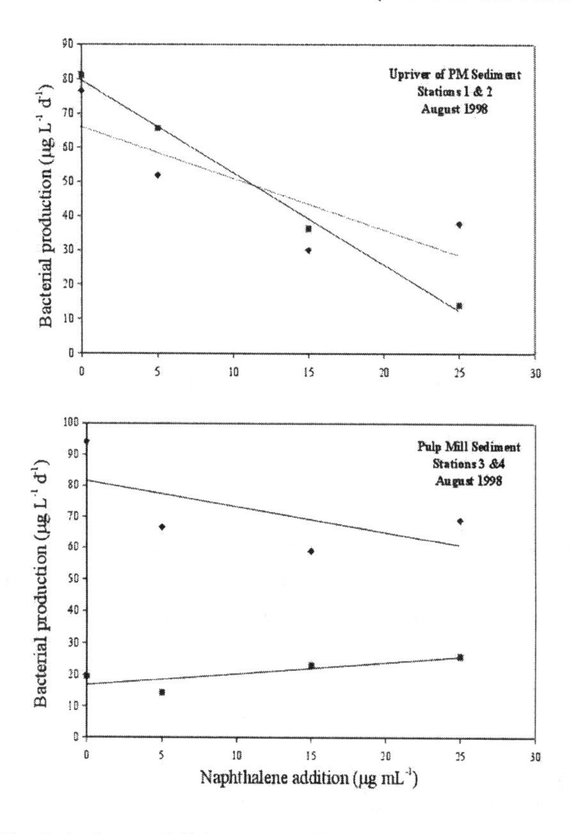

Figure 3. Naphthalene additions to sediment samples from upriver of the pulp mill (stations 1 & 2; top) inhibited bacterial production but there was little affect on stations by the pulp mill (stations 3 & 4; bottom).

rate of the heterotrophic assemblage may be inhibited by the elevated concentration of the PAHs. Bacterial production rates were also compared to PAH mineralization rates for those same stations. Some stations with PAH concentrations that were possibly inhibiting production, also had relatively high mineralization rates for certain PAHs (phenanthrene, Figure 2). Assuming that metabolic efficiency for degrading a given PAH did not vary between stations, PAH metabolism was a higher proportion of carbon which contributed to total heterotrophic production at these stations. We hypothesize that this is a result of an increase in the abundance of PAH-degrading bacteria amongst the natural assemblage. These changes may have been a community response to chronic contamination.

During the November 1997 and March 1998 cruises at Charleston, we found low heterotrophic production in the vicinity of the pulp mill (Station 4) in both surface waters and sediments. Here, we also found very rapid naphthalene mineralization rates suggesting the presence of a current PAH source (data not shown). Other PAHs were present in elevated concentrations, but naphthalene is of interest because it is relatively easily degraded by natural bacterial assemblages. During the August 1998 sampling, we added naphthalene (40-200 μM) to the sediment samples from the survey. Bacterial assemblages adapted to degrading naphthalene would be less sensitive to the additions. Naphthalene additions to sediment samples from upriver (Figure 3) and downriver of the pulp mill inhibited bacterial production. This suggests that a substantial proportion of the heterotrophic bacterial population at these sites is sensitive to the presence of naphthalene. Naphthalene additions to sediment samples from near the pulp mill did not affect bacterial production (Figure 4) suggesting that much of the bacterial assemblage is adapted to the presence of naphthalene. This experiment provides more evidence that certain areas were receiving recent, chronic inputs of PAHs and were adapted for intrinsic PAH biodegradation.

REFERENCES

Boyd, T. J., B. J. Spargo, and M. T. Montgomery. 1996. "Improved method for measuring biodegradation rates of hydrocarbons in natural water samples." In B. J. Spargo (Ed.), *In Situ Bioremediation and Efficacy Monitoring,* pp. 113-122. NRL/PU/6115--96-317,

Bååth, E. 1992. "Measurement of heavy metal tolerance of soil bacteria using thymidine incorporation into bacteria extracted after homogenization-centrifugation." *Soil Biol. Biochem. 24*: 1157-1165.

Bååth, E., M. Draz-Raviza, C. Frostegård, and C.D. Campbell. 1998. "Effect of metal-rich sludge amendments on the soil microbial community." *Appl. Environ. Microbiol. 64*: 238-245.

Bauer, J. E., and D. G. Capone. 1985. "Effects of four aromatic organic pollutants on microbial glucose metabolism and thymidine incorporation in marine sediments." *Appl. Environ. Microbiol. 49*(4): 828-835.

Carman, K.R., J. W. Fleeger, J. C. Means, S. M. Pamarico, and D. J. McMillin. 1995. "Experimental investigation of the effects of polynuclear aromatic hydrocarbons on estuarine sediment food web." *Mar. Environ. Res. 40*: 289-318.

Fuhrman, J. A., and F. Azam 1982. "Thymidine incorporation as a measure of heterotrophic bacterioplankton production in marine surface waters: evaluation and field results." *Mar. Biol. 66*: 109-120.

Harvey, R. W., R. L. Smith, and L. George. 1984. "Effect of organic contamination upon microbial distributions and heterotrophic uptake in a Cape Cod, Mass., aquifer." *Appl. Environ. Microbiol. 48*(6):1197-1202.

Huertas, M., E. Duque, S. Marques, J. L. Ramos. 1998. "Survival in soil of different toluene-degrading *Pseudomonas* strains after solvent shock." *Appl. Environ. Microbiol. 64*: 38-42.

Jensen, B. K. 1989. "ATP-related specific heterotrophic activity in petroleum contaminated and uncontaminated groundwaters." *Can. J. Microbiol. 35*(8): 814-818.

Kazumi, J., and D. G. Capone 1994. "Heterotrophic microbial activity in shallow aquifer sediments of Long Island, New York." *Microb. Ecol. 28*(1): 19-37.

Kirchman, D. L., E. K'Nees, and R. Hodson. 1985. "Leucine incorporation and its potential as a measure of protein synthesis by bacteria in natural aquatic systems." *Appl. Environ. Microbiol. 49*:599-607.

Leahy, J.G., and R.R. Colwell. 1990. "Microbial degradation of hydrocarbons in the environment." *Microbiological Rev. 54*: 305-315.

National Research Council (Eds.). 1993. *In situ bioremediation: when does it work?* National Academy Press, Washington, DC.

Renner, R. 1998. "Intrinsic remediation under the microscope: concerns about the use of 'natural attenuation' have led to a major review of the method's effectiveness." *Environ. Sci. Technol. 32*: 180A-182A.

Smith, D. C., and F. Azam. 1992. "A simple, economical method for measuring bacterial protein synthesis rates in seawater using ^{3}H-leucine." *Mar. Microb. Food Webs 6*(2):107-114.

FE(III) LIGANDS AS OXIDANTS TO ENHANCE BIOREMEDIATION

Björn Trepte, Stockholm University, Sweden.

ABSTRACT: Laboratory experiments showed that NTA and EDTA could enhance the bioremediation rate of PAH. Soil for the studies was collected at the gasworks site at Husarviken, Stockholm, with a total PAH concentration of about 130 mg/kg. The indigenous microflora was used to degrade PAH using PVC columns, with water circulating through the soil. Soil samples were analysed after 3-4 months treatment. With the addition of NTA the total PAH was reduced by 2/3, which was about 90% more efficient than in the control, where only water was circulated. 1 mM NTA in the water was enough to give maximum enhancement of biodegradation of PAH. Higher concentrations of NTA had a negative effect on PAH degradation. NTA with addition of Fe(III) has no additional effect on PAH degradation. The best PAH degradation was achieved with only dry yeast added, which indicates that it contains important nutrients and maybe even chelating agents. It was also observed that NTA was totally degraded and adsorbed to soil particles in 3-7 weeks.

INTRODUCTION

A majority of bacteria and other microorganisms in a soil use organic matter as a carbon source and obtain energy by degrading this organic matter, which may be anything from dead plants to an organic pollutant, like creosote. In the degradation process oxygen is consumed as electron acceptor. It is generally accepted by the scientific community that degradation takes place at a much slower rate during anoxic conditions than when the conditions are aerobic. The bacteria then have to use alternative electron acceptors, like nitrate, sulphate or iron(III) oxides. Generally, iron(III) oxides occur abundantly in the soil, but the rate of bacterially catalysed iron(III) reduction is very low, because iron(III) is present as a solid phase. By adding a chelate that binds to iron(III) and makes it soluble, the bioavailability of iron(III) can be greatly increased. This has been shown in laboratory experiments, where the biodegradation rate of benzene and toluene was enhanced when a chelate (NTA) was added (Lovley et al., 1994).

In this project it has been investigated if the addition of a chelate will enhance the bioremediation rate of more complex pollutants, like PAH, and in a complex matrix, like landfill soil. The chelating agents that were tested were NTA (nitrilotriacetic acid) and EDTA (ethylenediaminetetraacetic acid).

The gasworks site. The gasworks at Klara (1853-1922) was built to provide Stockholm with gas for heating and illumination. The city grew and prospered and the need for gas increased. When the capacity of Klara was too small to meet the demands, another gasworks was built (Värtaverket) at Husarviken. This was in 1893. During the following 79 years (until 1972) Värtaverket produced 7 billion m^3 of gas from 20 million tons of coal. A large amount of by-products was also produced, among others 700,000 tons of tar, 13 million tons of coke and more than 100,000 tons of benzene products. Due to spills, discharges and leakage, the gasworks site is today the largest contaminated site in Sweden (36 ha). The sediments in the water adjacent to the site are also heavily polluted, mainly by creosote (Ljungqvist Kärneryd, 1991).

MATERIALS AND METHODS

Column study #1. The aim of the first column study was to establish if the chelates NTA and EDTA could enhance the degradation of PAH. The soil that was used for the study can hardly be considered as natural, as it consisted to a great extent of slag and clinker. The creosote contamination was spread rather heterogeneously in the soil and analysis of four different samples showed results ranging from 86 to 196 mg/kg total PAH. The average of the four samples was 129 mg/kg.

10 columns were constructed of PVC pipes with an inner diameter of 10 cm and a length of 30 cm. After homogenisation and removal of stones and other particles larger than 2 cm, each column was filled with 2 L of soil, which weighed 1 kg.

For each column a 2-litre water container was supplied. From each container water was pumped to the top of the column. The initial flow rate was about 36 ml/h, but was finally lowered to 9 ml/h due to clogging in some of the columns.

After 97 days the column study was terminated and soil from each column was analysed. It was established that both NTA and EDTA did enhance the degradation of PAH, with slightly better results for NTA.

Column study #2. A second column study was started to test different concentrations and combinations of iron and NTA, with NTA additions up to 8 mmol/L, as the solubility of NTA at room temperature is about 7 mmol/L. This gives a saturated NTA solution. Dry baking yeast was added at 0.2 weight % to two of the containers. Baking yeast contains among other things vitamin B_{12}, which might be a limiting factor for bacterial growth. The same columns were used as in column study #1, but this time larger water containers (5 L) were used and column tops and containers were covered to prevent evaporation.

Before the water pump was started pH was adjusted to 6.2, which is the original pH of the soil. After 121 days the study was terminated and soil samples from the columns were analysed.

RESULTS AND DISCUSSION

The PAH concentration in the soil was reduced from the initial 129 mg/kg to 33-83 mg/kg in all columns. The results show that with only water circulating through the soil the PAH concentration is reduced by 36 %. Increased oxygen supply and mobilisation of microorganisms and contaminants explain this reduction. Results from column study #1 and #2 are presented in table 1.

TABLE 1. This table shows the PAH content (mg/kg) in the soil before and after treatment. The soil from column study #1 was treated during 97 days and the soil from column study #2 during 121 days. The control is an average of the controls from column study #1 and #2.

	Orig. soil	Con-trol	COLUMN STUDY #1			COLUMN STUDY #2			
NTA (mM)	0	0	0.43 mM NO_3	1 mM EDTA	1	2	4	8	0
Fe (mM)	0	0	0	1	1	0	2	2	0
Yeast (w%)	0	0	0	0	0	0	0	0	0.2
2+3 rings	31	25	28	16	18	13	20	12	11
4 rings	57	35	37	22	17	17	24	20	14
5+6 rings	41	23	17	12	7	12	24	25	8
Total PAH	**129**	**83**	**82**	**50**	**42**	**42**	**68**	**57**	**33**

High concentration of NTA (4-8 mmol/l) had a negative effect on the degradation of the heavier PAH compounds (5+6 rings), especially in combination with iron addition, where the degradation was lower than in the control. One can establish that an increase in NTA concentration does not enhance the degradation of PAH. In fact, it seems like the highest NTA concentrations have not been as efficient as the lower ones. An explanation to this could be that higher NTA concentrations lead to more extensive mobilisation of heavy metals, which in turn leads to higher toxicity and inhibition of bacterial activity. Another explanation is that high concentrations of NTA could be directly toxic to microorganisms.

With nitrate the effect is the opposite; the lighter PAH compounds (2-4 rings) were reduced to a lesser extent than in the control, while the heavier ones (5+6 rings) were reduced better, while the overall reduction compared to the control is negligible. The advantages with nitrate are that it is completely soluble in water and can function both as a nutrient and as electron acceptor during biodegradation. However, it is also connected with some difficulties, such as accumulation of nitrite, which is toxic to nitrate-reducing bacteria and can therefore become an inhibiting factor (Arvin et al., 1994).

The best PAH reduction (74%) was achieved with only dry yeast added, which indicates that it contains important nutrients and maybe even chelating agents, besides vitamin B_{12}, which in itself is a good chelator. All ring sizes were reduced 65-80 %.

With EDTA the effect is similar to that of NTA. The total PAH reduction was 61%, with a slightly better reduction for the heavy PAH compounds.

In table 2 one can follow the degradation of single PAHs in the control from column study #2. After 121 days all PAHs have been reduced and in the following 156 days all except two either have been reduced or stay constant. The reason for two of the PAHs to show an increase during the last 156 days is probably due to the problem of taking homogenous soil samples from the columns. The substantial reduction of the 6-ring benzo(ghi)perylene is most likely correct, as it has proven to undergo extensive reduction during earlier column studies (Trepte, 1997).

TABLE 2. Shows the concentration of individual PAHs in the soil after 0, 121 and 277 days respectively. Data are from the control in column study #2.

PAH (mg/kg) days	0	121	277
Naphtalene	1,4	0,6	0,6
Acenaphtylene	0,2	0,1	0,0
Acenaphtene	0,6	0,2	0,2
Fluorene	0,6	0,2	0,2
Phenantrene	10,3	3,6	3,2
Anthracene	2,1	0,8	0,7
Fluoanthene	16,2	8,9	6,5
Pyrene	13,2	5,7	5,2
Benzo(a)anthracene	10,4	5,0	3,6
Chrysene	10,2	3,8	4,2
Benzo(b)fluoranthene	13,9	5,9	5,4
Benzo(k)fluoranthene	9,3	5,3	3,5
Benzo(a)pyrene	12,8	5,7	4,9
Indeno(c,d)pyrene	11,0	6,2	2,8
Dibenz(a,c)anthracene	3,3	1,5	3,7
Benzo(g,h,i)perylene	14,1	8,7	0,8
Total	**129,3**	**62,3**	**45,4**

Figure 1 shows the variation of organic carbon in the percolating water of column study #2, where different concentrations of NTA were tested.

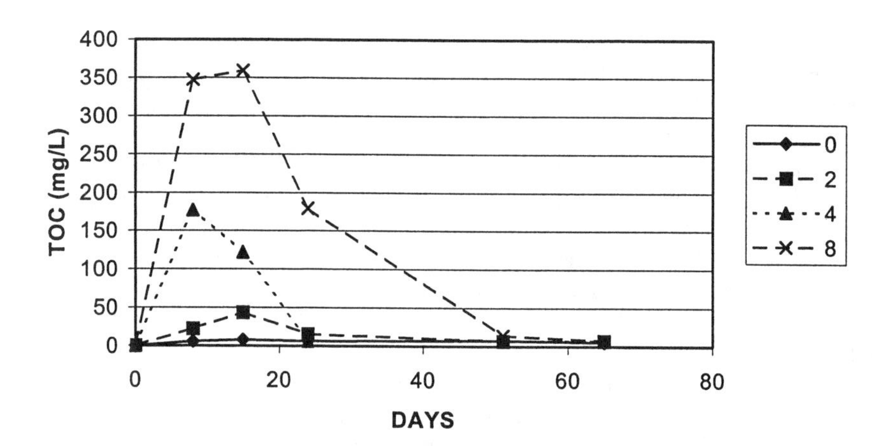

FIGURE 1. Variation of total organic carbon (TOC) in percolating water in column study #2. The numbers in the legend show the concentration of NTA (mM) added to the water.

The TOC concentration in the control (with no NTA) is constantly 5-8 mg/L. Calculations showed that the increase in organic carbon where NTA was added can be attributed to NTA itself, as it contains 37.7 % carbon. What figure 1 shows is that microorganisms start degrading NTA after 1-2 weeks and that NTA is adsorbed to soil particles. After 3-7 weeks all of the added NTA was degraded and adsorbed. An explanation for the lower PAH reduction where the highest NTA concentration was used could be that the bacteria preferred NTA as carbon source instead of PAH.

CONCLUSIONS

- 1 mmol/L NTA was enough to give maximum enhancement of biodegradation of PAH. Higher concentrations of NTA are more likely to have a negative effect on PAH degradation.
- NTA with addition of Fe(III) has no additional effect on PAH degradation.
- NTA is totally degraded and adsorbed to soil particles in 3-7 weeks.
- The best PAH degradation was achieved with only dry yeast added, which indicates that it contains important nutrients and maybe even chelating agents, besides vitamin B_{12}.

ACKNOWLEDGEMENTS

Thanks to Staffan Lundstedt, Umeå University, for help with PAH-analysis.

REFERENCES

Arvin E, Jensen B.K, Jörgensen C & Mortensen E (1994) "Biodegradation of toluene by a denitrifying enrichment culture." In *In situ bioreclamation*, ed. Hinchee R.E & Olfenbuttel R.F, pp. 480-487. Columbus, Ohio: Battelle Memorial Institute.

Ljungqvist Kärneryd A. (1991) "Gasverksprojektet, slutrapport. Sammanfattning av delrapporterna." *Stockholm Energi AB.*

Lovley D.R, Woodward J.C & Chapelle F.H (1994) "Stimulated anoxic biodegradation of aromatic hydrocarbons using Fe(III) ligands." *Nature* **370**, 128-131.

Trepte, B (1997) "Fe(III)ligands as oxidants to enhance bioremediation." *Progress report, August 97.* COLDREM report.

EFFECTS OF AGING ON THE MOBILITY OF PAHs IN SOILS

Borhane Mahjoub and Rémy Gourdon
(INSA, Villeurbanne, France)

ABSTRACT: The goal of this study is to generate scientific knowledge and data that may take into account the major mechanisms of aging in the predictive models of of organic contaminant behavior in soils. The effects of aging on the mobility of PAHs have been investigated using laboratory batch and column assays. Naphthalene and phenanthrene have been used as model contaminants because they are the most soluble PAHs. Experiments were conducted with four soil matrices: a sand, a clay, and two soils with different organic contents. Aging was conducted in water saturated and non-saturated conditions, under different conditions of temperature (constant temperature (20°C) or freeze/thaw cycles). Experimental results of the batch assays showed a reduction of the pollutant mobility with increasing aging times. Freeze/thaw cycles were found to reduce this effect significantly and increase the extractability of naphthalene. The observed effects of aging may be related to possible evolutions of pollutant/sorbent interactions, mechanical alterations of the soil micro-aggregates, and slow diffusion of naphthalene or phenanthrene in micro-pores.

INTRODUCTION

Soil and underground water pollution by polycyclic aromatic hydrocarbons (PAHs) is an important environmental problem, in France as well as in most industrialized countries. PAHs, especially naphthalene and phenanthrene, are the most frequently detected organic pollutants in underground water. Predicting the fate of this type of pollutant in soils to determine the risk of pollutant dispersion or to optimize the various treatment techniques of polluted soils is one of the crucial issues confronting researchers and engineers working in this area.

Organic pollutants in the soil are likely to be transformed, either through chemical reactions or biological reactions due to the presence of microorganisms. The capacity of bacteria to degrade xenobiotic hydrophobic compounds varies according to the state and location of the pollutants in the soil (Hatzinger et al., 1995). It has been shown that adsorbed naphthalene has a low bioavailability for degradation. In the same way, a PAH present in the micro-and nanopores of the soil particles or in the liquid or solid organic phase is much less available for biodegradation by micro-organisms. The transfer from the internal micro-and nano-porosity or from the organic phase constitutes a kinetic limit to mineralization of pollutants (Nam et al., 1998). The success of a biological treatment of a polluted site therefore depends on the mobility and bioavailability of pollutants in the soil.

It has often been observed that mineralization of hydrophobic organic pollutants declined when the contact time with the soil increased (Erickson et al., 1993). The pollutants therefore become more resistant in time to biodegradation and extraction (Hatzinger and Alexander, 1997). The set of phenomena observed concerning the evolution of polluted soil characteristics generally is referred to as aging. Apart from certain recent studies, the importance of aging to the fate of organic pollutants such as PAHs is a largely unexplored area of study. The mechanisms responsible for the observed effects are not well known and there is little data available quantifying the effects of aging on the mobility and bioavailability of pollutants in soils. It therefore appears fundamental to assess the differences in behavior between freshly added pollutants and aged pollutants (which have been in contact with the soil for a long time).

The general aim of this study is to promote knowledge and generate scientific data to allow development of transport and fate models of organic pollutants in soils, taking into account the aging phenomena which either have slow kinetics or are irreversible.

MATERIALS AND METHODS

Pollutants. The assays were carried out on 2 organic molecules selected as model pollutants : naphthalene and phenanthrene, which are the most soluble PAHs. The results presented here only concern naphthalene.

Soils and soil fractions. Four matrices were selected in order to assess the effect of aging according to the characteristics of the adsorbent. Two of the matrices are relatively simple : a sand (Fontainebleau sand) and a silty clay containing 0.7% of organic carbon. The adsorption/desorption assays on this type of material act as controls with non-porous particles (sand) and allow assessment of the role of the mainly mineral micro-porosity (clay) in the long term evolution of the mobility of the organic pollutants tested. The other two materials used were natural soils, which are representative of a wide range of European soils and which differ only in organic carbon content and the nature of organic material present (cultivated soil CSAC and a meadow soil CSAP from the Côte Saint-André, Isère, France).

Aging of pollutants. The study of the effect of aging on the mobility of organic pollutants in the soil was carried out using two distinct experimental methodologies: batch assays and column assays. Only the experimental results from batch assays are presented here.

The batch assays were carried out in 25-ml airtight centrifugation glass tubes with Teflon lined silicone stoppers to avoid adsorption of the pollutant. All the assays were carried out in triplicate. The mass ratio between the liquid and solid phases (L/S) was 3. For all the assays, 6 g of material (soil or soil fraction) were suspended in 18 ml of solution to limit the headspace (<1 ml) thus reducing losses through volatilization. To limit biodegradation, mercuric chloride was added (400 mg/L) to the aqueous solution. The $HgCl_2$ was added under sterile conditions and the tubes containing the soil were sterilized by two successive autoclavings (20 min at 120°C). The tubes were then shaken vertically at room temperature using a rotary shaker. After 24 hrs, the tubes either were centrifuged to carry out successive desorptions (by replacing the supernatant with water) or to be aged in unsaturated water conditions, or the tubes were stored vertically in a chamber at constant temperature away from the light to be aged in saturated water conditions. The mixtures therefore were aged in saturated and unsaturated conditions and at different temperatures (T constant (20°C), freeze (-18°C)/thaw (20°C) cycles).

At specified times (0,1,3,6, or 12 months), the solid and liquid phases were separated by centrifugation at 4,000g for 15 minutes (Prolabo SR 2002 centrifuge). The supernatant was sampled with a glass syringe for high-performance liquid chromatorgraphy (HPLC) analysis, to determine the quantities adsorbed by the solid matrix. Weighing of the supernatant after centrifugation allowed determination of the residual volume of solution present in the pellet, and therefore the quantity of pollutant present in the interstitial solution of the soil. The supernatant then was replaced by the same volume of water to study desorption. Second, third, and fourth desorptions were carried out using the same protocol.

After 4 desorptions with water, extraction with methanol and then an extraction with butanol-1 under sonic treatment were carried out. The extractions were carried out following the same protocol as for desorption in water, except that the supernatant was replaced by the same volume of methanol in the first stage, then by butanol-1. Sonic treatment was used to partially desegregate the adsorbent medium. The pellets in suspension in butanol were subjected to sonication (50W for 10 minutes). The tubes were then shaken for 24 hours, sonicated for another 10 minutes, and centrifuged.

RESULTS AND DISCUSSION

Silty clay. The assays on silty clay show that the role of this matrix in pollutant mobility is far from negligible (Figure 1). In fact, the clay used has a significantly retention capacity, not only the result of the presence of organic matter at relatively high concentration, but also from physical entrapping of molecules between the clay layers. This latter process is more pronounced for longer times of contact. An example of results is illustrated in Figure 1, which represents the percentage of non-restituted naphthalene after each extraction with water or organic solvent.

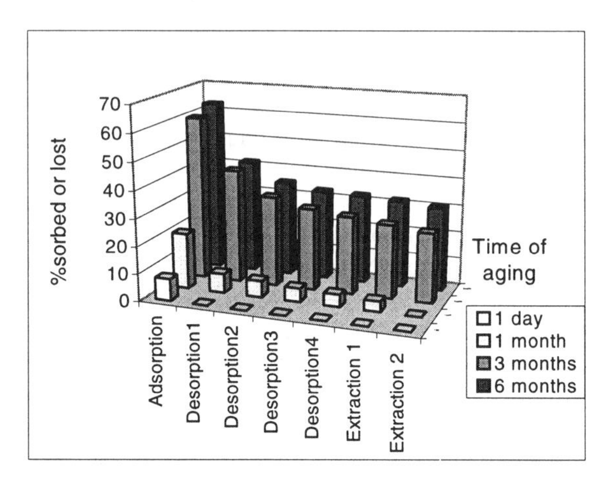

FIGURE 1: Effect of contact time on naphthalene retention in Clarsol silty clay in saturated conditions (data at 20°C).

The assays carried out with CSAC (Figures 2 and 3) and CSAP (results not shown) confirm that naphthalene retention (Figure 2) and phenanthrene retention (results not shown) take place in two steps: a rapid adsorption phase followed by a slow phase. The longer the contact of the pollutant with the soil, the greater its retention in the soil. Furthermore, whether aging takes place in saturated (Figure 2) or unsaturated conditions (Figure 3), the extractability of naphthalene in water or organic solvents decreases significantly when the contact time increases. The non-extracted fraction of naphthalene also may be partially explained by losses through volatilization. Similar results were obtained with phenanthrene (not shown).

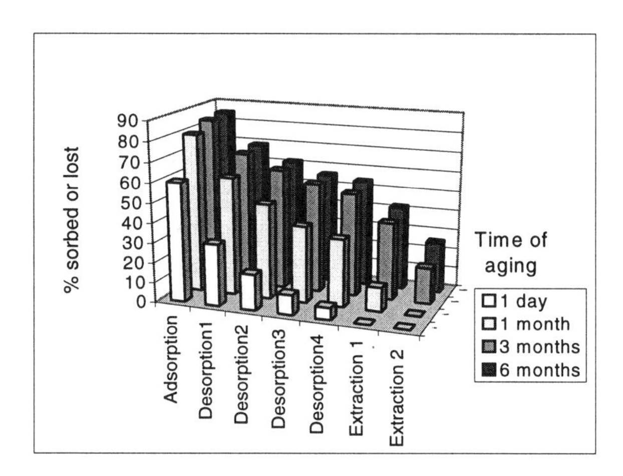

FIGURE 2: Effect of contact time on naphthalene retention in CSAC soil in saturated conditions (data at 20°C).

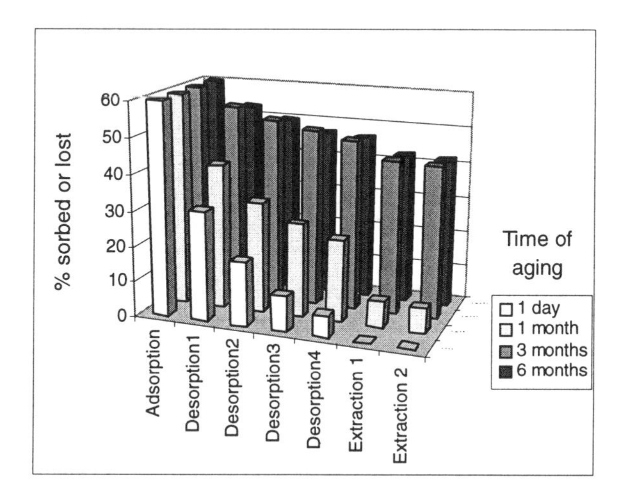

FIGURE 3: Effect of contact time on naphthalene retention in SCAC soil in unsaturated conditions (data at 20°C).

The succession of freeze/thaw cycles during aging of CSAC or CSAP soils has no observable effect on phenanthrene extractability (results not shown). However, for naphthalene, these cycles tended to limit the decrease in extractability induced by aging (Figure 4). In this case, it can be suggested that the mechanical effect of freeze/thaw cycles causes desegregation of certain soil microstructures, and improves accessibility of pollutants to extraction solvents. A fraction of the pollutants also can be extracted as adsorbed to the micro-particles thus formed. Experiments have been scheduled to quantify these possible phenomena on the last samples after one year of aging.

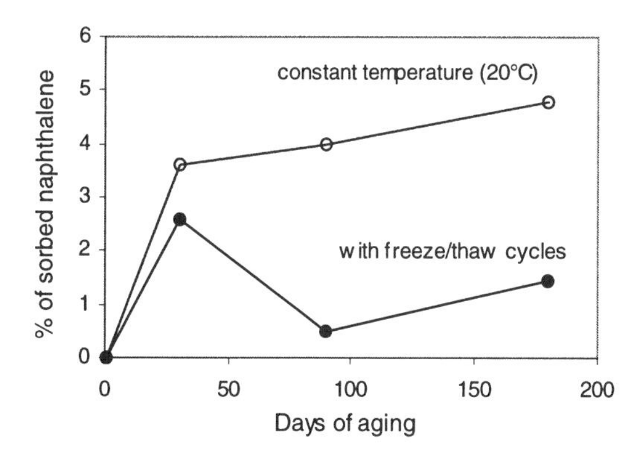

FIGURE 4: Levels of naphthalene and phenanthrene retained by Clarsol clay after 2 desorptions with water, versus contact time and aging conditions.

CONCLUSION

In a general way, the results of this study show that pollutants aged in the soil or silty clay are more significantly sequestered and less extractable than freshly added pollutants. These results suggest that pollutant sequestration by the soil or clay is probably the result of a process related to slow molecular diffusion within the micro-and nanopores of the aggregates or within the adsorbent matrix (organic matter)

Therefore, soils containing freshly added pollutants behave differently from soils having remained in contact with the pollutants for a long time. The results of laboratory assays carried out with freshly added pollutants can not therefore be used directly to predict the behavior of pollutants present in the soil over a number of years. Future research should focus on an understanding of slow sorption processes and on kinetic limitations induced by these phenomena, particularly in the case of bioremediation.

ACKNOWLEDGEMENTS

The authors gratefully acknowledge the financial support by Association RE.CO.R.D. (Réseau Coopératif de Recherche sur les Déchets).

REFERENCES

Erickson, D.C., R.C. Loehr, and E.F. Neuhauser. 1993. "PAH Loss during Bioremediation of Manufactured Gas Plant Site Soils". *Wat. Res. 27*(5): 911-919.

Hatzinger, B., and M. Alexander. 1997. "Biodegradation of Organic Compounds Sequestered in Organic Solids or in Nanopores within Silica Particles". *Environmental Toxicology and Chemistry. 16*(11): 2215-2221.

Hatzinger, P.B. et al. 1995. "Effect of Aging of Chemicals in Soil on Their Biodegradability and Extractability". *Environ. Sci. Technol. 29*(2): 537-545.

Nam, K., N. Chung, and M. Alexander. 1998. "Relationship between Organic Matter Content of Soil and the Sequestration of Phenanthrene". *Environ. Sci. Technol. 32*(23): 3785-3788.

PREDICTION OF PAH BIOAVAILABILITY IN SOILS AND SEDIMENTS BY PERSULFATE OXIDATION

Michiel P. Cuypers, J. T. C. Grotenhuis, and W. H. Rulkens
(Wageningen Agricultural University, Wageningen, The Netherlands)

ABSTRACT: To predict PAH bioavailability in soils and sediments a fast characterization test was developed, based on the chemical oxidation of soil organic matter by persulfate. The effect of oxidation time, oxidation temperature, and persulfate to organic matter ratio on the extent of PAH oxidation was investigated, yielding a characterization method that was validated both for a sediment with low and for a sediment with high PAH bioavailability. Validation showed that the amount of PAH that was removed by persulfate oxidation was a good measure of the amount of PAH that was available to micro-organisms.

INTRODUCTION

Biological treatment is an increasingly popular alternative for the decontamination of PAH-contaminated soils and sediments. Field practice, however, has shown that biological treatment is not always very successful. Often, high residual PAH concentrations are found. In many cases these high residual concentrations are caused by the limited availability of PAH for micro-organisms. To be able to predict PAH bioavailability before an actual bioremediation process is started, a characterization test is needed.

Existing characterization tests are generally based on microbiological conversion of PAH. Although reliable, these tests are long and laborious. Therefore, alternative tests were and are developed, most of them based on the desorption of well-available PAH in the water phase, followed either by immobilization of solubilized PAH on a solid sorbent (Cornelissen et al., 1998; Macrae and Hall, 1998), transport of solubilized PAH over a membrane (Macrae and Hall, 1998), or absorption of solubilized PAH in a soluble molecule (Reid et al., 1998). The main objective is to keep the water phase empty to allow maximal desorption of well-available PAHs. Although faster than a biological characterization, these tests still take at least one day of desorption time. For the prediction of long-term availability, longer desorption times are needed. A faster prediction of bioavailability may be achieved by solvent extraction (Kelsey et al., 1997). However, the use of solvents will likely change the structure of the material investigated and therefore this material differs from the material as present during bioremediation. Also, none of the solvent extractions described in literature has been extensively validated.

In our research we studied the potential of a fast chemical oxidation method for the prediction of PAH bioavailability. This method is based on the selective chemical oxidation of a part of the soil organic matter by persulfate ($S_2O_8^{2-}$). Persulfate, when heated, decomposes and forms sulfate radicals ($SO_4^{\bullet-}$)

or, in some cases, hydroxyl radicals ($OH^{\bullet -}$) (House, 1962). These radicals can oxidize 30-40% of soil humic acids, disrupting the weaker bonds present in the humic acids (Martin and Gonzalez-Vila, 1984).

In literature soil organic matter has been described as a material consisting of expanded and condensed regions which exhibit different sorption behavior for hydrophobic organic contaminants (Young and Weber, 1995). We assume that expanded organic matter, which is thought to contain predominantly well-available PAHs, is oxidized by persulfate at elevated temperatures. Condensed organic matter, which is thought to contain PAHs that are not easily available to micro-organisms, is assumed not to be oxidized by persulfate. Experiments by Young and Weber (1995) support this hypothesis as they found considerably higher K_{oc} values for phenanthrene after oxidation of soils with persulfate.

In the present study we developed a persulfate oxidation method for the prediction of PAH bioavailability. We investigated the effect of oxidation temperature, oxidation time, and persulfate to organic matter ratio on the extent of PAH oxidation in a sediment with high PAH bioavailability. On the basis of the results a method was selected which was briefly validated, both with a sediment with high PAH bioavailability and with a sediment with low PAH bioavailability.

MATERIALS AND METHODS

Sediments. Two sediments were used. The first sediment was dredged from the Amsterdam Petroleum Harbor (PH sediment). This sediment is contaminated with large quantities of oil, PAHs, and various other pollutants. Typical concentration ranges are 1000 to 10000 mg oil per kg and 500 to 3000 mg PAH per kg (dry matter). Biological experiments showed that the PAHs are well-available to micro-organisms. Typically, 70-90% of the PAHs can be biodegraded. The dry matter content of the sediment was 44 %, the organic matter content was 11%.

The second sediment was dredged from the Overschie in Rotterdam (OR sediment). This sediment is contaminated by 150 to 250 mg PAH per kg. Typically, 10% of the PAHs can be degraded microbiologically. The dry matter content of the sediment sample was 55 %, the organic matter content was 16 %.

Method Development. To determine the effect of oxidation temperature and oxidation time on the PAH oxidation by persulfate, 250 ml Schott bottles were filled with 5.0 g of wet PH sediment and 80 ml of demineralized water. Next, 4.0 g of potassium persulfate ($K_2S_2O_8$, Merck) was added and bottles were placed in a water bath at 50, 70, 90, or 120 °C. The slurry was shaken end-to-end (except at 120°C where bottles were placed in a pressure cooker) and after 0, 1, 3, 8, and 24 hours three bottles were sacrificed for PAH analysis. The slurry was filtered and solids were extracted with 1-methyl-2-pyrrolidinone (Noordkamp et al., 1997). Extracts were centrifuged (5 min, 13000 rpm) and analyzed for 16 EPA PAHs by HPLC. PAHs were separated over a Vydac 5 C18 reverse phase column (250x4.6 mm) with external guard column and concentrations were determined by UV absorbance at 254, 264, 287, 300, and 335 nm on a Waters 991 PDA detector.

To determine the effect of the persulfate to organic matter ratio on the PAH oxidation, experiments were carried out at 3, 6, 12, and 29 g $S_2O_8^{2-}$ per g organic matter (12 g/g in 1st experiment). Twenty ml water was added per g of $K_2S_2O_8$ and the temperature was set at 70°C. At 0, 3, and 24 h three samples were sacrificed for PAH analysis.

Method Validation. The oxidation method derived from the above experiments was validated using PH and OR sediment. Each sediment was both oxidized and treated biologically. Residual PAH concentrations after oxidation were compared to concentrations after biological treatment. Additionally, biologically treated material was oxidized by persulfate. This was done to study whether PAHs could be removed by oxidation after bioremediation had taken place.

Persulfate oxidation was carried out at 70°C, at a persulfate ($S_2O_8^{2-}$) to organic matter ratio of 12 g/g. Per g of $K_2S_2O_8$ 20 ml water was added. Oxidation time was 3 h and experiments were carried out in quadruple as described above.

Biodegradation experiments were carried out following a method described by Bonten et al. (1999). Sediment was mixed with a mineral medium and inoculated with an active enrichment culture. The slurry was incubated at 30°C and mixed on a rotary tumbler. On days 0 and 21 four flasks were sacrificed for PAH analysis.

RESULTS AND DISCUSSION

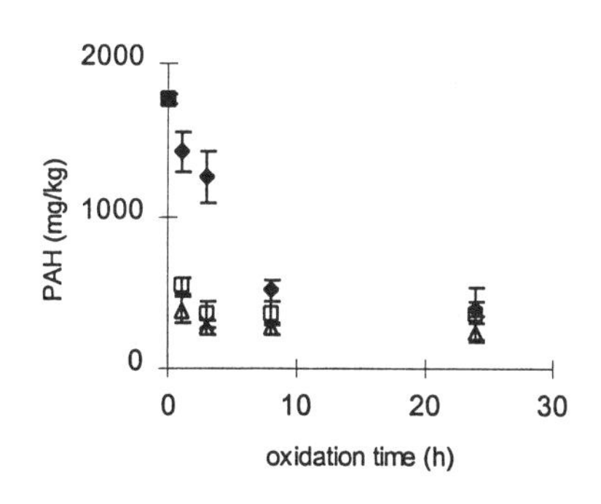

FIGURE 1. PAH concentrations (16 EPA) in PH sediment after persulfate oxidation at 50 (◆), 70 (□), 90 (Δ), and 120 (x) °C.

Method Development. The residual PAH concentrations after persulfate oxidations at different temperatures are presented in figure 1. In the temperature range 50-70°C persulfate oxidation becomes faster with increasing temperature. At 70°C and higher, PAH oxidation is hardly influenced by temperature and no further oxidation of PAH is observed after 3 hours. For the oxidations of 3 h and longer, at temperatures of 70°C and higher, the differences between the residual PAH concentrations are not significant (t-test, 95%). Given the fact that high temperature may increase bioavailability of PAHs by change of organic matter structure (Bonten et al.,1999) it is preferable to perform oxidations at the lowest temperature possible, within the shortest time possible. The results show that a 3 h oxidation at 70°C is optimal.

The persulfate to organic matter ratio does not influence the extent of PAH oxidation within the range studied (data not shown). Residual concentrations were not significantly different. Also the repeated addition of 12 g persulfate per g organic matter did not yield lower PAH concentrations. Therefore, it may be concluded that a persulfate to organic matter ratio of 12 g/g guarantees a satisfactory excess of persulfate.

Method Validation. In figure 2 the PAH concentrations in fresh, biologically treated, and chemically oxidized PH and OR sediment are presented.

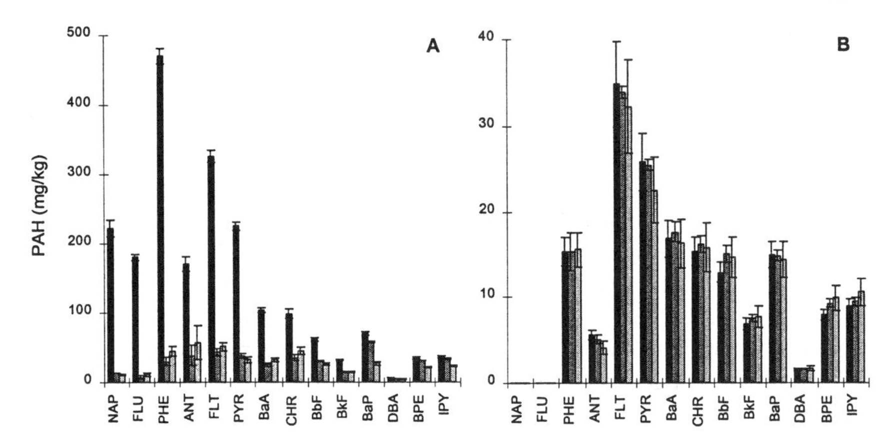

FIGURE 2. PAH concentrations in PH (A) and OR sediment (B) (16 EPA PAHs, without acenaphthalene and acenaphthene). Black: fresh sediment; Gray: after biodegradation; Light gray: after persulfate oxidation.

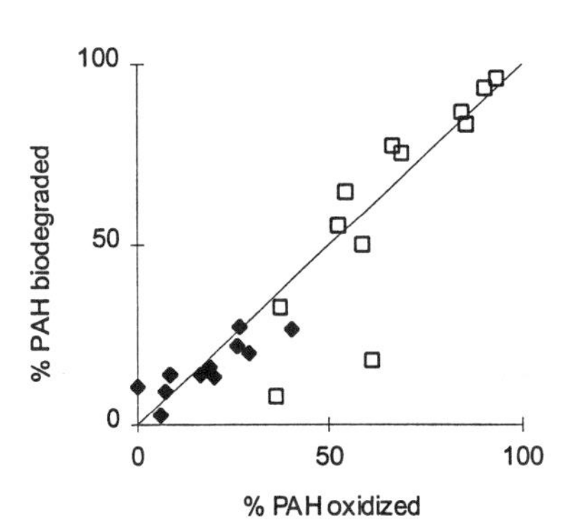

FIGURE 3. Percentage of individual PAHs biodegraded vs percentage oxidized; PH (□), OR (◆) sediment.

In PH sediment the microbiological PAH removal strongly differed for the individual PAHs. Low molecular weight PAHs were degraded efficiently, with removal efficiencies up to 96%, whereas high molecular weight PAHs were hardly degraded. The residual PAH concentrations after persulfate oxidation follow this pattern. Despite that, the PAH concentrations after biodegradation and after oxidation differ significantly for 9 out of 14 PAHs (t-test, 95%). This may be caused by the high removal efficiencies, leaving very low residual PAH concentrations compared to the initial concentrations. As can be seen in figure 3, the removal percentages of the individual PAHs are fairly similar, except for benzo[a]pyrene, benzo[ghi]perylene, and indeno[123-cd]pyrene. These high molecular weight PAHs show better removal by oxidation than by biological degradation. This most likely is caused by biological limitations rather than by limited availability. Similar observations have previously been reported by Cornelissen et al. (1998) and Macrae et al. (1998).

In OR sediment hardly any PAHs were removed by biological treatment or oxidation. The small differences between the residual concentrations after oxidation and biodegradation are not significant (t-test, 95%).

Figure 2 and 3 show that the residual PAH concentrations after bioremediation could be well predicted by persulfate oxidation, both for a sediment with high and for a sediment with low biological PAH removal. In case of high PAH removal the exact residual concentrations are more difficult to predict. However, in that case the percentual PAH removal can be fairly accurately predicted.

Figure 4 shows that no more PAHs could be removed from PH and OR sediment by persulfate oxidation after biological treatment. The differences between the PAH concentrations before and after oxidation are not significant (t-test, 95%), except for fluorene, which could not be detected in oxidized PH sediment. The results indicate that the fraction of PAHs that is removed by persulfate oxidation is the same as the fraction removed by biological treatment. This indicates that the organic matter that is affected by persulfate oxidation most likely is the same organic matter that contains PAHs which can easily desorb.

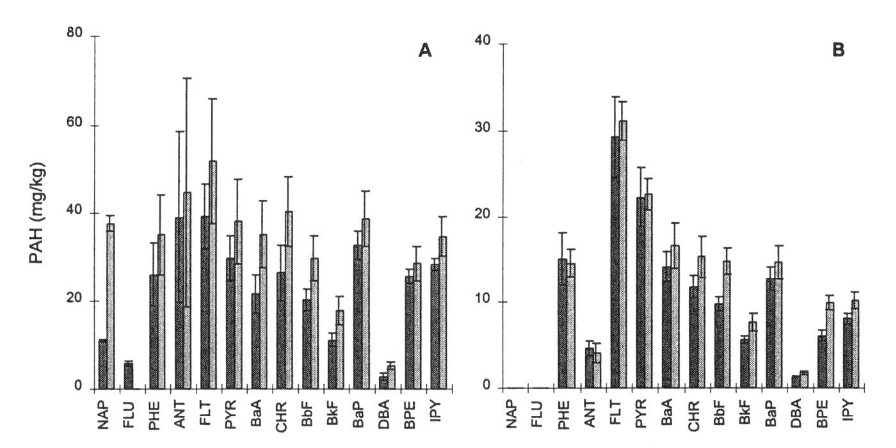

FIGURE 4. Residual PAH concentrations in PH (A) and OR sediment (B). Gray: after biodegradation; Light gray: after biodegradation and oxidation.

CONCLUSIONS

The best process conditions for persulfate oxidation are an oxidation time of 3 h, an oxidation temperature of 70 °C, and a persulfate to organic matter ratio of 12 g/g. The amount of PAH that was removed from PH and OR sediment under these conditions proved to be a good measure of the amount of PAH that is available to micro-organisms.

To establish the general applicability of the persulfate oxidation method for the prediction of PAH bioavailability a more extensive validation is currently performed.

ACKNOWLEDGEMENTS

This work was financially supported by the Dutch Organization for Applied Scientific Research (TNO). We thank Jelle Botma and Bart van Gelder for their technical assistance.

REFERENCES

Bonten, L. T. C., T. C. Grotenhuis, and W. H. Rulkens. 1999. "Enhancement of PAH Biodegradation in Soil by Physicochemical Pretreatment." *Chemosphere,* accepted.

Cornelissen, G., H. Rigterink, M. M. A. Ferdinandy, and P. C. M. van Noort. 1998. "Rapidly Desorbing Fractions of PAHs in Contaminated Sediments as a Predictor of the Extent of Bioremediation". *Environ. Sci. Technol. 32*(7): 966-970.

House, D. A. 1962. "Kinetics and Mechanisms of Oxidations by Peroxydisulfate." *Chem. Rev. 62*(3): 185-203.

Kelsey, J. W., B. D. Kottler, and M. Alexander. 1997. "Selective Chemical Extractants to Predict Bioavailability of Soil-aged Organic Chemicals." *Environ. Sci Technol. 31*(1): 214-217.

Macrae, J. D. and K. J. Hall. 1998. "Comparison of Methods Used to Determine the Availability of Polycyclic Aromatic Hydrocarbons in Marine Sediment." *Environ. Sci. Technol. 32*(23): 3809-3815.

Martin, F. and F. J. Gonzalez-Vila. 1984. "Persulfate Oxidation of Humic Acids Extracted from Three Different Soils." *Soil Biol. Biochem. 16*(3): 207-210.

Noordkamp, E. R., J. T. C. Grotenhuis, and W.H. Rulkens. 1997. "Selection of an Efficient Extraction Method for the Determination of Polycyclic Aromatic Hydrocarbons in Soil and Sediment." *Chemosphere. 35*(9): 1907-1917.

Reid, B. J., K. T. Semple, and K.C. Jones. 1998. "Prediction of Bioavailability of Persistent Organic Pollutants by a Novel Extraction Technique." In *Contaminated Soil '98*, pp. 889-890. Thomas Telford, London.

Young, T. M. and W. J. Weber Jr. 1995. "A Distributed Reactivity Model for Sorption by Soils and Sediments. 3. Effects of Diagenetic Processes on Sorption Energetics." *Environ. Sci. Technol. 29*(1): 92-97.

TEMPORAL CHANGES IN THE BIOAVAILABILITY OF PAHs IN SOILS

K.T.Semple, J. Stokes, B.J. Reid
(Lancaster University, Lancaster, U.K.).

ABSTRACT: A study was conducted to investigate the ageing of phenanthrene in soil, and in particular how the ageing process was influenced by the presence and concentration of hydroxypropyl-β-cyclodextrin (HPCD). [9-^{14}C]Phenanthrene-spiked soil was aged in the presence and absence of HPCD. In general, an increase in the soil-phenanthrene contact time resulted in decreased extractability and mineralization. It was noted that the highest concentration of HPCD resulted in enhanced loss from the system and there was an observed increase in the recalcitrant fraction of the soil-associated PAH. Lower HPCD concentrations retarded ageing with respect to both bioavailability and mineralization with only small changes observed between time 0 and 84 d. Additionally, this study showed a high degree of agreement between HPCD extraction of phenanthrene with its mineralization by catabolically active microorganisms. In terms of the assessment of the bioavailability of phenanthrene, extraction of contaminated soils with 50 mM HPCD can predict the bioavailability of the PAH to degrading microorganisms. Ultimately, this predictive tool may prove useful for determining the success or failure of bioremediation strategies in the field.

INTRODUCTION

Persistent organic pollutants (POPs), such as polycyclic aromatic hydrocarbons (PAHs), enter soils via atmospheric deposition and industrial and agricultural activities. These compounds can associate with/partition strongly onto soils and national inventories show that soils represent the major environmental repository for these contaminants (Wild and Jones, 1995). Because of the long half-lives of POPs, they are accumulating in soils nationally. This raises critical questions as to whether soil-bound POPs can be considered as immobilized and thus unlikely to enter food chains or whether they have the potential - perhaps in the long-term - to remobilize and be subject to uptake or biodegradation.

Previous studies have observed that as the length of time an organic compound remains in contact with soil increases, both the rate and extent of degradation of that compound by microorganisms decreases (Hatzinger and Alexander, 1995). Similarly, the solvent extractability of the compound decreases as compound/soil contact time increases. This decrease in compound availability, with time, has been termed "*ageing*" (Hatzinger and Alexander, 1995). The nature and extent of ageing is putatively dependent upon many factors including (i) soil structure - particle size and organic matter content; (ii) compound's physicochemical properties - molecular size, aqueous solubility, vapour pressure and octanol:water partition coefficient (K_{ow}), and (iii) intra-soil processes - sorption onto soil particles, diffusion into spatially remote areas, such as soil macro and micro pores and the entrapment within soil organic matter (Pignatello and Xing, 1996).

It is evident that the fate and behaviour of organic compounds within the soil environment is dependent on a complex array of processes. Processes which ultimately govern bioavailability, and thereby dictate the feasibility of bioremediation strategies. In terms of bioremediation, can these ageing processes be retarded which would result in greater levels of degradation in the soil? Previous studies have highlighted cyclodextrins as a class of chemicals that can be used to enhance the degradation of non-polar pollutants in liquid culture (Wang et al., 1998) and also increase the concentration of POPs in the aqueous phase (Reid et al., 1998a).

This project the aimed to characterise the fate of PAHs in soil in the presence of a range of concentrations of hydroxpropyl-β-cyclodextrin (HPCD); to assess the extractability with dichloromethane and HPCD after specific soil-PAH contact times and to characterise the biodegradation of PAHs in aged and non-aged soils.

MATERIALS AND METHODS

Spiking of soils with PAHs. Soil selected for previous studies (Reid et al., 1998b) will be used in this study. The soils will be air-dried and sieved. One half the soil will be rehydrated in the presence of a [9-^{14}C]phenanthrene (Reid et al., 1998b) and the other half of the soil will be rehydrated with a solution of HPCD over a concentration range (0, 2, 10 and 20%). The two halves will then be thoroughly mixed together to provide HPCD contents of 0, 1, 5 and 10% at a phenanthrene concentration of 25 mg kg^{-1} and placed into sealed jars with minimal headspace for the duration of the experiment. The soils were sampled at 0 d and 84 d. Each soil treatment was produced in triplicate.

Assessment of Total Activity: The spiked soils were left for 1 d, to allow the compound to become incorporated/associated with the soil. After this conditioning period 6 samples (1 g) were removed for analysis and packed into paper combustion cones. The samples were combusted using a Packard 307 sample oxidiser. The combustion process was conducted over 3 min, aided by the addition of Combust-aid™ (100 μL) injected onto the samples prior to oxidation. The resultant $^{14}CO_2$ trapped in Carbosorb-E and eluted solutions were counted using a Canberra Packard Tri-Carb 2250CA liquid scintillation analyzer

Dichloromethane (DCM) Extraction: Soil (2.5 g) was ground using a pestle and mortar with oven dried sodium sulphate (10 g). The resultant free-flowing powder was transferred to a cellulose extraction thimble. Extraction was achieved using a Tecator, Soxtech System HT (1043 extraction unit). A 30 min boil followed 150 min rinse cycle was used. The resultant extrtacts were made up to volume and sampled (10 ml) into Ultima Gold XR liquid scintillation fluid and counted as before. Extractions were run in triplicate.

HPCD Extraction: Freshly spiked and aged soils (1.25 g) were weighed in triplicate into Teflon centrifuge tubes (35 ml capacity) and HPCD solution (50 mM) added (25 ml). Samples containing unspiked soil were also prepared to provide analytial blanks. The tubes were sealed and placed on their sides on an orbital shaker and shaken at 150 rev min^{-1} for 20 h. The tubes were then centrifuged at 27,000 x *g* using a Beckman JA 21/2 centrifuge. The supernatants were then sampled (6 ml) and added to Ultima Gold XR scintillation fluid (14 ml). The resultant solutions were counted using a Canberra Packard Tri-Carb 2250CA liquid scintillation analyzer for 10 min.

Bioavailability assessment: The mineralization of [9-^{14}C]phenanthrene was assessed using respirometers (modified Erlenmeyer flasks) (Semple and Fermor, 1997). To the respirometers soil (10 g) was added along with 30 ml of a minimal basal salts solution containing 10^6-10^7 cell ml^{-1}. $^{14}CO_2$ liberated by the mineralization of the phenanthrene was trapped using 1M KOH (1 ml) present in a glass scintillation vial (solvent washed) suspended from the top of the respirometer by a stainless steel clip. The vials were removed and replaced at regular intervals and analysed by liquid scintillation counting after mixing with Ultima Gold scintillation fluid. Respirometers were run in triplicate.

RESULTS AND DISCUSSION

Immediately after spiking (time 0) the soils and after 84 d ageing, the soils were oxidized to quantify the total [9-^{14}C]phenanthrene present (Table 1). In each of the sets of soil incubations, there were losses of 5%, 6.4%, 11.1% and 35.5% for soils containing 0%, 1%, 5% and 10% HPCD, respectively, after 84 d ageing (Table 1).

DCM-soxhlet extractions were carried out at time 0 and again after 84 d ageing. At time 0, 97.7%, 98.4%, 98.5% and 98.8% of the [9-^{14}C]phenanthrene was measured after extraction of soils containing 0%, 1%, 5% and 10% HPCD, respectively (Table 1). However, after 84 d ageing, the amount extractable had been reduced to 94.5%, 94.5%, 94.4% and 85% for the soils, respectively (Table 1). This suggests that there is very little change with ageing in the amount of phenanthrene which can be extracted from the soils containing 0%, 1% and 5% HPCD. However, there is a substantial decrease in the amount of the PAH extractable in the soils containing 10% HPCD. Coupled with the increased loss measured after sample oxidation at this high HPCD concentration, the data may be suggesting enhanced removal of labile soil-associated phenanthrene with ageing.

Soil samples were extracted using a 50 mM HPCD solution at time 0. This aqueous solvent did not extract as much of the PAH as the DCM, with values of 74.4%, 76%, 77.9, and 80.3% for soils containing 0%, 1%, 5% and 10% HPCD, respectively. Interestingly, the amount of extractable PAH increased with increasing HPCD concentration in the soils, similar to DCM, but more marked. After 84 d, in all the soils there was a decrease in the amount of extractable phenanthrene relative to the time 0 values, with 58%, 71.1%, 72.7% and 60.4% for soils containing 0%, 1%, 5% and 10% HPCD, respectively (Table 1).

Table 1. Comparisons between residual [9-^{14}C]phenanthrene (%) in soils using different extraction techniques and bacterial mineralization

HPCD (mM)	Residual [9-^{14}C]phenanthrene[1] (%) in soils at 0 d and 84 d							
	Sample Oxidation		DCM Extraction		HPCD Extraction		Mineralization	
	0 d	84 d	0 d	84 d	0 d	84 d	0 d	84 d
0	100	95.0	97.7	94.5	74.4	58.0	75.6	60.3
1	100	93.6	98.4	94.5	76.0	71.1	60.3	58.1
5	100	88.9	98.5	94.4	77.9	72.7	62.5	60.7
10	100	64.5	98.8	85.0	80.3	60.4	62.1	41.8

[1]DCM extractions, HPCD extractions and mineralization values were normalized to the amount of ^{14}C-activity measured after sample oxidation at 0 and 84 d. Each value is the arithmetic mean of three separate measurements.

Bioavailability of the [9-^{14}C]phenanthrene was quantified by measuring the mineralization of the PAH by a known microbial degrader. At time 0, the soils which had not been spiked with HPCD mineralized the greatest amount of the PAH at 75.6% (Table 1). The soils which were spiked with increasing concentrations of HPCD, mineralized 60%-63% of the PAH (Table 1). After 84 d ageing, the values for mineralization were 60.3%, 58.1%, 60.7% and 41.8% for the soils containing 0%, 1%, 5% and 10% HPCD, respectively (Table 1). The only soils, which showed a marked decrease in mineralization with soil-PAH contact time, were those with 0% and 10% HPCD. In the 0% HPCD soils, the amount of phenanthrene which was HPCD-extractable for time 0 and 84 d ageing (74.4% and 58.0, respectively) was very similar to the amount of the PAH which was mineralized (75.6 and 60.3, respectively). This suggests, at least for soils with no HPCD present, that 50 mM HPCD extraction can account for the amount of phenanthrene which is available to degrading microorganisms. Interestingly, the values for 1% and 5% HPCD-spiked soils show very little difference in HPCD extractability or mineralization between time 0 and 84 d (approx. 5% and 2%, respectively). This suggests that, at lower concentrations, HPCD may retard the ageing process.

A problem that is often encountered with POPs is their hydrophobicity and the need to used organic solvents to extract/solubilize them. This is a problem because organic solvent extractions do not necessarily reflect compound bioavailability (Kelsey and Alexander, 1997). To address this, an aqueous solvent composed of hydroxpropyl-β-cyclodextrin (HPCD) has been used. The molecule has a high aqueous solubility but contains a hydrophobic cavity which has been shown to form a 1:1 inclusion complex with hydrophobic organic molecules (Brusseau et al., 1997). Contrary to excluding compounds from microorganisms, it has been shown that the formation of such inclusion complexes enhances the degradation of organic compounds, such as phenanthrene (Wang et al., 1998). This phenomenon is evident in the total amounts of quantified by sample oxidation after 84 d (Table 1).

CONCLUSIONS

This study has shown that there is a decrease in bioavailability and extractability of phenanthrene in soil-PAH contact time. Additionally, the data suggest that soils spiked with 1% and 5% HPCD can retard this ageing process, although the mechanisms are not known at this time. Conversely, soil spiked with 10% HPCD exhibits enhanced loss with ageing time. In terms of the assessment of the bioavailability of phenanthrene, extraction of contaminated soils with 50 mM HPCD can predict the bioavailability of the PAH to degrading microorganisms. Ultimately, this predictive tool may prove useful for determining the success or failure of bioremediation strategies in the field.

Acknowledgements

The authors would like to acknowledge the support for this project from The Nuffield Foundation and the Natural Environment Research Council (NERC) UK.

References

Brusseau, M.L., X. Wang, and W-Z. Wang, W-Z. 1997. "Simultaneous elution of heavy metals and organic compounds from soil by cyclodextrin." *Environ. Sci. Technol. 31:* 1087-1092.

Hatzinger, P.B. and M. Alexander. 1995. "Effect of ageing of chemicals in soil and their biodegradability and extractability." *Environ.l Sci. Technol. 29:* 537-545.

Kelsey, J.W. and M. Alexander. 1997. "Declining bioavailability and inappropriate estimation of risk of persistent pollutants." *Environ. Toxicol. Chem. 16:* 582-585.

Pignatello, J.J. and B. Xing. 1996. "Mechanisms of slow sorption of organic chemicals to natural particles." *Environ. Sci. Technol. 30*: 1-11.

Reid, B.J., K.T. Semple, C.J.A. Macleod, K.J. Weitz and G.I. Paton. 1998a. "Feasibility of using prokaryote biosensors to assess toxicity of hydrophobic pollutants using a novel solvent." *FEMS Microbiol. Lett. 169*: 227-233.

Reid, B.J., G.L. Northcott, K.C. Jones and K.T. Semple. 1998b. "Evaluation of spiking procedures for the introduction of poorly water soluble contaminants into soil." *Environ. Sci. Technol. 32*: 3224-3227.

Semple, K.T. and T.R. Fermor. 1997. "Enhanced mineralization of [UL-^{14}C]PCP in mushroom composts." *Res. Microbiol. 148*: 795-798.

Wang, J-M., E.M. Marlowe, R.M. Miller-Maier and M.L. Brusseau. 1998. "Cyclodextrin-enhanced biodegradation of phenanthrene." *Environ. Sci. Technol. 32*: 1907-1912.

Wild, S. R. and K.C. Jones. 1995. "Polynuclear aromatic hydrocarbons in the United Kingdom environment: a preliminary source inventory and budget." *Environ. Poll. 88*: 91-108.

CAN BIOAVAILABILITY OF PAHs BE ASSESSED BY A CHEMICAL MEANS?

Brian J. Reid; Kevin C. Jones; ***Kirk T. Semple***
(Lancaster University, Lancaster, UK).

ABSTRACT: A method is presented for extracting PAHs from contaminated soil using aqueous solutions of hydroxypropyl-β-cyclodextrin (HPCD). The procedure enables elucidation of ageing trends and illustrates kinetic restraints on compound release as ageing proceeds. A comparison is made between HPCD extracted fractions and the fractions obtained by traditionally established exhaustive methodologies i.e. dichloromethane (DCM) soxhlet extraction. Comparisons of the amount of soil associated ^{14}C-radiolabelled compound mineralised by degrading microorganisms with the amount of compound extractable into HPCD solutions indicates a strong linear correlation with slope of unity. The data presented here supports the hypothesis that an aqueous solution of cyclodextrin can provides a good prediction of organic compound bioavailability in soil.

INTRODUCTION

Traditionally, soil extraction techniques have been concerned with the determination of "total" compound concentrations. Thus soils have been exhaustively extracted by "harsh" techniques, such as soxhlet extraction (Brilis and Marsden., 1990). However, such methods may give very little information about proportion / amount of compound which is transportable / bioavailable in soil (Bosma et al., 1997; Kelsey et al., 1997). More recently less exhaustive techniques have been investigated with the aim of establishing a method which may mimic compound bioavailability (Cornelissen et al., 1997; Kelsey and Alexander 1997; Kelsey et al., 1997). The use of less exhaustive organic solvent based techniques, while potentially more appropriate in terms of risk assessment, still have very little relevance to the transfer mechanisms inherent to bioavailability. It is generally recognised that for microbial degradation an organic compound must be in (or transferable with) the aqueous phase (Bosma and Harms, 1996; Bosma et al., 1997). It is hypothesised that by using an aqueous solution of cyclodextrin to extract soil that molecules which can transfer with the aqueous phase will do so and then be encapsulated by the cyclodextrin macromolecule (Reid et al. 1998a). Conversely, strongly bound or sequestered molecules will not be readily transferred to the aqueous phase and thus will not be extracted. We therefore conducted the studies described in this paper to investigate whether extraction with an aqueous phase cyclodextrin solution could adequately mimic the trends in declining bioavailability during compound ageing.

MATERIALS AND METHODS

Soil collection and spiking: Subsurface soil was collected and spiked as described elsewhere (Reid et al., 1998b). A phenanthrene concentration of 10 mg $kg^{-1}_{wet\ soil}$ was used for method development. Studies of ageing trends were carried out at 10 mg $kg^{-1}_{wet\ soil}$ (kinetic release studies) and 25 mg $kg^{-1}_{wet\ soil}$ or 50 mg $kg^{-1}_{wet\ soil}$ (extractability and mineralisation assessment). ^{14}C-9-phenanthrene provided a radioactivity of between approximately 50 and 100 Bq $g^{-1}_{wet\ soil}$ depending on the nature of the experiment.

Assessment of total activity: The spiked soils were left for 1 d, to allow the compound to become incorporated/associated with the soil. After this conditioning period 6 samples (1 g) were removed for analysis and packed into paper combustion cones. The samples were combusted using a Packard 307 sample oxidiser. The combustion process was conducted over 3 min, aided by the addition of Combust-aid™ (100 µL) injected onto the samples prior to oxidation. The resultant $^{14}CO_2$ trapped in Carbosorb-E and eluted solutions were counted using a Canberra Packard Tri-Carb 2250CA liquid scintillation analyzer.

Optimisation of hydroxypropyl-β-cyclodextrin (HPCD) extraction concentration. HPCD solutions were prepared using Milli Q water to provide a range in concentration from 0 to 60 mM. Freshly spiked soil (1.25 g) was weighed in triplicate into Teflon centrifuge tubes (35 mL capacity) and HPCD solution added (25 mL). Samples containing unspiked soil were also prepared to provide analytical blanks. The tubes were sealed and placed on their sides on an orbital shaker and shaken at 150 rev min^{-1} for 20 h. The tubes were then centrifuged at 27,000 x *g* using a Beckman JA 21/2 centrifuge. The supernatants were then sampled (6 mL) and added to Ultima Gold XR scintillation fluid (14 mL). The resultant solutions were counted using a Canberra Packard Tri-Carb 2250CA liquid scintillation analyzer for 10 min. A mass balance was determined at the completion of the extraction by sample oxidation (see before) of the residual pellet after removal of the supernatant.

Optimisation of extraction time: Freshly spiked soil was extracted into a 50 mM HPCD aqueous solution for 3, 6, 18, and 48 h. In this way the optimum time for effective extraction of ^{14}C-9-phenantherene was determined.

Release kinetics: The experiment described above for establishing sufficient extraction time was repeated using a freshly spiked soil (1d ageing) and a soil which had been allowed to age for 117 d. Extractions were terminated after 3, 6, 12, 18, 48 and 112 h.

Chemical extractability and biodegradability: After successive ageing periods (1 d, 42 d and 84 d) soils were sampled, mixed with a spatula and sub-samples were either sample oxidised to determine the residual activity (see above), extracted with dichloromethane (DCM) or extracted with hydroxypropyl-β-cyclodextrin (HPCD). Biodegradation of soil-associated phenanthrene was determined by respirometry.

Dichloromethane Extraction: Soil (2.5 g) was ground using a pestle and mortar with oven dried sodium sulphate (10 g). The resultant free-flowing powder was transferred to a cellulose extraction thimble. Extraction was achieved using a Tecator, Soxtech System HT (1043 extraction unit). A 30 min boil followed 150 min rinse cycle was used. The resultant extracts were made up to volume and sampled (10 mL) into Ultima Gold XR liquid scintillation fluid and counted as before. Extractions were run in triplicate.

HPCD Extraction: The optimised procedure using 50 mM HPCD was used to extract soil over a 20 h period as described before. Extractions were run in triplicate.

Bioavailability assessment: The mineralization of ^{14}C-9-phenanthrene was assessed using respirometers (modified Erlenmeyer flasks) (Semple and Fermor, 1997). To the respirometers soil (10 g) was added along with 30 ml of a minimal basal salts solution. $^{14}CO_2$ liberated by the mineralization of the phenanthrene was trapped using 1M KOH (1 ml) present in a glass scintillation vial (solvent washed) suspended from the top of the respirometer by a stainless steel clip. The vials were removed and replaced at regular intervals and analysed by liquid scintillation counting after mixing with Ultima Gold scintillation fluid. Respirometers were run in triplicate.

RESULTS AND DISCUSSION

Optimisation of HPCD extraction concentration. With increasing concentrations of HPCD (0 to 40 mM), there is a corresponding increase in the amount of phenanthrene extracted from the soil (Figure 1A). Above a HPCD concentration of 40 mM there is no additional increase in extraction efficiency (Figure 1A). Sample oxidation of the residual pellets after extraction at excess concentrations of HPCD indicated that all of the introduced compound could be accounted for (data not shown).

Optimisation of extraction time: An increasing extraction time (0 to 6 h) resulted in a concurrent increase in the extraction efficiency (Figure 1B). However, after 6 h the extraction efficiency plateaued. Thus, the extraction time of 20 h employed in the other parts of this investigation was concluded to be appropriate for complete extraction.

Release kinetics: The release of phenanthrene from both the freshly spiked and aged soils was characterized by two modes of exchange. The freshly spiked soil exhibited an initial rate of exchange of 12.2% h^{-1} followed by a very slow subsequent rate of exchange 0.02% h^{-1}. The aged soil exhibited a reduced rapid rate of exchange 8.2% h^{-1} while the subsequent slower rate was greater than in the unaged samples, at 0.12% h^{-1}. The results suggest the formation with increased ageing time of a less rapidly exchangeable soil-associated fraction. This biphasic behaviour has been reported elsewhere (Wu and Gschwend. 1986)

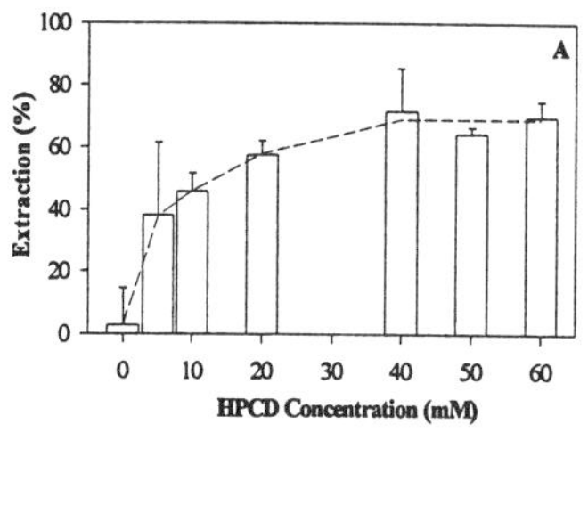

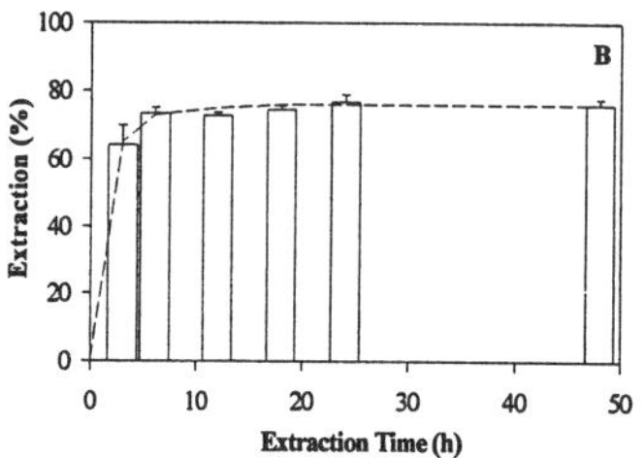

FIGURE 1
Optimisation of HPCD aqueous extraction in terms of: HPCD concentration (A) and extraction time (B).

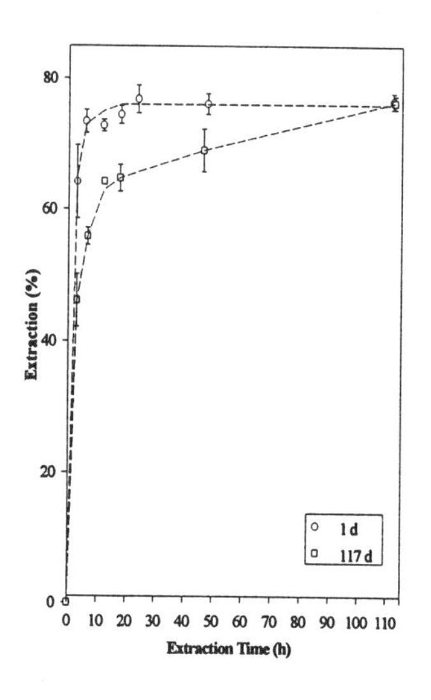

FIGURE 2
Kinetic extraction time-series of freshly spiked (1 d) and aged (117 d) phenanthrene from soil using an aqueous HPCD solution

Chemical extractability and biodegradability of phenanthrene: Where phenanthrene was introduced into the soil at a concentration of 25 mg kg^{-1} there was very little loss with time. Extractability of the residual compound with DCM was very close to 100% in all cases. Extractability with HPCD however decreased with increased ageing. The extractability with HPCD very closely related to the extent of mineralization after 240 h assay time (sufficient for degradation to be complete) at all ageing times. Where phenanthrene was introduced into the soil at a concentration of 50 mg kg^{-1} there was significant loss with time. This may be due to the induction of catabolically active microorganisms at this higher phenanthrene concentration. Extractability of the residual compound with DCM decreased with increased ageing time, indicating the formation of recalcitrant fraction. Extractability with HPCD also decreased with increased ageing. The extractability with HPCD most closely related to the extent of mineralization (after 240 h assay time), while the extractability with DCM typically over estimated availability by approximately 60% at all ageing times.

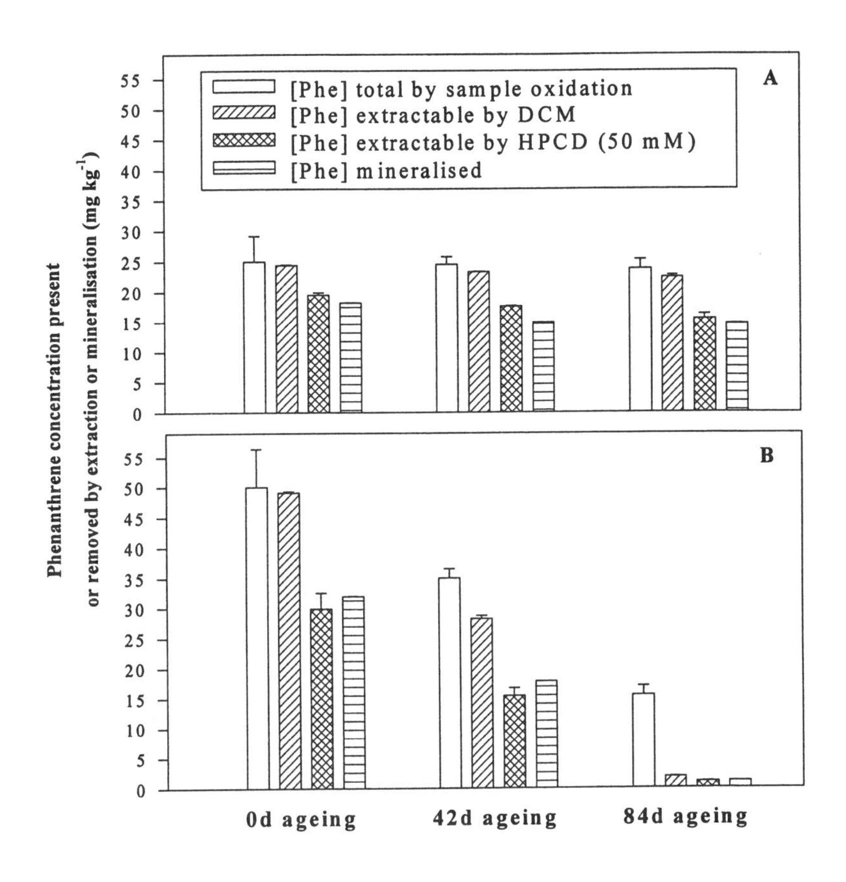

FIGURE 3 The influence of compound ageing on residual concentration, extractability with DCM or HPCD and mineralisation, at two initial phenanthrene concentrations: 25 mg kg^{-1} (A) and 50 mg kg^{-1} (B).

CONCLUSIONS

This investigation indicates that ageing occurs as compound contact time increases. The processes of ageing not only influence compound partitioning within the soil but also the subsequent release of compound and its bioavailability. Additionally, the fate of phenanthrene during ageing is phenanthrene concentration depend. Finally the data strongly supports the use of an aqueous based HPCD extraction to quantify the bioavailable fraction of soil associated compound.

ACKNOWLEDGEMENTS

The authors would like to acknowledge the support for this project from the Natural Environment Research Council (NERC) UK and the Nuffield Foundation.

REFERENCES

Brilis, G. M. and P. J. Marsden, 1990. "Comparative evaluation of soxhlet and sonication extraction in the determination of polynuclear aromatic hydrocarbons in soil". *Chemosphere. 21*: 91-98.

Bosma, T. and H. Harms. 1996. "Bioavailability of organic pollutants". *EAWAG News. 40*: 28-31.

Bosma T. N. P.; P. J. M. Middeldrop; G. Schraa. and A. J. B. Zehnder. 1997. "Mass transfer limitation of biotransformation: quantifying bioavailability". *Environ. Sci. Technol 31*: 248-252.

Cornellissen G.; P. C. M.van Noort.; H. A. J. Govers. 1997. "Desorption kinetics of chlorobenzenes, polycyclic aromatic hydrocarbons, and polychlorinated biphenyls: sediment extraction with Tenax® and effects of contact time and solute hydrophobisity". *Environ. Toxicol. Chem. 16*: 1351-1357.

Kelsey, J. W. and M. Alexander. 1997. "Declining bioavailability and inappropriate estimation of risk of persistent compounds". *Environ. Toxicol. Chem. 16*: 582-585.

Kelsey, J. W.; B. D. Kottler.and M. Alexander. 1997. "Selective chemical extractants to predict bioavaibility of soil-aged organic chemicals". *Environ. Sci. Technol. 31*: 214-217.

Reid, B.J.; K.T. Semple. and K.C. Jones. 1998a. "Prediction of bioavailability of persistent organic pollutants by a novel extraction technique". *Sixth International FZK/TNO Conference on Contaminated Soil. Vol 2:* 889-890.

Reid, B.J.; G.L. Northcott.; K.C. Jones. and K.T. Semple. 1998b. "Evaluation of spiking procedures for the introduction of poorly water soluble contaminants (PAHs) into soil". *Environ. Sci. Technol. 32:* 3224-3227.

Semple, K.T. and T.R. Fermor. 1997. "Enhanced mineralization of [$U^{14}C$]PCP in mushroom composts". *Res. Microbiol. 148*: 795-798.

Wu S. C. and P. M. Gschwend. 1986. "Sorption kinetics of hydrophobic oganic compounds to natural sediments and soils". *Environ. Sci. Technol. 20:* 717-725.

BIOAVAILABILITY OF CABLE INSULATING OIL TO SOIL BIOTA

Christopher J.A. MacLeod, Philip H. Lee, Brian J. Reid, Kafui Dzirasa and ***Kirk T. Semple*** (Lancaster University, Lancaster, UK)
D. Patel and S.G. Swingler (The National Grid Company, plc., Leatherhead, UK)

ABSTRACT: This study investigated the interactions between cable insulating oil and the soil microflora and macrofauna in freshly-spiked and in aged soils in the laboratory. Studies were carried out at two levels: (i) The impact of varying concentrations (1-10% w/w) of the non-aqueous phase liquid (NAPL) on soil microbial respiration using [1-^{14}C]glucose and (ii) The impact of the NAPL on the survival of earthworms (*Lumbricus terrestris*) in soil. In terms of soil-microbial respiration, [1-^{14}C]glucose mineralization was found to be enhanced by 32% over a range of oil-soil concentrations and contact times. In freshly spiked soils, all of the earthworms died after 14 d. However, in the 300 d aged soils, survival was significantly increased at lower concentrations of NAPL. If low concentrations ($\leq$ 1%) of NAPL are released to the soil environment from underground power cables it is postulated that a shift in the balance of the soil microflora from autochthonous to zymogenous microflora may occur. Additionally, the bioavailability of the NAPL to the soil macrofauna decreased with increased soil-NAPL contact time.

INTRODUCTION

The fate and behaviour of oil and oil-related compounds in the environment have been well researched in recent years using chemical extraction techniques, followed by an appropriate analytical methodology. Interest has also focused on assessing the ecotoxicity of xenobiotics in contaminated soils. Bio-indicators of toxicity may range from microorganisms to eukaryotes, such as earthworms.

Substrate-induced respiration (SIR) is an established method for measuring microbial activity after the addition of an easily utilizable carbon source, such as glucose or acetate. Living cells rapidly use glucose and produce CO_2 that can be easily measured (Anderson and Domsch, 1978). SIR has been used to assess the impact of pollutants on the respiratory activity of soil biomass (Van Beelen *et al.*, 1990). *Eisenia fetida* (Savigny) and *E. andrei* (Bouché) were the first soil invertebrates to be used in ecotoxicological studies (OECD, 1984). One recent study showed that freshly oiled soils were initially toxic to *Eisenia* sp. (Salanitro *et al*, 1997).

In the United Kingdom, Europe and North America, high voltage electricity transmission includes a small proportion of underground power cables. Dodecylbenzene either on its own or in combination with mineral oils is extensively used to insulate underground electrical cables. Due to the movement of these cables over a period of many years, leaks can occurs at joints, resulting in leakage of cable insulating oil to the surrounding soil environment. Leakage from such cables has in a number of cases been identified and contained. Further leaks may occur which will be difficult to locate and repair before oil is released into the soil environment. This investigation assessed the impact of cable insulating oil at various concentrations, and over increasing oil-soil contact times on soil microbial respiration and the earthworm *Lumbricus terrestris*, a widespread species in the UK (Sims and Gerard, 1985).

MATERIALS AND METHODS

Soil. Subsurface soil (5-15 cm) was collected from a rural hillside environment (Lancaster University, Hazelrigg, Field Station, UK - O.S. sheet 97, [493578]). The soil was passed through a 10 mm gauge sieve, air dried for 14 d and subsequently passed through a 1.7 mm gauge sieve to remove roots. Prior to use, the soil was rehydrated using Milli-Q™ water to its original moisture content (35% soil dry wt). The organic content of the soil (mass loss on ignition) was determined to be 7% of the soil dry weight.

Chemicals. Cable insulating oil was provided, as a mixture (50:50) of dodecylbenzene and mineral oil, by The National Grid Company, plc, UK. [1-^{14}C]Glucose was obtained from Sigma Aldrich Co. Ltd., UK. The liquid scintillation cocktail, Ultima Gold, was obtained from Canberra Packard, UK. Potassium hydroxide used in CO_2 respirometer traps was obtained from Fisons, Ltd, UK.

Soil spiking. Soil was spiked with oil concentrations of 0% 1%, 5%, and 10% oil (w/w dry wt of soil). This was achieved by adding oil to wet soil (1.7 kg) contained within a 2 l Kilner jar and vigorously mixed using a high speed drill with a stainless steel paddle attachment. The jars had air tight rubber seals and were opened only for sampling. Triplicate incubations were produced for each concentration.

Substrate Induced Respiration. The mineralization of [1-^{14}C]glucose was used to assess the respirometric activity of the microbial community present in the soil. Samples (30 g) of soil were placed in respirometers (modified Erlenmeyer flasks) (Semple and Fermor, 1997). To each sample, 15 ml of 0.3 mM [1-^{14}C]glucose (total activity 770 Bq) was introduced and the $^{14}CO_2$ produced was trapped using 1 M KOH (1 ml) present in the centre well of the respirometers. The KOH was removed and replaced at regular intervals and analysed by liquid scintillation counting (Canberra Packard Tri-carb 2250CA) after mixing with Ultima Gold scintillation fluid. Soil samples were removed from the microcosms after 0, 9, 21 and 300 d oil-soil contact time. By sampling from each microcosm a triplicate set of glucose mineralization results were obtained for each oil concentration.

Toxicity of cable insulating oil on *Lumbricus terrestris* L. Four specimens of *L. terrestris* were introduced to freshly prepared oil-soil incubations in triplicate (containing 1.7 kg of wet soil) of 0%, 1%, 5% and 10% oil (w/w soil dry wt). After contact times of 7 d and 14 d the earthworms were removed from the incubations and their mortality was assessed.

RESULTS AND DISCUSSION

Substrate induced respiration in freshly spiked and aged soils. The results indicate that the presence of the cable insulating oil has an impact on the overall extent of [1-^{14}C]glucose mineralization (Table 1).

Table 1. Mineralization of [1-^{14}C]glucose in the presence of cable insulating oil (1, 5 and 10%) with increasing soil-oil contact time.

Oil-Soil Contact Time (Days)	Mineralization[1] (%) of [1-^{14}C]glucose in the presence of Cable insulating oil (w/w soil dry wt)			
	0%	1%	5%	10%
0	43.7	45.4 (4%)	46.6 (7%)	46.0 (5%)
9	45.8	52.9 (16%)	56.0 (22%)	58.4 (28%)
21	39.1	93.5 (139%)	62.5 (60%)	57.8 (48%)
300	40.1	59.2 (48%)	61.9 (54%)	54.9 (37%)

[1]The mineralization value was determined over an incubation period of 48 h. Each value is the arithmetic mean of three separate measurements. Values in the brackets are the percentage increase in mineralization compared to the control soil incubation (0% oil).

In all the oil-soil incubations there was an increase in glucose mineralization (median value of 32%), however this increase was not proportional to increasing oil concentrations. After the minimum oil-soil contact time (0 d), all of the oiled soils showed an increase (arithmetic mean =5%) in mineralization relative to the control. The extents of mineralization after the 48 h assay time were approximately 44%, 45%, 47% and 46% for the control, 1%, 5% and 10% oil concentrations, respectively.

After 9 d oil-soil contact time, all of the oiled soils showed an increase (arithmetic mean =22%) in mineralization relative to the control. The extents of mineralization after the 48 h assay time were approximately 46%, 53%, 56% and 58% for the control, 1%, 5% and 10% oil concentrations, respectively.

After 21 d oil-soil contact time, all of the oiled soils showed an increase (arithmetic mean =82%) in mineralization relative to the control. The extents of mineralization after the 48 h assay time were approximately 39%, 94%, 62% and 58% for the control, 1%, 5% and 10% oil concentrations, respectively. It is postulated that lower oil concentrations (up to and including 1% oil) may shift the balance of the soil microbial community from autochthonous to zymogenous microflora, similar to phenomena described by Shen and Bartha (1996a and b). In these studies, the authors suggested that the increase in the extent of glucose mineralization was commensurate with an increase in the number of (zymogenous) microorganisms which had the ability to utilise oil for growth. Incubations with 0.75% oil concentration show a similar increase in the extent of glucose mineralization (data not shown). A similar shift in community structure may also be occurring at higher oil (5% and 10%) concentrations but its effect less apparent due to a putatively toxic effect at these higher oil concentrations. None the less, there is also an increase in respirometric activity with time at these concentrations.

After 300 d oil-soil contact time (chronic exposure), all of the oiled soils showed an increase (arithmetic mean =46%) in mineralization relative to the control. The extents of mineralization after the 48 h assay time were approximately 40%, 59%, 62% and 55% for the control, 1%, 5% and 10% oil concentrations, respectively. With the exception of the 1% oiled soils, the extents of glucose mineralization were similar to that of 21 d oil-soil contact time. This suggests that the mineralization of glucose is enhanced for at least 300 d after the addition of a NAPL to the soil environment. Interestingly, the trend in the elevation of glucose mineralization was similar for all the oil-soil concentrations, with the largest increase occurring after 21 d oil-soil contact time.

Toxicity of cable insulating oil to *Lumbricus. terrestris* in freshly spiked and aged soils. *L. terrestris* was incubated in soils spiked with cable insulating oil (0%, 1%, 5% and 10% w/w soil dry wt).

Table 2. The mortality of *L. terrestris* in soil freshly spiked and 300 day aged with cable insulation oil.

Oil-Soil Contact Time(d)	Mortality (%) of *L. terrestris* in the presence of cable insulating oil at various concentrations (w/w soil dry wt) after 7 and 14 d incubation time							
	0%		1%		5%		10%	
	7 d	14 d	7 d	14 d	7 d	14 d	7 d	14 d
0	0	0	44	100	100	100	100	100
300	0	0	11	44	78	89	100	100

After 7 d, 44%, 100% and 100% of the earthworms had died in 1%, 5% and 10% oiled soils, respectively (Table 2). Further, there was 100% earthworm mortality in all the oiled soils after 14 days incubation. These oiled soils were allowed to age for 300 d and were then again inoculated with *L. terrestris*. Table 2 shows that there was a decrease in the mortality of the earthworms as compared to the freshly spiked oiled soils. After 7 and 14 d incubations there was a decrease in mortality in the 1% (75% and 12%, respectively) and 5% (56% and 11%, respectively) oiled soils. However, in the 10% oiled soils, all the earthworms died in the aged systems, as well as the freshly spiked soils.

Traditionally, *Eisenia fetida* and *E. andrei* have been used in ecotoxicological studies (OECD, 1984). However, *L. terrestris* was used in this investigation because it is a species common to the UK., found in a wide range of soil types (Sims and Gerard, 1985). Also, as a deep burrowing species that consumes substantial quantities of mineral soils as well as organic matter in various states of decay, it represents a more realistic subject for this sort of study. *Eisenia fetida* and *andrei* are only abundant in soils with very high organic matter contents or compost environments (Sims and Gerard, 1985). The results of this study agree with those of Salanitro *et al.* (1997), where the freshly spiked soils were more toxic than the aged (300 d) soils. It has been observed that, as the length of time an organic compound remains in contact with soil increases, the bioavailability, and therefore the toxicity of the compound decreases (Hatzinger and Alexander, 1995).

CONCLUSIONS

This study shows that the presence of a NAPL enhanced the mineralization of glucose with a median increase of 32% over a range of oil-soil concentrations and contact times. The data indicates that the oil-contaminated soils contain increased microbial activity which possibly possesses oil-degrading capabilities. However, the oiled soils have been shown to be toxic to earthworms, although at lower concentrations the toxicity diminishes with increased soil-oil contact time.

ACKNOWLEDGEMENTS

Funding for this project was provided by The National Grid Company plc. We thank the Natural Environment Research Council for additional support.

REFERENCES

Anderson, J. P. E., and K. H. Domsch. 1978. "A physiological method for the quantitative measurement of microbial biomass in soils." *Soil Biology. Biochemistry*. 10: 215-221.

Hatzinger, P. B., and M. Alexander. 1995. "Effect of Ageing of chemicals in Soil on Their Biodegradability and Extractability." *Environmental Science and Technology*. 29: 537-545.

Huesemann, M. H., and K. O. Moore. 1993. "Compositional changes during landfarming of weathered Michigan oil-contaminated soil." *Journal of Soil Contamination*. 2: 245-264.

Llanos, C., and A. Kjoller. 1976. "Changes in the flora of soil fungi following oil waste application." *Oikos*. 27: 377-382.

Piphonen, R., and V. Huhta. 1984. "Petroleum fractions in soil: effects on populations of nematoda, enchytraeidae and microarthrododa." *Soil Biology and Biochemistry*. 16: 347-350.

Organization for Economic Co-operation and Development 1984. "Earthworm acute toxicity tests." *Guideline for Testing Chemicals*. No. 207.

Salanitro, J. P., P. B. Dorn, M. H. Huesemann, K. O. Moore, I. A. Rhodes, L. M. Rice Jackson, T. E. Vipond, M. M. Western, and H. L. Wisniewski. 1997. "Crude oil hydrocarbon bioremediation and soil ecotoxicity assessment." *Environmental Science and Technology*. 31: 1769-1776.

Semple, K. T., and T. R. Fermor. 1997. "Enhanced mineralisation of UL-^{14}C-pentachlorophenol by mushroom composts." *Research in Microbiology*. 148: 795-798.

Shen, J., and R. Bartha. 1996a "Priming effect of substrate addition in soil-based biodegradation tests." *Applied and Environmental Microbiology*. *62*: 1428-1430.

Shen, J., and R. Bartha. 1996b "Metabolic efficiency and turnover of soil microbial communities in biodegradation tests." *Applied and Environmental Microbiology 62*: 2411-2415.

Sims, R. W., and B. M. Gerard. 1985. "Earthworms. Keys and notes for the identification and study of the species." *Synopses of the British Fauna (New Series)*. 31: 1-171.

Torstensson, L. 1997. "Microbial Assays in Soils." In J. Tarradellas, G. Bitton and D. Rossel (Eds.),*Soil Ecotoxicology*, pp. 207-234. CRC Press, Inc. Boca Raton, FL.

Van Beelen, P., Fleuren-Kemilä, A.K., Huys, M.P.A., van Mil, A.C.H.A.M. and van Vlaardingen, P.L.A. 1990. "Toxic effect of pollutants on the mineralisation of substrates at low experimental concentrations in soils, subsoils and sediments." In F. Arendt, M. Hinseveld and H.J. van den Brink (Eds.),*Contaminated Soil*, pp. 431-438. Kluwer Academic Publishers, Netherlands.

ENHANCED BIODEGRADATION OF PAH CONTAMINANTS BY THERMAL PRETREATMENT

Luc T.C. Bonten, Tim C. Grotenhuis, Wim H. Rulkens
(Wageningen Agricultural University, Wageningen, The Netherlands)

ABSTRACT: Biodegradation of hydrophobic contaminants in soil is often limited by slow desorption from the soil. The fraction of the contaminants that show slow desorption can be decreased by heating the soil (65 - 100 °C) for a short period of time (10 min. - 24 h). Desorption and biodegradation experiments have been performed to investigate which process parameters determine the decrease of the slow desorbing fraction. This fraction decreased more with higher temperatures and longer treatment times. This decrease is probably caused by rearrangement of the soil organic matter in which the contaminants are sorbed.

INTRODUCTION

Biological decontamination of soils polluted with hydrophobic organic compounds (e.g. PAH, mineral oil) has not always been successful due to low degradation rates and high residual concentrations that do not meet the legal clean-up guidelines (Rulkens and Honders, 1996). This problem, which is also called 'limited bioavailability', is mainly caused by slow desorption of the contaminants from soil particles to micro-organisms and is, in general, not due to slow degradation by microorganisms. It often has been shown that desorption of hydrophobic compounds from soil particles takes place in two phases: a phase of fast desorption followed by one of slow desorption (LeBoeuf and Weber, 1997; Kan et al., 1997). The distribution of the contaminants between the two desorption phases is influenced by temperature (Xing and Pignatello, 1997), in such a way that at higher temperatures more contaminants show fast desorption. In previous research it was shown that also after the soil had cooled to room temperature the fast desorbing fraction of the contaminants was still increased (Bonten et al., 1999).

The objective of the research presented here is to show which process parameters determine the increase in bioavailability of hydrophobic contaminants after a thermal pretreatment of soil. These factors will give more insight in the mechanisms governing the increase in bioavailability. Two types of experiments were used: a physical desorption experiment with an adsorbent as a sink for desorbed contaminants and a biological degradation experiment.

MATERIALS AND METHODS

Soil. For both the desorption and the biological experiments sludge from a soil washing plant has been used. The original soil was obtained from a former gas plant site (Kralingen, The Netherlands). The soil was separated at a cut-off diameter of 63 μm using hydrocyclones. The sludge residue - i.e. the fraction containing the smallest particles - was air dried and sieved at 2 mm to remove coarse, light material that could not be separated by hydrocyclonage. The total PAH concentration (16 US-EPA) was 115 mg/kg dry soil. The sludge further contained 300 mg mineral oil/kg and 9 mg cyanide/kg. The organic matter content of the sludge was 12.3 % determined by loss on ignition at 550 °C. The mineral part of the sludge consisted of 27 % clay, 46 % silt and 27 % sand.

Desorption experiments. The method used for determining desorption kinetics was adopted from Cornelissen et al. (1997) and Caroll et al. (1994). In this method a polymer adsorbent is used to extract desorbed compounds from the waterphase. In this way the contaminant concentration in the waterphase is kept very low and the maximal desorption rate will be achieved.

Before the desorption experiment, 2 grams of air dried sludge were brought in a serum flask that was closed with a Teflon lined screw cap. The serumflask was heated in a waterbath at a set temperature and during a set timespan. Then the sludge was transferred into a 50 ml separation funnel and 40 ml of 0.01 M $CaCl_2$ solution, 0.5 g of Tenax-TA (Chrompack, mesh 20-35) as adsorbent, and 10 mg NaN_3 to prevent microbial degradation were added. The funnel was put on a end-over-end mixer (22 rpm, 30 °C). At set times the soil suspension was separated from the Tenax and fresh Tenax was added. The residual Tenax was extracted with 20 ml acetone on a shaker for 15 minutes. PAH concentrations in the acetone were determined by HPLC as described in Bonten et al. (1999). At the end of the experiment, residual PAH concentrations in the soil were determined by extracting the soil with 10 ml 1-methyl-2-pyrrolidinone (99 %, Acros) in a microwave oven (MDS 2100, CEM corp.) at 130 °C for 1 hour (Noordkamp, 1997). After extraction the pyrrolidinone was analyzed by HPLC. The desorption experiments were carried out in duplicate.

Biological experiments. Before biodegradation, the soil was thermally pretreated under several conditions which are shown in Table 1. To determine the biological degradation rates of PAH compounds in soil, biological experiments were performed by thoroughly mixing of a slurry of pretreated soil and nutrient solution together with a microbial inoculum. At set times the PAH concentrations of the soil were determined. The method for the biological experiments has been more extensively described in Bonten et al. (1999).

TABLE 1. Conditions of thermal pretreatment before biodegradation

# exp.	T (°C)	t (h)	water content (ml/g dry soil)
1 (control)	-	-	-
2	65	2	1.5
3	80	2	1.5
4	95	2	1.5
5	95	0.5	1.5
6	95	24	1.5
7	95	2	0
8	95	2	0.5

Data interpretation. The desorption kinetics were described with a two compartment model consisting of a slow desorbing and a fast desorbing compartment. Desorption from each compartment is assumed to be first order:

$$\frac{S_t}{S_0} = F_{slow} e^{-k_{slow} t} + F_{fast} e^{-k_{fast} t} \qquad (1)$$

where S_t and S_0 are the concentrations of PAH in the soil at time t and at the start of the experiment, respectively; F_{slow} and F_{fast} are the initial fractions of the contaminants in the slow and the fast desorbing compartment, respectively (N.B. F_{slow} = 1-F_{fast}); k_{slow} and k_{fast} are the first-order rate constants of slow and fast desorption, respectively. Values for F_{slow}, F_{fast}, k_{slow} and k_{fast} were determined by calculating the least sum of squared differences between the calculated and experimental values of S_t/S_0.

RESULTS AND DISCUSSION

Values for F_{fast}, k_{fast} and k_{slow} of phenanthrene, pyrene and chrysene in the classification sludge were derived from the desorption experiments. No effect of a thermal treatment on k_{fast} and k_{slow} could be found (data not shown). Values for F_{fast} are presented in Figure 1.

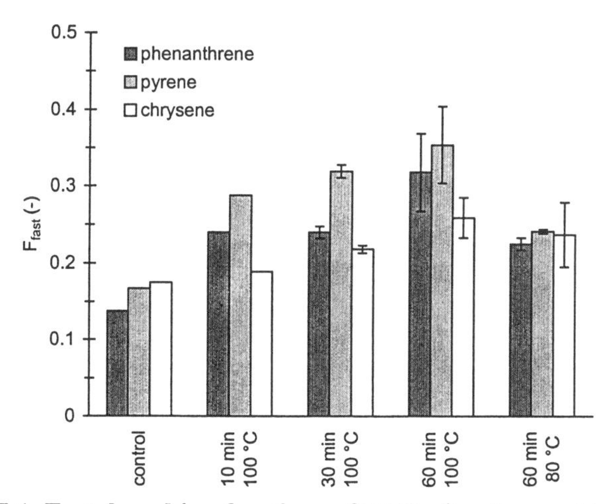

FIGURE 1. Fast desorbing fractions of PAH after thermal treatment and in control experiment; error bars indicate 1 standard deviation; if no error bars are shown, no duplicate experiment was performed

Figure 1 shows that the fraction of the sorbed PAH compounds that showed fast desorption increased after a thermal treatment. A longer treatment time and a higher treatment temperature led to a larger increase of F_{fast}. The fractions of three PAH compounds that were biodegraded in Kralingen classification sludge in 21 days after a thermal treatment under various conditions are shown in Figure 2.

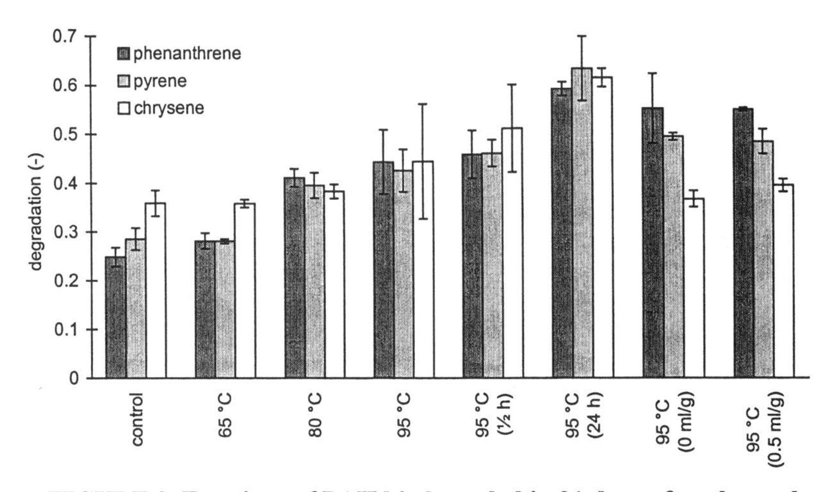

FIGURE 2. Fractions of PAH biodegraded in 21 days after thermal treatment during 2 h with 1.5 ml H_2O/g soil (unless indicated otherwise); errors bars indicate 1 standard deviation

Figure 2 shows that a larger fraction of the PAHs sorbed in the classification sludge was degraded after a thermal treatment. A pretreatment time of 24 hours led to a larger fraction that was degraded than a pretreatment time of 0.5 and 2 hours. The fractions that were degraded also increased with increasing treatment temperature. The amount of water showed no significant effect on the degradation (t-test, 2-sided, $\alpha<0.05$) except for the degradation of phenanthrene which was improved.

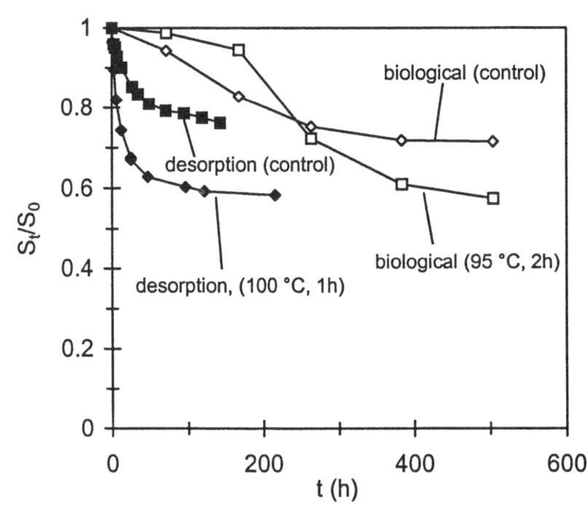

FIGURE 3. Desorption and biodegradation of pyrene in soil; thermally treated and non-treated soil

In Figure 3 both the desorption and the biodegradation of pyrene are presented in a control experiment and after a thermal treatment. This figure clearly shows that the residual concentrations for desorption and biodegradation were comparable, for both the control experiment and the experiment with thermal treatment. This indicates that residual concentrations after biodegradation are determined by the slow desorbing fractions. Notice that the fast desorbing fractions in Figure 1 are smaller than the fractions degraded in Figure 2. This is caused by the long timespan of the degradation experiments in which a small part of the slow desorbing fraction was also degraded. Further, Figure 3 shows that the initial desorption rates are much larger than the initial biodegradation rates. This means that biodegradation is initially biologically limited by biological factors instead of by desorption.

LeBoeuf and Weber (1997) and Xing and Pignatello (1997) showed that sorption isotherms of hydrophobic organics to soil became more linear with increasing temperature. This was attributed to the elimination of micro-voids, in which contaminants should be physically incorporated. The elimination of these voids resulted from the conversion of glassy soil organic matter to rubbery soil organic matter. However, glass/rubber transition as found in synthetic polymers is a reversible process (Billmeyer, 1970). In contrast to this, our experimental results showed that after a thermal treatment the fast desorbing fractions were still larger than before thermal treatment. Kan et al. (1997) and Schlebaum et al. (1998) stated that the formation of slow desorbing fractions resulted from incorporation of the contaminants by rearrangement of soil organic matter. This rearrangement might be (partially) undone by a thermal treatment. A temperature raise leads to a larger thermal movement of both the contaminants and the soil organic matter in such a way that contaminants are probably less tightly bound after a thermal treatment.

CONCLUSIONS

Bioavailability of hydrophobic organic contaminants in soil is increased after a short thermal treatment. The treatment temperature and the treatment time are the most important parameters that determine bioavailability increase. This increase is probably caused by a rearrangement of the soil organic matter.

REFERENCES

Billmeyer Jr., F. W. (Ed.). 1970. *Textbook of Polymer Science.* 2nd ed., John Wiley and Sons, New York, NY

Bonten, L. T. C., J. T. C. Grotenhuis, and W. H. Rulkens. 1999. "Enhancement of Biological Degradation of PAH in Soil by Physicochemical Pretreatments." *Chemosphere.* accepted.

Carroll, K. M., M. R. Harkness, A. A. Bracco, and R. R. Balcarcel. 1994. "Application of a Permeant/Polymer Diffusional Model to the Desorption of Polychlorinated Biphenyls from Hudson River Sediments." *Environ. Sci. Technol.* 28: 253-258.

Cornelissen, G., P. C. M. van Noort, J. R. Parsons, and H. A. J. Govers. 1997. "The Temperature Dependence of Slow Adsorption and Desorption Kinetics of Organic Compounds in Sediments." *Environ. Sci. Technol.* 31: 454-460.

Kan, A. T., G. Fu, M. A. Hunter, and M. B. Tomson. 1997. "Irreversible Adsorption of Naphthalene and Tetrachlorobiphenyl to Lula and Surrogate Sediments." *Environ. Sci. Technol.* 31: 2176-2185.

LeBoeuf, E. J., and W. J. Weber. 1997. "A Distributed Reactivity Model for Sorption by Soils and Sediments. 8. Sorbent Organic Domains: Discovery of a Humic Acid Glass Transition and an Argument for a Polymer based Model." *Environ. Sci. Technol.* 31: 1697-1702.

Noordkamp, E. R., J. T. C. Grotenhuis, and W. H. Rulkens. 1997. "Selection of an Efficient Extraction Method for the Determination of Polycyclic Aromatic Hydrocarbons in Contaminated Soil and Sediment.", *Chemosphere.* 35: 1907-1917.

Rulkens, W. H., and A. Honders. 1996. "Clean-up of Contaminated Sites: Experiences in the Netherlands.", *Wat. Sci. Tech.* 34: 293-301.

Schlebaum, W., A. Badora, G. Schraa, and W. H. van Riemsdijk. 1998. "Interaction between a Hydrophobic Organic Chemical and Natural Organic Matter: Equilibrium and Kinetic Studies." *Environ. Sci. Technol.* 32: 2273-2277.

Xing, B., and J. Pignatello. 1997. "Dual-Mode Sorption of Low-Polarity Compounds in Glassy Poly(Vinyl Chloride) and Soil Organic Matter." *Environ. Sci. Technol.* 31: 792-799.

SEQUESTRATION OF PAHs IN SIZE- AND DENSITY FRACTIONATED ESTUARINE SEDIMENTS

Karl J. Rockne (Chemical Engineering, Rutgers University, Piscataway, NJ)
Gary L. Taghon (Institute of Marine and Coastal Sciences, Rutgers University, New Brunswick, NJ)
Lily Y. Young (Biotechnology Center for Agriculture and the Environment, Rutgers University, New Brunswick, NJ)
David S. Kosson (Chemical Engineering, Rutgers University, Piscataway, NJ)

ABSTRACT: The degree of sediment or soil contamination typically is quantified based on the bulk contaminant concentration or normalized to the bulk fraction of organic carbon. Rather than consider contaminant levels in the bulk sediment, our approach has focused on the examination of different phases or fractions of particles, in particular size- and density-fractions. Our initial hypothesis was that the majority of nonionic, hydrophobic contaminants in sediments would be associated with a small minority of the total particle mass in particular, with low-density particles. We have found that there is a distinctive pattern to this association for sediment from Piles Creek, one of our field sites in the NY/NJ Harbor Estuary. The majority of polycyclic aromatic hydrocarbons (PAHs) are associated with low-density particles (<1.9 g/mL) for larger size classes of sediment particles (generally particles >90 to 300 μm). These large, low-density particles are primarily remains of terrestrial and marine plant tissues that have accumulated in the sediment. As particle size decreases, however, there is a reversal in these associations. For the smallest particles, the majority of the PAHs are associated with the high-density particles, representing mineral phases. This association is not explainable by equilibrium partitioning theory based on organic carbon normalization of PAH concentrations. An analysis of the size distribution of sediment-associated PAHs at Newtown Creek (also in the NY/NJ Harbor Estuary) demonstrates that sequestration is both a function of sediment density and PAH structure. For example, three-ring PAHs, while comprising 5% of the total PAHs in both the bulk sediment and high density phase (ρ >1.9 g/mL), account for over 35% of the total PAHs in the lightest particles (ρ <1.3 g/cm^3).

INTRODUCTION

The quantification of contaminant levels in environmental samples is of paramount importance for the determination of exposure levels, risk assessment, contaminant fate, and bioavailability. Quantification of highly hydrophobic compounds such as polycyclic aromatic hydrocarbons (PAHs) in sediment can be complicated by the tight binding of these compounds in the organic phase and/or micropores of the sediment particles, resulting in poor quantification of the bioavailable fraction.

If contaminants sequester within the sediment solid phases, the thermodynamic partitioning of contaminants between the phases, diffusion within each phase, and inter-phase mass transfer will control the flux to bacteria or uptake by benthic animals. Contaminant fluxes from particles will be a strong function of particle characteristics such as size and nature and content of organic matter. The goal of this project is to determine the bioavailability of sediment-associated contaminants to microbes and benthic animals. As a first step to understanding and quantifying the relationship between the physical and chemical characteristics of sediments, we characterized the distribution of PAHs in field-contaminated sediments. Specifically, we characterized the sequestration behavior of sediment-associated PAHs in size- and density-fractionated sediments from two sites in the NY/NJ harbor estuary with different contamination levels, physical characteristics, and natural organic carbon content.

MATERIALS AND METHODS

Sediment from two PAH-contaminated sites in the NY/NJ harbor estuary were used for these experiments. Piles creek is an inter-tidal creek that runs through a marshy area with abundant emergent vegetation. Newtown creek is a sub-tidal industrial waterway in Queens NY. Piles creek sediment was sampled by surface grabs at low tide as described previously (Tso and Taghon, 1997). Sediment from several grabs was pooled together and homogenized with an impeller (15 min) and stored at 4° C until use. Newtown creek was sampled with a Polnow dredge and homogenized as described for Piles creek sediment.

Sediment was size fractionated by wet sieving in sequential sieves from 1000 μm down to 38 μm. Sediment was separated into density fractions by floatation in CsCl solution of varying densities as described (Mayer et al., 1993).

Physical Characterization. Surface area was measured by nitrogen adsorption at 77 K using a Micromeritics model ASAP 2010 accelerated surface area and porisimetry system. Sediment was dried (110° C, 24 hrs) and placed under high vacuum to degas the sediment as per manufacturers instructions (Micromeritics, 1996). Nitrogen was allowed to adsorb and desorb as per manufacturers instructions. Specific surface area was calculated using the BET isotherm transformation software supplied with the unit. Particle size analysis was performed by both wet sieving and x-ray sedimentation analysis using a Micromeritics model 5100 particle size analyzer (particles <300 μm) as per manufacturers instructions (Micromeritics, 1994).

Chemical Characterization. Total organic carbon was measured using a Carlo Erba elemental analyzer as described previously (Mayer et al., 1993). PAHs were extracted from the sediment using a modification of the extraction procedure of Leeming and Maher (1994). Sediment was air dried in a fume hood for several days. Sediment (1 g dry wt.) was transferred into a Teflon-lined centrifuge tube (50 mL) and methylene chloride (20 mL) was added. The extraction tubes were placed in a heated sonicator water bath (40° C) for 30 min. The tubes were centrifuged (8000 x g, 15 min) and the methylene chloride was decanted into a Rotovap flask (50 mL). The sediment was then re-extracted twice with methylene chloride and the three extracts were pooled in the Rotovap flask and concentrated to less than 5 mL. The concentrated extract was then brought up to 5 mL with methylene chloride, run through an activated normal-phase silica gel cartridge (Waters), and concentrated under a purified N_2 stream (less than 1 mL). Acetonitrile (5 mL) was added and then concentrated under N_2 to 1 mL to remove residual methylene chloride. The extract was then brought up to 5 mL with acetonitrile, run through an activated reverse-phase C-18 column (Waters), concentrated to 0.5 mL under N_2, and transferred to HPLC autosampler vials. PAHs were analyzed by HPLC (Hewlett Packard model 1050) with digital diode array detection and identification.

RESULTS AND DISCUSSION

Physical and Chemical Characterization. Piles creek sediment differed significantly from Newtown creek sediment in particle size distribution, surface area, and organic carbon content. Piles creek sediment was composed primarily of sand- and silt-sized particles whereas Newtown creek had almost no sand sized particles and was dominated by silt-sized particles (Table 1). Both sediments were comprised of similar fractions of clay-sized particles. Newtown creek sediment also had significantly greater organic carbon and specific surface area; 30% and 70% greater, respectively.

Table 1. Physical and Chemical Properties of test sediments. Mean ± SEM.

Sediment	TOC (mg/g)	Surface area[†] (m^2/g)	% Sand	% Silt	% Clay
Newtown	64.4 ± 2.2	10.6 ± 0.06	1.8 ± 1.0	73.4 ± 3.3	24.3 ± 3.5
Piles	48.5 ± 4.4	6.3 ± 0.09	42.9 ± 1.6	35.7 ± 1.3	20.8 ± 0.8

[†]Determined by BET transformation of nitrogen adsorption data.

PAH Distribution. Fractionation of Piles creek sediment demonstrated a particle size and sediment density phase dependent PAH distribution. For particles greater than 125 μm, the vast majority of PAHs were in the low density (<1.9 g/mL) phase, even though the low density phase represented only 2-4% of the total sediment mass (Figure 1). The low density phase was comprised predominantly of detrital remains of terrestrial and marine plant tissues. In contrast, most of the PAH mass was present in the high density mineral phase for smaller particles in the silt and clay fractions.

PAH concentrations (normalized to sediment dry weight) were also greatly concentrated in the low density fractions of Newtown creek sediment, as shown by the >10:1 ratio of low density phase to bulk sediment phase PAH concentration (Figure 2a). There is a general trend with the highest PAH concentrations in lower density fractions of sediment, with sediment <1.3 g/mL having PAH concentrations 10-50 times higher than either the bulk or high-density phase. If PAH concentrations were equally distributed throughout the sediment fractions, we would expect a 1:1 ratio of PAH concentration across phases, as predicted by equilibrium partitioning theory (Di Toro et al., 1991). It is apparent that the high density fraction is virtually identical in PAH concentration with the bulk sediment, as shown by the 1:1 relationship in PAH distribution (Figure 2a).

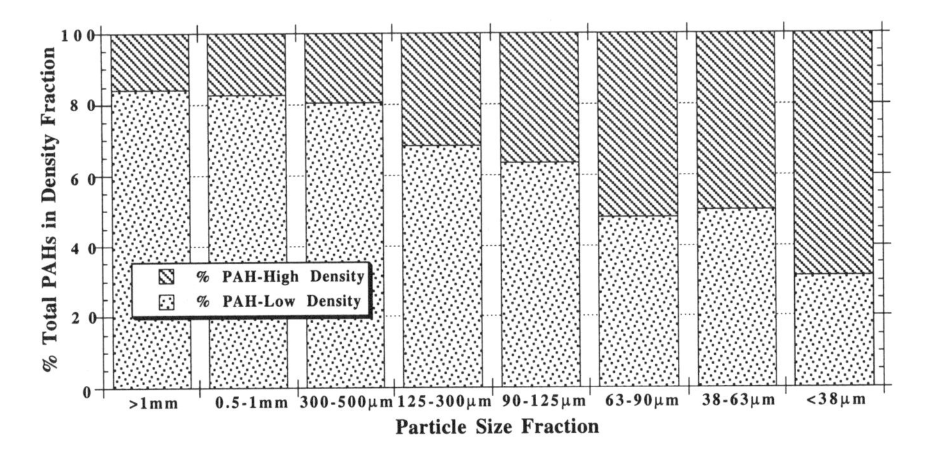

Figure 1. Total PAH distribution in low- and high-density phases as a function of particle size fractions of Piles creek sediment.

PAH concentration data was normalized to organic carbon because the low-density phase was greatly enriched in organic carbon. Organic carbon content has been shown to control the concentration of many hydrophobic contaminants in soils and sediments (Di Toro et al., 1991). Although the ratios of low density to bulk phase PAH concentrations were not as large as when normalized to dry weight, carbon-referenced PAH concentrations were still enriched up to a factor of 20:1 (Figure 2b).

There was also a demonstrated dependence of PAH structure on sequestration behavior in the low and high density phases. Carbon-referenced PAH concentrations were increased by the greatest amount for smaller PAHs; a factor of 24 for 2- and 3-ring PAHs. Conversely, carbon-referenced 4-, 5-, and 6-ring PAH concentrations were only increased by a factor of three in the <1.3 g/mL fraction, demonstrating that most of the increased distribution of PAH in the low density fraction was due to partitioning of the smallest PAHs.

All of these results demonstrate that PAHs preferentially sequester in the low density particles at levels not predictable by equilibrium partitioning theory. The results with Piles creek sediment also demonstrate that the larger low-density particles differ significantly in PAH sequestration potential from the smaller low density particles. A likely explanation for the larger distribution of PAHs in the mineral phase for smaller particles is that these particles have much larger specific surface areas which can harbor large amounts of natural organic matter. The overall particle density will still likely be >1.9 g/mL. The organic matter sorbed to these high surface area particles then acts as a sorbent for PAHs.

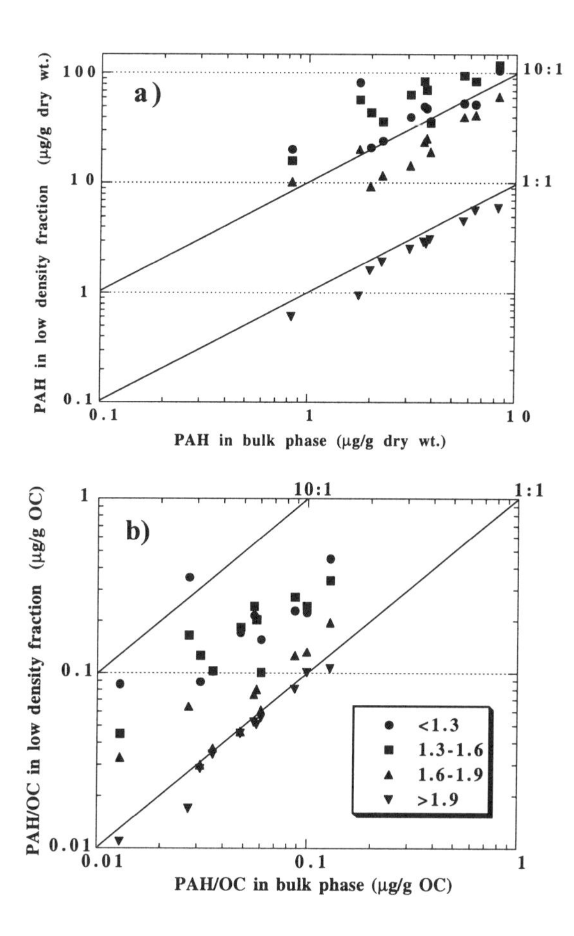

Figure 2. Comparison of a) PAH and b) carbon referenced-PAH concentration in bulk versus several low-density phase fractions of Newtown creek sediment.

CONCLUSIONS

Understanding and quantifying the relationship between the physical and chemical characteristics of sediments and fluxes of contaminants to microbial and animal communities is essential for prudent risk-based decision making. The selectivity of benthic animal feeding towards low density, high organic matter content particles is well established (Mayer et al., 1993). The contrast in sequestration behavior between bulk and low density sediments (the particles preferentially selected by detritivores) demonstrated in this study suggests that bulk contaminant concentration alone may be a poor predictor of the potential for PAH bioaccumulation in higher animals feeding on contaminated sediments.

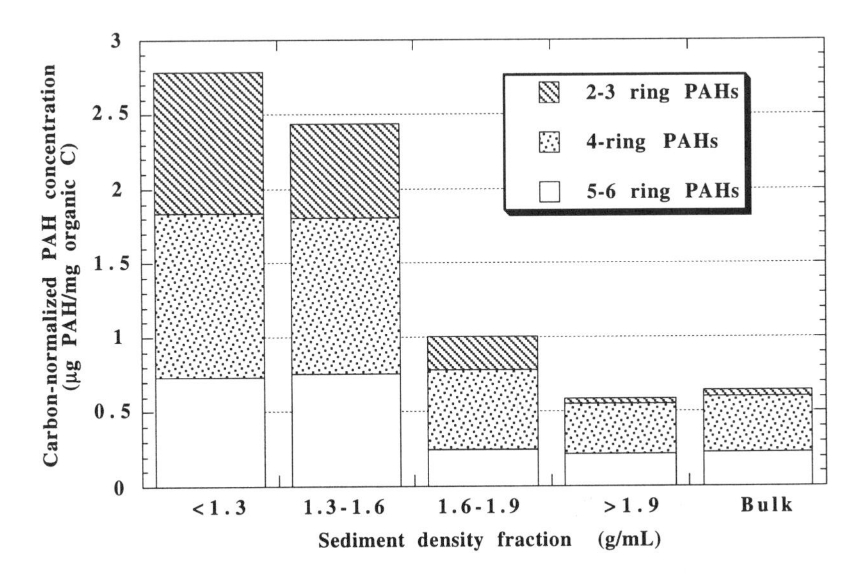

Figure 3. PAH structure-dependent sequestration behavior in bulk and several density fractions of Newtown creek sediment.

ACKNOWLEDGMENTS

This work was supported by grant R825303 from the Environmental Protection Agency.

REFERENCES

Di Toro, D. M., C. S. Zarba, et al. 1991. "Technical basis for establishing sediment quality criteria for nonionic organic chemicals using equilibrium partitioning." *Environ. Toxicol. Chem.* 10:1541-1583.

Leeming, R. and W. Maher. 1994. "Sources of polycyclic aromatic hydrocarbons in Lake Burley Griffin, Australia." *Org. Geochem.* 18(5):647-655.

Mayer, L. M., P. A. Jumars, G. L. Taghon, S. A. Macko, and S. Trumbore, (1993). "Low-density particles as potential nitrogenous foods for benthos." *Journal of Marine Research* 51: 373-389.

Micromeritics. 1994. *Sedigraph 5100 Particle Size Analysis System Operators Manual. V3.07*. Micromeritics Instrument Corporation. Norcross, GA.

Micromeritics. 1997. *ASAP 2010 Accelerated Surface Area and Porisimetry System Operators Manual.* V3.01. Micromeritics Instrument Corporation. Norcross, GA.

Tso S. F. and G. L. Taghon. 1997. "Enumeration of protozoa and bacteria in muddy sediment." *Microbial Ecology*. 33:144-148.

FIELD-SCALE BIOREMEDIATION MONITORING UTILIZING BIOLUMINESCENT GENETICALLY ENGINEERED MICROORGANISMS

Steven Ripp, David E. Nivens, Robert Burlage, and Gary S. Sayler, Center for Environmental Biotechnology, University of Tennessee, Knoxville, Tennessee 37996 and Oak Ridge National Laboratory, Oak Ridge, Tennessee 37831

ABSTRACT: An intermediate-scale field release of a genetically engineered microorganism was initiated in October, 1996 to demonstrate the ecological fate and capabilities of a bioluminescent bioreporter in the monitoring and control of polyaromatic hydrocarbon (PAH) degradation during a subsurface soil bioremediation process. The release occurred in intermediate-scale, semi-contained lysimeters containing soil contaminated with a mixture of transformer oil, naphthalene, anthracene, and phenanthrene. The engineered microbe, *Pseudomonas fluorescens* HK44, contains a gene for bioluminesce (*lux*) incorporated into a catabolic pathway for naphthalene degradation. Thus, HK44 bioluminesces as it biodegrades, providing an on-line, real-time detection of naphthalene bioavailability and biodegradative capacity. A portable photomultiplier tube and fiber optic-based biosensors were incorporated into the lysimeters for measuring the bioluminescent response. Both types of light detection devices were successful in monitoring HK44 bioluminescence, and could be used to survey overall naphthalene bioavailability within the soil ecosystem during the two-year study period. Thus, bioreporter organisms are capable of providing, in real-time, a general assessment of the effectiveness of a bioremediation process.

INTRODUCTION

Bioremediation takes advantage of the biodegradative potentials of indigenous microorganisms for the removal of hazardous chemicals from contaminated sites. Central to this process is the need for rapid on-site monitoring of localized microbial processes to determine whether conditions are favorable for bioremediation to occur. A promising strategy is to use genetically engineered bioreporter microorganisms for continuous on-line monitoring of contaminant bioavailability during the bioremediation process. In this study, an engineered *P. fluorescens* strain designated HK44 was released into a PAH-contaminated field site to assess its capabilities as a bioreporter. *P. fluorescens* HK44 was derived from a soil-borne manufactured gas plant microbe that was capable of degrading naphthalene, a common constituent of PAHs. The *lux* gene from *Vibrio fischeri* was incorporated into the strain's catabolic pathway for naphthalene degradation, thus producing a bacterial cell that emits light in response to naphthalene catabolic gene expression (Heitzer, et al., 1992; King, et al., 1990). Thus, utilizing light-sensing devices such as photomultiplier tubes and fiber optic cables, a rapid, on-site, on-line, and real-time assessment of biodegradative activities can be achieved.

MATERIALS AND METHODS

The lysimeter facility was situated in a one acre clearing at Oak Ridge National Laboratory, Oak Ridge, Tennessee. Each of the six lysimeters utilized consisted of an epoxy-coated corrugated steel culvert, 4 m deep by 2.5 m in diameter, set into concrete foundations 3 m below ground level (Fig. 1). Each of the lysimeters was loaded with a stratified bed consisting of a 31 cm layer of gravel, a 61 cm layer of sand, a 92 cm layer of clean soil, a 92 cm layer of treated soil, and a 61 cm cap of clean soil. All soil was a Huntington Loam blend made up of 42% sand, 40% silt, 18% clay, and 1.3% organic carbon. A stainless steel lid covered each lysimeter. Sensors for measuring soil temperature, moisture content, and oxygen and carbon dioxide concentrations were incorporated into the lysimeters.

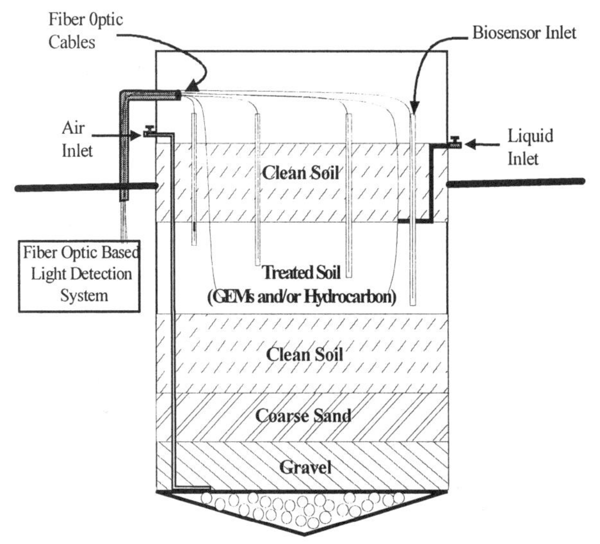

FIGURE 1. Schematic representation of a lysimeter.

The treated layer of soil consisted of either chemically contaminated or clean soil, with or without a *P. fluorescens* HK44 inoculum (Table 1). Chemically contaminated soil was artificially contaminated with naphthalene at 1000 mg/kg soil and anthracene and phenanthrene each at 10 mg/kg soil. Soil within the treatment layer was added in approximate 10 cm increments, each being compacted to a dry density of 1300 kg/m^3. Between each 10 cm increment was sprayed 4 L of a *P. fluorescens* HK44 inoculum in saline, producing initial cell densities of approximately 1 x 10^6 cfu/g soil. Lysimeter 6 served as a control and did not receive an HK44 inoculum.

TABLE 1. Lysimeter treatments.

Lysimeter	Chemically Contaminated	*P. fluorescens* HK44 Inoculum	Biosensors	Fiber Optic Cables	Portable PMT
1	+	+	+	+	+
2	+	+	+	+	+
3	-	+	+	+	+
4	+	+	-	-	+
5	-	+	-	-	+
6	+	-	-	-	+

Due to a delay in receiving regulatory agency permission to initiate loading of the lysimeters, the artificially contaminated soil underwent an extended storage period during which a significant amount of naphthalene was lost through volatilization and natural biodegradation. Therefore, an additional contaminant mixture was added 135 days after the lysimeters had been loaded with soil. Into each of lysimeters 1, 2, 4, and 6 was pumped 208 L of a transformer oil (Exxon Univolt 60) into which was dissolved 6 kg naphthalene and 0.5 kg anthracene. Seven days later a minimal salts nutrient medium was also added to these lysimeters as a means of stimulating *P. fluorescens* HK44 population growth.

Bioluminescence Detection Methods. A goal of this study was to utilize the bioluminescent response of *P. fluorescens* HK44 to monitor naphthalene bioavailability during the bioremediation process. Biosensors, a fiber optic system, and a portable photomultiplier tube (PMT) probe were incorporated into the lysimeters to provide both continuous and intermittent bioluminescent detection capabilities (Sayler, et al., 1998) (Table 1).

Biosensors. The biosensors were utilized for the detection of soil-borne vapor phase naphthalene (Webb, et al., 1997). Lysimeters 1, 2, and 3 each contained four biosensors placed at varying depths within the treated soil layer. Biosensors consisted of alginate-encapsulated *P. fluorescens* HK44 cells placed into a porous, light-tight stainless steel tube (14 cm long by 2.4 cm diameter). A fiber optic cable was inserted into each biosensor to transfer bioluminescent signals to a photon counting PMT/computer-based light monitoring system.

Fiber optic based system. One millimeter diameter fiber optic cables were buried throughout the treatment zone of lysimeters 1, 2, and 3. This allowed for bioluminescence emitted by HK44 cells residing directly in the soil to be transferred on a continuous basis to the light monitoring equipment.

Portable PMT probe. The portable PMT consisted of a light-tight housing containing a photon counting PMT module that could be lowered into the treatment zone to monitor light emanating from within the soil matrix. The portable PMT served much the same function as the buried fiber optic cables but afforded greater sensitivity and lower detection limits resulting from a larger area of detection and the elimination of signal transfer via fiber optic cables.

Soil Analysis. Soil cores were routinely removed from the lysimeters to obtain microbiological and chemical data for assessing the bioremediation process over the two-year study period. Cores typically consisted of soil representing a transverse section through the 92 cm treatment layer as well as a 30 cm section below the treatment layer. Total heterotrophic and *P. fluorescens* HK44 microbial populations were analyzed using non-selective and selective spread plate methods. Presumptive HK44 cells growing on selective plates were genotypically verified through colony hybridization protocols. Additionally, a most-probable-number technique based on bioluminescence was also developed for rapidly estimating HK44 numbers. Soil contaminant concentrations were determined using a solvent extraction and separation procedure followed by gas chromatograph-mass spectrometer analysis.

RESULTS AND DISCUSSION

Bioluminescent Response of *P. fluorescens* HK44. Monitoring of bioluminescence from encapsulated HK44 cells placed into biosensors was successful in the detection of volatile PAHs. Bioluminescent signals were produced from biosensors placed in chemically contaminated lysimeters 1 and 2 (Fig. 2). Biosensors in uncontaminated lysimeter 3 produced only background levels of bioluminescence due to basal levels of *lux* gene expression. Bioluminescent responses typically persisted for approximately five days, after which the alginate encapsulation matrix became too dehydrated to adequately support continued cell growth and maintenance.

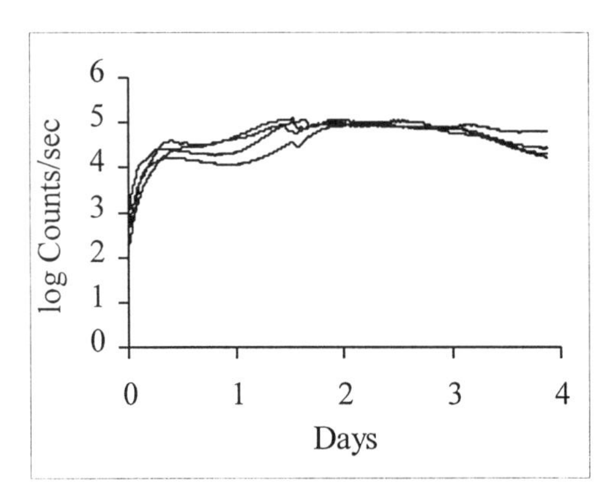

FIGURE 2. Bioluminescent responses generated from four biosensors located within the soil treatment layer of lysimeter 2

The network of fiber optic cables buried within the soil treatment zone were ineffective at detecting bioluminescent signals, likely due to their small 1 mm diameter area of detection. The portable PMT probe, however, did successfully detect, directly from soil, HK44 cells bioluminescing in response to naphthalene amendments. Figure 3 illustrates portable PMT-based light data acquisition from within the treatment layer of lysimeter 2. Due to the utilization of the portable PMT only within year two of the study, most of the naphthalene had been depleted from the lysimeter beds. Consequently, a supplementary amendment of 100 mg naphthalene and 100 mL minimal salts solution was placed within the general area of insertion of the portable PMT in order to restimulate HK44 popula-

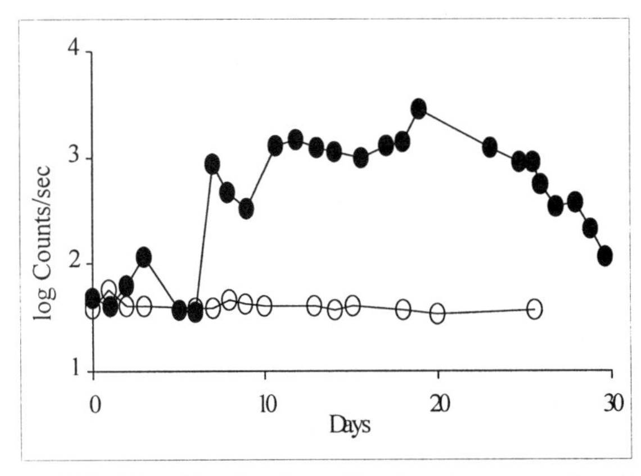

FIGURE 3. Monitoring of bioluminescent signals using the portable PMT in chemically contaminated lysimeter soil containing *P. fluorescens* HK44 (●) and in soil absent of HK44 (○).

tions. As shown in Figure 3, *P. fluorescens* HK44 had sufficiently recovered to produce a bioluminescent signal that persisted for 30 days, exhibiting photon counts well above background levels.

Survival and Maintenance of *P. fluorescens* HK44. *P. fluorescens* HK44 populations were assessed through selective plate counts utilizing a tetracycline resistance marker incorporated into the strain's genetic construct. Growth on selective plates, however, was not conclusive for HK44 enumeration. A background population of indigenous tetracycline resistant microorganisms was present at an average concentration of 2.0 (± 2.9) x 10^4 cfu/g soil. Therefore, colony hybridizations using a ^{32}P-labeled *luxA* gene probe were used to genetically verify which cells growing on selective plates were *P. fluorescens* HK44. Resulting population dynamics are shown in Figure 4. Overall, HK44 populations declined from an initial average of 1.5 (± 0.46) x 10^6 cfu/g soil to a final average concentration of 1.7 (± 1.5) x 10^3 cfu/g soil over a total sampling period of 656 days. At no time did HK44 numbers approach extinction. Population assessments of *P. fluorescens* HK44 were also monitored using a modified *lux* most-probable-number assay. This assay was significantly faster than plate count methods (24 hr versus 4 days) and produced HK44 population estimates that paralled those of selective plate counts, albeit at a one order of magnitude lower cell density (data not shown).

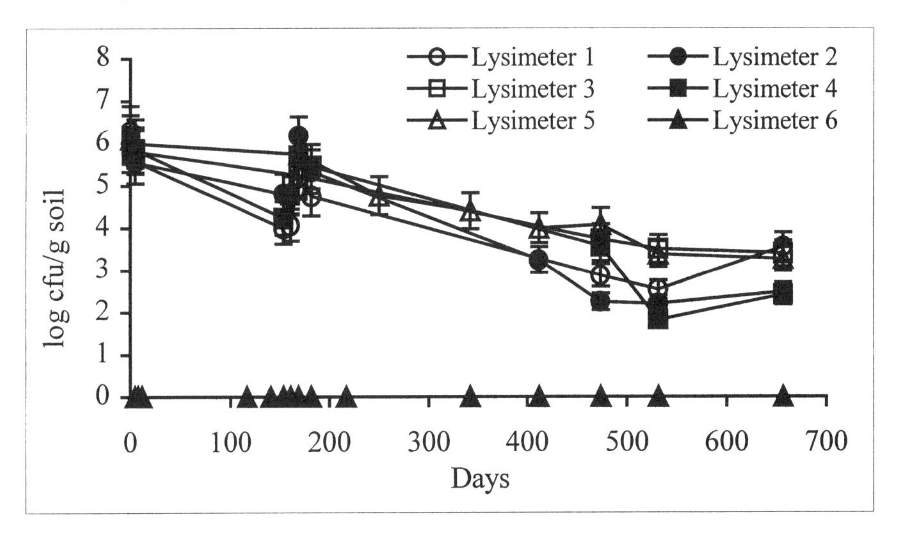

FIGURE 4. *P. fluorescens* HK44 population dynamics within each lysimeter.

Ecological considerations of *P. fluorescens* HK44 release. The introduction of an engineered microorganism into a natural ecosystem always carries with it the concern of recombinant DNA dissemination within the indigenous microbial community. Based on colony hybridization analyses utilizing four genetic markers specific to *P. fluorescens* HK44, no evidence of gene transfer events within the lysimeter microbial populations was observed.

CONCLUSIONS

Assessment of soil remediation processes usually employs gas chromatography/mass spectrometry-based detection protocols that are typically time consuming and expensive. Moreover, the results obtained are often of the 'below detection limit' variety. By implementing bioreporter organisms into the monitoring program, a general overview of bioremediation activities can be rapidly generated, on-line and in real time, thereby avoiding unnecessary sampling regimens. This study has demonstrated that *lux*-based bioluminescent bioreporters such as *P. fluorescens* HK44 can be utilized for such purposes, serving as competent and effective tools for bioremediation process monitoring and control.

ACKNOWLEDGMENTS

Research sponsored by the Office of Biological and Environmental Research, U.S. Department of Energy. Oak Ridge National Laboratory is managed by Lockheed Martin Energy Research Corp. for the U.S. Department of Energy under contract number DE-AC05-96OR22464.

S. Ripp was supported by an appointment to the Alexander Hollaender Distinguished Postdoctoral Fellowship Program sponsored by the U.S. Department of Energy, Office of Health and Environmental Research, and administered by the Oak Ridge Institute for Science and Education.

REFERENCES

Heitzer, A., O. F. Webb, J. E. Thonnard, and G. S. Sayler. 1992. "Specific and Quantitative Assessment of Naphthalene and Salicylate Bioavailability by using a Bioluminescent Catabolic Reporter Bacterium." *Appl. Environ. Microbiol. 58*: 1839-1846.

King, J. M. H., P. M. DiGrazia, B. Applegate, R. Burlage, J. Sanseverino, P. Dunbar, F. Larimer, and G. S. Sayler. 1990. "Rapid, Sensitive Bioluminescence Reporter Technology for Naphthalene Exposure and Biodegradation." *Science. 249*: 778-781.

Sayler, G. S., C. D. Cox, R. Burlage, S. Ripp, D. E. Nivens, C. Werner, Y. Ahn, and U. Matrubutham. 1998. "Field Application of a Genetically Engineered Microorganism for PAH Bioremediation Process Monitoring and Control." In R. Y. Flashner and S. Reuveny (Eds.), *Novel Approaches for Bioremediation of Organic Pollution.* 42nd OHOLO Conference, Eilat, Israel.

Webb, O. F., P. R. Bienkowski, U. Matrubutham, F. A. Evans, A. Heitzer, and G. S. Sayler. 1997. "Kinetics and Response of a *Pseudomonas fluorescens* HK44 Biosensor." *Biotech. Bioeng. 54*: 491-502.

IMPACT OF ADSORPTION AND DESORPTION ON BIOREMEDIATION OF PAH CONTAMINATED ESTUARIAN SEDIMENTS

J.E.Antia[1], ***H.H.Tabak***[2] and M.T.Suidan[1]
([1]University of Cincinnati, Cincinnati, Ohio 45221-0071, USA)
([2]U.S. Environmental Protection Agency, Cincinnati, Ohio 45268, USA)

ABSTRACT: The extent of adsorption and desorption of organic pollutants in sediments has a major influence on their fate and transport in the environment. In this work, experimentally obtained adsorption/desorption isotherms for 2, 3 and 4-ring polycyclic aromatic hydrocarbons (PAHs), a major class of carcinogenic environmental contaminants, are reported. The possible predictive nature of their calculated octanol-water partition coefficients for determining sorption capacities of sediments for PAHs having more than 4 carbon rings, is also explored.

INTRODUCTION

Widespread contamination of aqueous sediments by PAHs has created a need for cost-effective remediation processes. In practice, restoration of sediments contaminated with persistent organic chemicals practically involves either mechanical removal of the sediments with long term containment and, when appropriate, *ex-situ* treatment or use of natural attenuation/recovery including natural sediment burial by clean sediment (U.S.EPA, 1997). Even after mechanical removal considerable contaminant residuals may remain and also require monitored natural attenuation/recovery for acceptable restoration. Unfortunately natural attenuation/ recovery has uncertainties and limitations, including possible sediment resuspension, transport and distribution over wide areas, that can produce substantial risk to human health and the environment (Bishop, 1998). Thus, affordable and practical *in-situ* biotreatment is clearly needed for restoration of sediments contaminated with persistent organics.

PAHs are typical organic contaminants deposited in sediments and exhibit high octanol/water partition coefficients (K_{ow}), thereby resulting in the accumulation of these compounds in fatty tissues with subsequent biomagnification in the food chain (Safe, 1998). Also, PAHs dissolve only sparingly in water and are taken up readily by suspended particles containing organic matter in aquatic environments. As a result of particle settlement, sediments tend to be a major sink for PAHs in stream, lakes, estuaries and oceans.

The fate and distribution of PAHs is influenced by interactions with sediment. As a first step in the treatment of sediments contaminated with these PAHs, it is necessary to quantify the amounts of contaminants that are adsorbed and desorbed by sediments under abiotic equilibrium conditions with the surrounding aqueous phase. In turn, these data may be used to quantify bioavailability of PAHs in contaminated sediments.

Objective. The main purpose of this phase of the study is to quantify the abiotic adsorption and desorption amounts of 2, 3 and 4 ring PAHs in estuarian sediments. Data from adsorption/desorption isotherms are required to calculate PAH adsorption/desorption rates in sediment. Further these data are also useful in determining diffusion and mass transfer coefficients of PAHs in sediments.

MATERIALS AND METHODS

Samples of uncontaminated New York Harbor sediment and overlaying natural water were used in this work. Sediment samples were prepared by powdering and sieving air-dried sediments. Saturated natural water solutions of ^{14}C radiolabeled naphthalene, phenanthrene and pyrene (representative of 2, 3 and 4-ring PAHs respectively, from Sigma Chemicals) were prepared. Around 5 g. of air-dried sediment and 57-59mL of individual ^{14}C radiolabeled PAH-contaminated natural water were sealed into 60mL teflon-capped glass bottles. Each bottle was also spiked with a 1mL solution of sodium azide (10mM) and sodium molybdate (5mM) to inhibit aerobic/anaerobic bioactivity. The filled bottles were axially tumbled for 14 -28 days to allow the PAH to achieve equilibrium between the sediment and aqueous phases. Desorption experiments were conducted by successive substitution of 25 mL of the contaminated aqueous phase with clean natural water.

Liquid samples are withdrawn from each bottle, centrifuged, and filtered. The filtrate is analyzed on a liquid scintillation counter (LSC) to measure the PAH concentration in the aqueous phase. Equilibrium is assumed to have been reached when scintillation counts between any two ensuing time intervals are within 10% of each other. Table 1 presents PAH solubilities and other relevant experimental data. The amount of PAH sorbed in the sediment was estimated indirectly by subtracting the measured final PAH amount of the aqueous phase from the initial amount PAH in aqueous solution.

Table 1. Radiolabeled PAHs used in this study

PAH	No. Carbon rings	Radioactivity (mCi/mmole)	Solubility (μg/L)
Naphthalene	2	8.1	22,000
Phenanthrene	3	12.4	1,100
Pyrene	4	58.7	150

RESULTS AND DISCUSSION

Clean sediment samples was analyzed to quantify the amounts of PAH present. Adsorption isotherms, obtained for the three PAHs used in this work, are employed to estimate partition coefficients and possibly predict K_{ow} values for higher PAHs (> 5 carbon rings).

Uncontaminated Sediment Characterization. Preliminary characterization ofNew York Harbor sediment was performed before adsorption/desorption experiments were begun. In particular, initial PAH concentration in the sediment was obtained by extraction with methylene chloride/acetone mixture and analyzed by both, gas chromatography and gas chromatography/mass spectroscopy. Only trace quantities of PAHs were observed to be present in the sediment: Naphthalene (0.1μg/g), phenanthrene (0.2 μg/g) and pyrene (0.6μg/g). Sediment pH was measured at 6.9, BET surface area estimated at 21.4 m^2/g and organic carbon content was found to be 10.8%.

Adsorption/Desorption Experiments. Isotherms for both, adsorption and desorption of the model PAHs are presented in Figures 1 to 3, where the PAH amounts per weight of dry sediment are plotted against equilibrium aqueous PAH concentrations. For the three PAHs used here, the adsorption isotherms are essentially linear. Similar behavior was observed by other researchers (Chiou et al.,1998, Fu et al.,1994). At low loading rates, equilibrium adsorption was observed to occur within a few hours; while at higher loading rates, equilibrium takes upto two days.

Desorption experiments were conducted by successive aqueous phase dilutions on sediments contaminated with individual PAHs during adsorption. Preliminary results from two desorption experiments are presented in Figures 1 to 3, and follow the hysteresis (irreversible adsorption) behavior reported by others (Fu et al., 1994).

PAH sorption capacity. The calculated values of the sediment-water distribution coefficient K_p obtained here for the three PAHs are reported in Table 2 and are of the same order of magnitude reported by other researchers for Massachusetts Bay sediment (Chiou et al., 1998). The value of K_p is strongly dependent on the sediment organic carbon content (f_{oc}). Hence, the organic carbon partition coefficient ($K_{oc} = K_p/f_{oc}$) of the individual PAHs between sediment organic matter and water are presented below in Table 1. The octanol-water coefficient (K_{ow}) is calculated using the following correlation by Karickhoff (Lyman, 1990): $\log K_{oc} = 1.00 \log K_{ow} - 0.21$

Table 2. Partition coefficients for the model PAHs

PAH	K_p (mL/g)	K_{oc}	log K_{ow}
Naphthalene	12.8	118.52	2.2837
Phenanthrene	535.2	4955.56	3.9051
Pyrene	1195.7	11071.30	4.2542

The log K_{ow} values for PAHs studied in this work lie within the range reported (log K_{ow} = 2.11 - 6.34) for river sediments (Lyman, 1990).

Figure 4 shows that a plot of log K_{ow} versus log S is linear. This plot may be used to predict the octanol-water partition coefficients for PAHs having more than 5 carbon rings in their molecular structure. These larger PAHs have very low aqueous solubilities, thereby making them difficult to experimentally quantitate in the aqueous phase.

Conclusion and Research Direction. The extent of PAH sorption in sediments is dependent on contaminant hydrophobicity and most of the PAHs are irreversibly

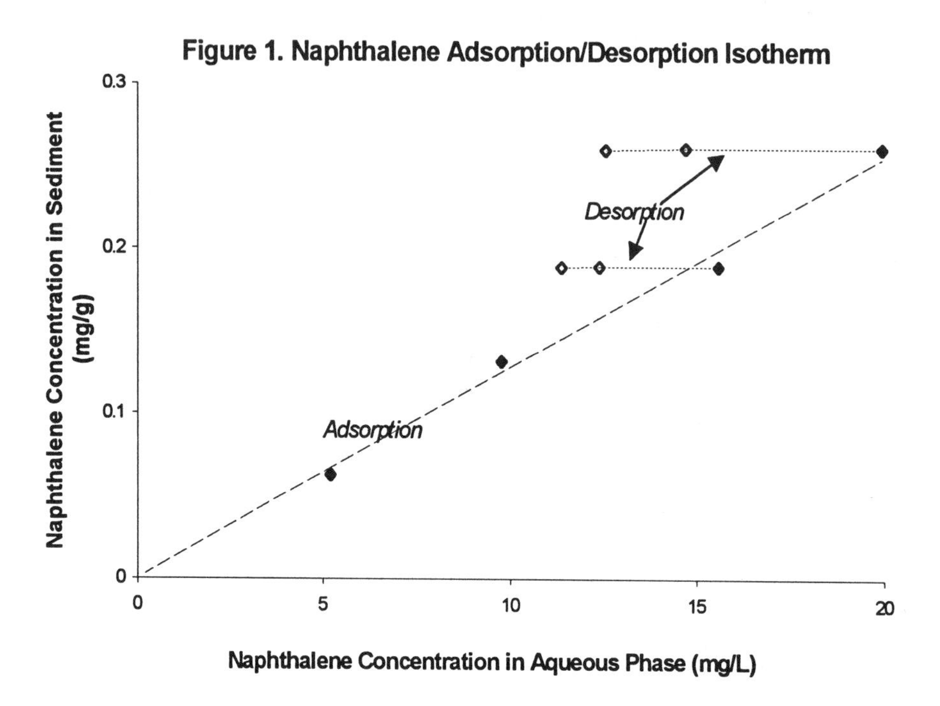
Figure 1. Naphthalene Adsorption/Desorption Isotherm
Naphthalene Concentration in Sediment (mg/g)
0.3
0.2
0.1
0
Desorption
Adsorption
0
5
10
15
20
Naphthalene Concentration in Aqueous Phase (mg/L)

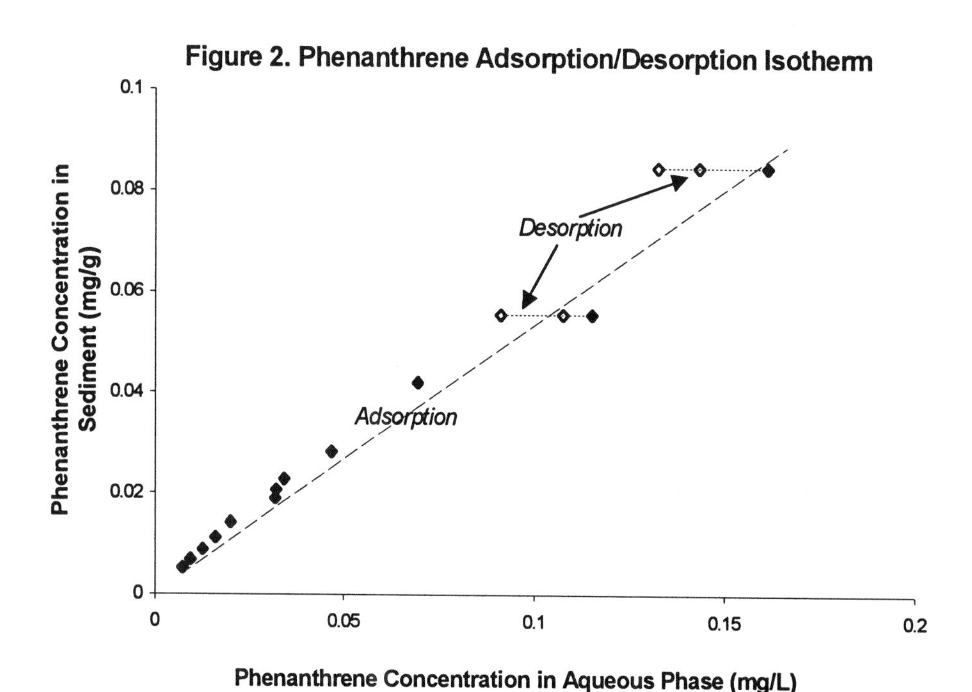
Figure 2. Phenanthrene Adsorption/Desorption Isotherm
Phenanthrene Concentration in Sediment (mg/g)
0.1
0.08
0.06
0.04
0.02
0
Desorption
Adsorption
0
0.05
0.1
0.15
0.2
Phenanthrene Concentration in Aqueous Phase (mg/L)

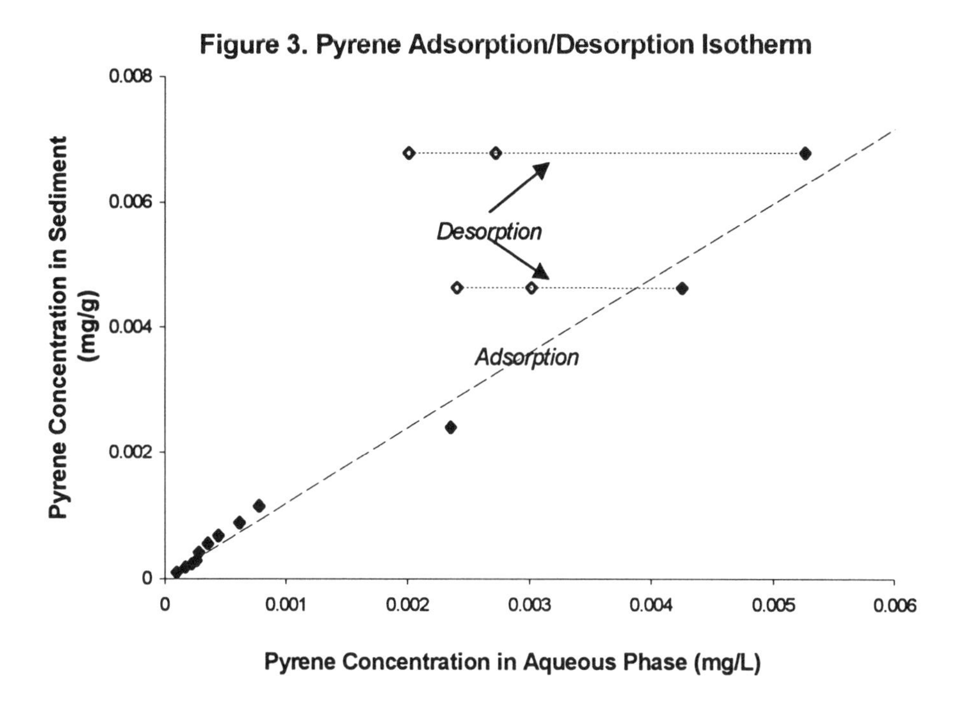
Figure 3. Pyrene Adsorption/Desorption Isotherm
Pyrene Concentration in Sediment (mg/g)
0.008
0.006
0.004
0.002
0
Desorption
Adsorption
0
0.001
0.002
0.003
0.004
0.005
0.006
Pyrene Concentration in Aqueous Phase (mg/L)

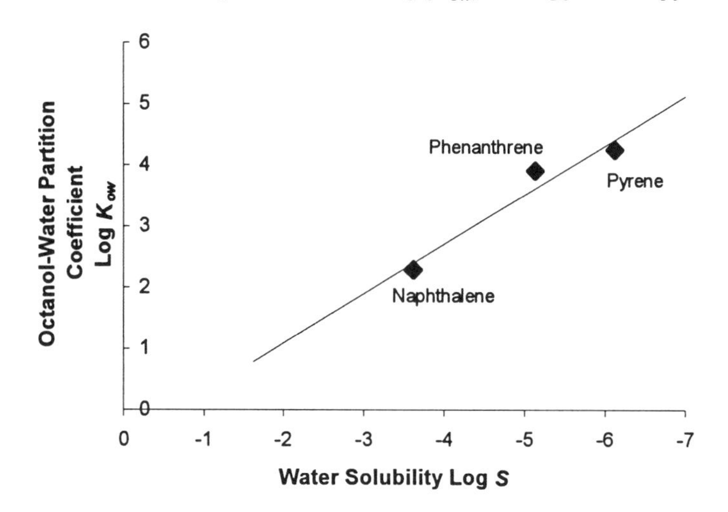
Figure 4. Plot of log (K_{ow}) vs Log(Solubility)
Octanol-Water Partition Coefficient Log K_{ow}
6
5
4
3
2
1
0
Phenanthrene
Pyrene
Naphthalene
0
-1
-2
-3
-4
-5
-6
-7
Water Solubility Log S

bound to the sediment particles. However, the effects of intra-particle diffusion (slow sorption) are poorly understood and may be critical in contaminant transport, bioavailability and remediation (Pignatello and Xing, 1996). Further factors which impede the development of bioremediation processes include the effect of complex PAH mixtures on adsorption, as well as, their bioavailability (Hughes et al., 1998). Also, the addition of nutrients and surfactants may enhance PAH bioremediation (Hughes et al.,1998; Venosa, et al.,1996). Future work on this project involves quantifying PAH desorption from contaminated sediments, establishing contaminant bioavailability, and studies on natural and enhanced *in-situ* biodegradation rates.

REFERENCES

U.S.EPA. 1997. *EPA's contaminated sediment management strategy*. Office of Water and Solid Waste, U.S. Environmental Protection Agency, Washington, DC.

Bishop, D.F. 1998. *Perspective on remediation and natural recovery of contaminated sediments.* National Risk Management Research Laboratory, U.S. Environmental Protection Agency, Cincinnati. In press.

Safe, S. 1998. "Metabolism, uptake, storage, and bioaccumulation." In R. Kimbrough (Ed), *Halogenated biphenyls, naphthalenes, dibenzodioxins, and related products.* pp.81-107. Elsevier/North Holland.

Fu, G., A.T.Kan and M.B.Tomson. 1994. "Adsorption and desorption hysteresis of PAHs in surface sediment." *Environ.Toxicol.Chem. 13*(10):1559-68.

Chiou, C.T., S.E.McGrody and D.E.Kile. 1998. "Partition characteristics of polycyclic aromatic hydrocarbons on soils and sediments." *Environ.Sci.Technol. 32*(2): 264-69.

Lyman, W.J. 1990. "Adsorption coefficient for soils and sediment." In W.J. Lyman, W.F. Reehl, D.H. Rosenblatt (Eds). *Handbook Of Chemical Property Estimation Methods : Environmental Behavior Of Organic Compounds.* Chapter 4. American Chemical Society.

Pignatello, J.J., and B.Xing. 1996. "Mechanisms of slow sorption of organic chemicals to natural particles." *Environ.Sci.Technol. 30*(1): 1-11.

Hughes, J.B., D.M.Beckles, S.D.Chandra and C.H.Ward. 1998. "Utilization of bioremediation processes for the treatment of PAH-contaminated sediment." *J.Ind. Microbiol.Biotechnol. 18*(2-3): 152-60.

Venosa, A.D., M.T. Suidan, B.A.Wrenn, K.L.Strohmeir, J.R.Haines, E.L.Eberhart, D.King and E.Holder. 1996. " Bioremediation of an experimental oil spill on the shoreline of Delaware Bay." *Environ.Sci.Technol. 30*(5): 1764-75.

MICROSCALE CHARACTERIZATION OF PAH SEQUESTRATION ON SEDIMENTS

Upal Ghosh and ***Richard G. Luthy*** (Carnegie Mellon University, Pittsburgh, PA)
J. Seb Gillette and Richard N. Zare (Stanford University, Stanford, CA)
Jeffery W. Talley (USACE Waterways Experiment Station, Vicksburg, MS)

ABSTRACT: Unique complementary spectroscopic and spectrometric techniques are employed to provide more direct information at the microscale on the sequestration of polycyclic aromatic hydrocarbon (PAH) contaminants in sediments. This information is used to assess where PAHs are bound to sediment particles and the association of such PAHs with organic matter. PAH measurements at a spatial resolution of 40 μm and organic matter identification at a spatial resolution of 10 μm have been performed on sediment particles, including both external surfaces and interior regions. A microprobe two step laser desorption laser ionization mass spectrometer is used for PAH measurement, wavelength dispersive X-ray microanalysis and infrared microspectroscopy for carbon measurement, and a cryo-microtome sectioning procedure for cross-sectional investigations. A strong correlation of organic matter location with PAH location is observed for Milwaukee harbor sediments. PAH levels on the black carbonaceous particles are two orders of magnitude higher than on the white siliceous particles. Additionally, most PAHs are found to be associated with the external surface regions of sediment carbonaceous particles indicating near surface sorption mechanisms.

INTRODUCTION

Hydrophobic organic compounds (HOCs) such as polycyclic aromatic hydrocarbon (PAH) compounds are known to sorb strongly to soils and sediments. This binding or sequestration poses an obstacle to remediation (Luthy *et al.*, 1997; NRC, 1994) and challenges our concepts about cleanup standards and risks with regard to bioavailability and toxicity of soils and sediments (Alexander, 1995; NRC, 1997). A basic understanding of where and how PAHs are bound on soils and sediments is required to address these challenges. Presently, there are no direct observational data revealing the sub-particle scale locations in which HOCs accumulate when they associate with soils and sediments. As a result, researchers must rely on inferences from macroscopic experimental observations that capture overall behavior and provide empirical evidence for deducing sorbent/sorbate mechanistic models (e.g., Pignatello and Xing, 1996; Weber and Young, 1997; Wu and Gschwend, 1986; Brusseau *et al.*, 1991).

In order to address these shortcomings, the primary goal of this research is to use more direct microscale measurement approaches to investigate where and how PAHs are sorbed or sequestered on sediments and the use of this information to gain insights to possible sorption mechanisms. A very powerful analytical tool used in this research is the microprobe laser desorption/laser ionization mass spectrometer that allows investigation at the microscale on the location of organic contaminants. Additionally, valuable microscale information on the mineral and organic matter environment is made possible using scanning electron microscopy (SEM) with wavelength dispersive X-ray microanalysis (WDX) and infrared microspectroscopy. The use of these analyses in conjunction with the particle sectioning techniques developed in this work makes it possible to directly visualize the location and associations of bound PAHs at a sub-particle level. Results of the direct investigation of PAH locations and particle characteristics of sediments from Milwaukee harbor confined disposal facility are presented here.

MATERIALS AND METHODS

Sediment obtained from the Milwaukee Harbor Confined Disposal Facility (CDF) operated by the US Army Corps of Engineers were used in this study. These sediments originated from the Milwaukee harbor during the process of dredging to maintain navigability. Concerns have been raised about the potential for release of contaminants (notably PAHs) from these CDF sites and about closure requirements, as discussed by Bowman *et al.* (1996). Sediment samples obtained from the CDF were separated into size fractions by wet sieving prior to analysis.

Microscale PAH measurement. A new analytical technique of microprobe two-step laser desorption/laser ionization mass spectrometry (μL^2MS) was used to identify and characterize the trace distribution of PAHs on sediment particles. The first step in the μL^2MS analysis involves desorption of constituent molecules on a sediment particle with a pulsed infrared (IR) laser beam focused to a ~ 40 μm spot or less. In the second step, the desorbed molecules are selectively ionized with a pulsed ultraviolet (UV) laser and the resulting ions are extracted into a reflectron time-of-flight (RTOF) mass spectrometer. Detailed description on the application of the μL^2MS technique for soil and sediment is provided in Gillette *et al.* (1999).

Elemental analysis. Elemental analysis of the particle surfaces was conducted and compared with PAH location and abundance. Measurement of microscale elemental composition of sediment particles was performed using scanning electron microscopy equipped with wavelength dispersive X-ray analysis (Phillips XL-FEG40).

Microscale organic carbon measurement. Fourier transform infrared (FTIR) micro-spectroscopy was used to discern the predominant types of sorbent carbon. Infrared microspectroscopy combines the spatial resolution of a microscope with infrared spectral analysis making it possible to analyze for IR absorbance down to a diffraction-limited spatial resolution of approximately 10 microns. Use of synchrotron infrared radiation provides nearly two-orders of magnitude brightness advantage over standard laboratory globar IR sources. This work was carried out at the National Synchrotron Light Source (NSLS), Brookhaven National Laboratory, using Spectra Tech IRμs microspectrometer connected to infrared beamline U10B. Further details of the performance of the microspectrometer are available in Carr *et al.* (1995).

Particle sectioning. Sediment particles were sectioned by cryo-microtomy (Leica UltracutR) to expose the internal surface for analysis. Sediment particles were frozen at -150 °C and cut using a diamond blade. Sectioning using cryo-microtomy was selected because unlike other methods of embedding in a resin matrix, cryo-microtomy does not involve any contact of the specimen with organic materials and solvents that may adversely affect the integrity of the sample. After sectioning, the particles were thawed in a dessicator and mounted on brass stubs for further analysis.

RESULTS AND DISCUSSION

Interparticle heterogenity of PAHs and organic carbon. The predominant class of particles in the sediment comprised silica having silicon and oxygen as the most abundant elements. The second most abundant particle class was black particles with high carbon abundance. An image of the Milwaukee sediment particles showing the black and silica particles along with their respective coincident IR absorbance spectrum and PAH concentration measurements are shown in Figure 1. FTIR microspectroscopy revealed strong C-H stretching vibrations in the IR spectrum with peak absorbances at 2928 and 2870 cm^{-1} indicating that the carbonaceous black particles were organic in nature. Additionally, the measured PAH mass signal on the black and white particles presented in Figure 1 indicate that PAHs are primarily associated with organic-carbon-rich particles.

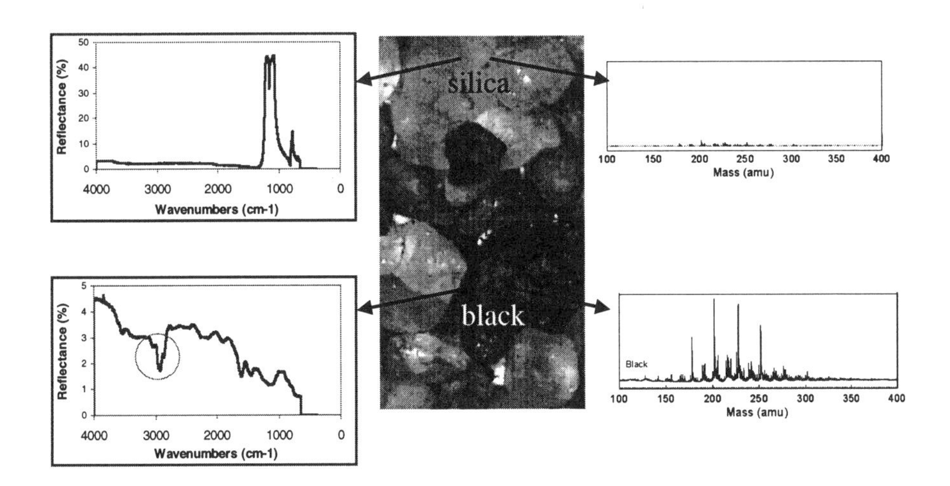

FIGURE 1. Infrared absorbance spectrum and PAH abundance measurements of silica and black Milwaukee CDF sediment particles.
(ref. Gillette *et al.*, 1999; Ghosh *et al.*, 1999)

The most abundant PAHs measured were mass numbers: 178(anthracene, phenanthrene), 202(fluoranthene, pyrene), 228(benzo(a)anthracene, chrysene), 252(benzo(b)fluoranthene, benzo(k)fluoranthene, benzo(a)pyrene), 276(benzo(ghi)perylene, indeno(1,2,3-c,d)pyrene) and their methylated derivatives. The distribution of PAHs measured on individual sediment particles using μL^2MS match closely with the PAH distribution observed from bulk extraction and analysis of the sediment. These direct observations at the microscale support the well-accepted theory that hydrophobic organic compounds are primarily associated with organic carbon present in soils and sediments. However, the importance of our observations is that much different levels and associations are evident for the different particles comprising the sediment mixture. These types of observations will be important for interpreting results of biological treatment of contaminated sediments. For example, certain particle classes or regions on particles may be more amenable to remediation due to faster release of PAHs. Further, findings of this nature have significant implications for possible new remediation technologies such as particle classification to separate the contamination by class of particles. In this case, if the black particles can be separated from the bulk sediments by an economical particle classification technology, most of the contamination may be removed by a physical separation process alone.

Measurement of PAH and organic matter within sediment particles. There has been no evidence in the literature to date on direct observation of HOCs on the interior of natural or anthropogenic particles. We report here an analytical methodology that can be used to make direct observation of PAHs inside sediment particles. Observations made from such analyses can then be used to assess degradation by biological treatment, toxicity or uptake of PAHs, as well as the applicability of pore diffusion or polymer diffusion models for PAH release for a given soil or sediment. Results are presented for the analysis of particle sections of the two most abundant types of particles present in the Milwaukee sediment: organic-carbon-rich black particles, and white siliceous particles.

Several samples each of the black and white Milwaukee sediment particles approximately 300-600 μm in size were sectioned and analyzed. A light microscope image of a typical black Milwaukee sediment particle after sectioning using cryomicrotomy is shown in Figure 2. Electron microscopy with WDX analysis revealed that the sectioned surface was abundant in elemental carbon throughout the interior. Mapping of organic carbon using FTIR microspectroscopy also revealed strong infrared C-H stretching absorbance throughout the particle interior indicating a more or less homogeneous substrate composed of organic matter. The sectioned sediment particle was analyzed for spot PAH measurement across the cut surface following two tracks (*a* and *b*) as shown in Figure 2 to reveal any interior location of PAH. Tracks *a* and *b* shown in Figure 2 start from outside the sectioned region, traverse the cut section, and terminate outside the cut section. Total PAH signal intensity measured along these two tracks indicates two orders of magnitude higher PAH concentrations on the exterior regions of this particle compared to the interior regions. Analyses such as that represented in Figure 2 have been conducted on multiple particles of the black Milwaukee sediment and the observations of PAH profiles across the cross sections have been found to be very similar, i.e., two orders of magnitude higher concentration of PAHs on the near surface regions compared to the interior regions. Therefore, it is possible to infer that a dominant sequestration mechanism for these sediment particles entails near-surface sorption processes, perhaps with shallow penetration. This may be the dominant phenomenon controlling PAH sequestration on these particles, as compared to incorporation deep within a porous substrate or within organic matter interior regions. White silica particles found in the Milwaukee sediment also were sectioned and analyzed for PAHs and organic carbon. Elemental SEM-WDX analysis revealed a primarily silicon and oxygen rich interior with regions of clay and traces of carbon on the outer surface. PAH concentrations were measured only on the outer surface and very little signal was detected inside the cut surface. These observations are similar to the observations made on the sectioned black particles.

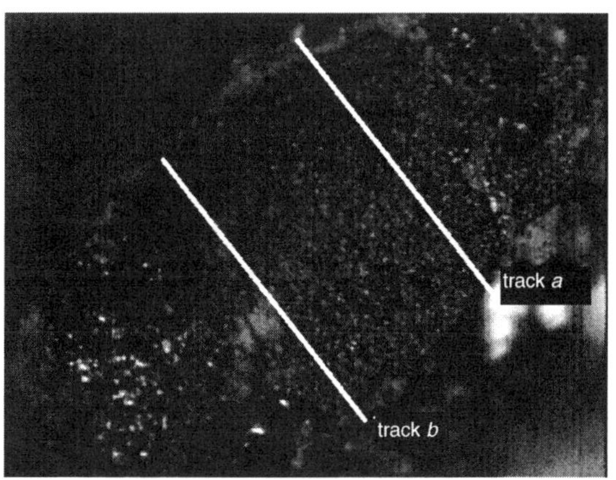

FIGURE 2. Image of black sediment particle showing sectioned surface and two tracks along which PAH measurements were performed.

These direct observations of the physical locations of PAHs indicate that near-surface sorption phenomena on organic matter particles are the likely mechanisms for PAH sequestration for these sediments. Near-surface sorption processes, perhaps with shallow penetration, appear to be much more significant for PAH sorption than incorporation deep within a porous substrate or organic matter interior regions. Individual carbonaceous particles contained two orders of magnitude more PAHs than individual silica particles. The methodologies presented in this work can be used to provide more direct evidence of sequestration mechanisms of HOCs on soils and sediments and thereby help interpret results from biological treatment and toxicity studies as well as help resolve some of the contentious issues involved in the understanding of sequestration processes. Causes of low bioavailability that plague efficient operation of biotreatment processes can be investigated using these micro-scale analysis methods by comparing particles before and after biotreatment. Direct physical evidence of sequestration mechanisms may be used to support the concept of reduced availability and risk of biotreated sediment.

REFERENCES

Alexander, M. 1995. "How Toxic are Toxic Chemicals in Soil?" *Environ. Sci. Technol., 29*, 2713-2717.

Bowman, D. W., J. M. Brannon, and S.A. Batterman. 1996. "Evaluation of Polychlorinated Biphenyl and Polycyclic Aromatic Hydrocarbon Concentrations in Two Great Lakes Dredged material Disposal Facilities." *Proceedings of the 11th seminar: Water Quality '96*, Seattle, Washington.

Brusseau, M.L., R. E. Jessup, and P.S.C. Rao. 1991. "Nonequilibrium Sorption of Organic Chemicals: Elucidation of Rate Limiting Processes." *Environ. Sci. Technol.* 25, 134-142.

Carr, G.L., J.A. Reffner, and G.P. Williams. 1995. "Performance of an Infrared Microspectrometer at the NSLS." *American Institute of Physics*. 66, 1490-1492.

Ghosh, U., J.S. Gillette, R.G. Luthy, R.N. Zare. 1999. "Investigation of PAH Location and Association in Contaminated Sediments Using Microscale Analysis." (in review)

Gillette, J.S., R.G. Luthy, S.J. Clemett, and R.N. Zare. 1999. "Direct Observation of Polycyclic Aromatic Hydrocarbons on Geosorbents at the Sub-Particle Scale." *Environ. Sci. Technol. in* press.

Luthy, R.G., G.R. Aiken, M.L. Brusseau, S.D. Cunningham, P.M. Gschwend, J.J. Pignatello, M. Reinhard, S.J. Traina, W.J. Jr. Weber, and J.C. Westall. 1997. "Sequestration of Hydrophobic Organic Compounds by Geosorbents." *Environ. Sci. Technol.* 31, 3341-3347.

NRC, 1994. *Alternatives for Groundwater Cleanup*. National Research Council Report, National Academy Press, Washington, D.C.

NRC, 1997. *Contaminated Sediments in Ports and Waterways, Cleanup Strategies and Technologies*. National Research Council Report, National Academy Press, Washington D.C.

Pignatello, J.J., and B. Xing. 1996. "Mechanisms of Slow Sorption of Organic Chemicals to Natural Particles (Critical Review)." *Environ. Sci. Technol.* 30, 1-11.

Weber, W.J., Jr., and T. M. Young. 1997. "A Distributed Reactivity Model for Sorption by Soils & Sediments. 6. Mechanistic Implications of Desorption under Supercritical Fluid Conditions." *Environ. Sci. Technol.* 31, 1686-1691.

Wu, S., and P. M. Gschwend. 1986. "Sorption Kinetics of Hydrophobic Organic Compounds to Natural Sediments and Soils." *Environ. Sci. Technol.* 20, 717-725.

PAH DISTRIBUTION AND BIODEGRADATION N THE DELAWARE AND SCHUYLKILL RIVERS

Thomas J. Boyd, Julia K. Steele, Michael T. Montgomery, and Barry J. Spargo
(US Naval Research Laboratory, Washington, DC)

ABSTRACT: Sediments and waters in the Philadelphia Naval Complex Reserve Basin and the Delaware and Schuylkill Rivers were sampled in December 1997. The primary goal was to determine the concentration, biodegradation, distribution and fate of contaminants within be sampling area. Forty-five sediment samples were analyzed for PAHs, hydrocarbons, metals, total organic carbon, PAH mineralization and heterotrophic bacterial productivity. PAHs, trace elements and total organic carbon were slightly elevated in the Reserve Basin and the Schuylkill River relative to the Delaware River. PAH biodegradation rates and bacterial productivity data suggested that active microbial communities were degrading organic contaminants in the Reserve Basin and in the associated watershed. Elevated trace metal concentrations appeared to inhibit PAH mineralization and bacterial productivity. PAH turnover times (removal times) ranged from weeks to several years at some stations within the Reserve Basin. We found correlation between total organic carbon and both PAH and metal concentrations in sediments, indicating that sediment organic::contaminant loads are similar throughout the watershed. There was no evidence to suggest that contaminants from the Reserve Basin are impacting adjacent sediments. Temperature and dissolved oxygen profiles showed little variation throughout the sampling area and most likely did not account for the differences observed in the biological activity between stations in the winter sampling. The major source for suspended particles entering the Reserve Basin is likely the Schuylkill River. We conclude that it is the primary vehicle for current contaminant input, particularly with respect to petroleum hydrocarbons. High levels of contaminants in a few stations within the Reserve Basin suggest historical contamination that was more significant than current input. Similarities in contaminant concentrations, locations and bioremediation suggest that the current source(s) of contaminants are similar for both the Schuylkill River and the Reserve Basin. Preliminary analysis from June and September 1998 sampling indicates dissolved oxygen may be a limiting factor for PAH mineralization. Mineralization rates did not appear to be affected by higher summer temperatures.

INTRODUCTION

Ecosystem dynamics involve complex transfer and cycling of metabolites though various levels of the food chain. Both natural sources of organic matter and contaminant hydrocarbons can be assimilated and respired by bacteria. The contaminant degradation rate is controlled by environmental factors that influence natural carbon and nutrient fluxes through microbial assemblages. Typical site evaluations focus on parent compound surveys and flask biotreatability studies and fail to determine factors that control contaminant metabolism as an integrated component of an ecosystem. As a result, subsequent remediation strategies often do not achieve cleanup goals and can even lead to unintended or adverse environmental consequences. Only through integrated study of intrinsic bioremediation rates and ecosystem dynamics affecting biological communities can it be determined if costly sediment treatment or removal actions will be the most ecologically viable long term strategy.

PAH compounds represent one of the most ubiquitous chemical contaminants in US sediments (Daskalakis and O'Connor, 1995). Petroleum refineries, wood treatment plants, and coke processing facilites are major point sources for PAH contaminantion (Cerniglia, 1984). In addition, non-point sources such as surface runoff and atmospheric deposition may also constitute significant watershed inputs (Dickhut and Gustafson, 1995). One of the most difficult tasks for both scientists and remediation project managers is to determine the intrinsic biodegradation rates of chemical contaminants in field settings (Fox, 1994; Sims and Sims, 1995). A number of methods have been used to assess bioremediation. Every method has limitations when soley employed. For this reason, a National Research Council panel has suggested a combination of testing methods for assessing microbial activities during field evaluations (National Research Council, 1993). We have measured physical parameters such as temperature, particle concentration, and transmissivity; chemical parameters such as dissolved oxygen concentration, total organic carbon, and PAH and trace element concentratons; and biological parameters such as bacterial productivity and PAH mineralization rates in order to determine the origin, fate and transport of contaminants in the Delaware and Schuylkill Rivers and the Philadelphia Naval Shipyard's Reserve Basin.

Site Description. The Reserve Basin (RB) within the Philadelphia Naval Complex (PNC) is a semi-enclosed body of water currently used primarily for ship storage. It extends from the west side of Broad Street to the Schuylkill River (Fig. 1). The Reserve Basin comprises approximately 109 acres and has a 25 ft average depth. The depth ranges from 20 to 31 ft (mean low water). Tidal action is the most significant mode of water inflow. Minor inputs come from National Pollutant and Discharge Elimination System (NPDES) discharges and stormwater drains. Past Naval activities included ship maintenance, construction and storage. Although current activity is not full-scale, periodic tugboat operations, ship movement and maintenance dredging still occur. Contaminants found on site include fuel hydrocarbons from spills (1964 and 1980) and upriver refinery activity, trace elements from the inactive Girard Point blasting grit disposal area and a rail loading facility, as well as organics and trace elements brought into the Reserve Basin through tidal action. Thirty stations were established in the Reserve Basin. Sixteen stations were established in the Delaware and Schuylkill Rivers (Fig. 1).

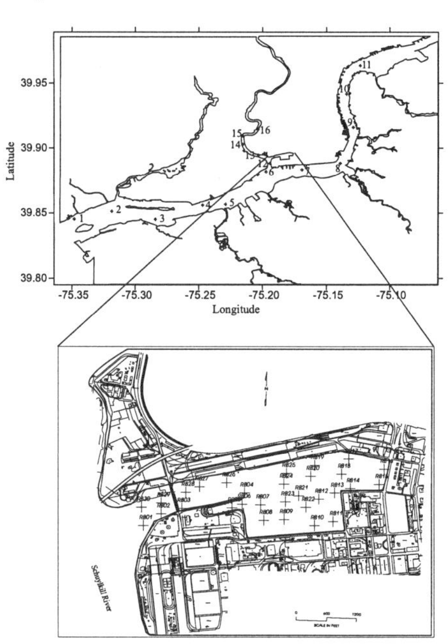

Figure 1.

MATERIALS AND METHODS

Each station was sampled using a petite Birge-Ekman or Ponar grab for sediments (Mudroch and MacKnight, 1994). A standard CTD rosette was used for water samples in the Delaware and Schuylkill Rivers aboard the *R/V Cape Henlopen* (Seabird Electronics). A Hydrolab portable multiprobe sensor package (Datasonde 4) system was deployed from an inflatable boat in the Reserve Basin. PAHs in sediments were extracted and analyzed using a modification of EPA SW 8270 (Fisher, 1997). Bacterial productivity was measured (via ^{3}H-leucine incorporation) by the method of Smith and Azam, 1992. PAH mineralization was measured by trapping $^{14}CO_2$ liberated from ^{14}C-labeled PAH substrates (Boyd et al., 1996) . Trace elements were analyzed using EPA method SW-846 (6010B and 3051). TOC was measured from sediment samples using EPA method 440.4. Particle concentration were measured gravimetrically using pre-tared filters. Nepheloid material (benthic boundary layer) was sampled using an inverted funnel attached to a teflon-lined pump.

RESULTS AND DISCUSSION

Total PAH concentrations were highest on average in the December 1997 sampling (although standard devations were very high). Stations within the Schuylkill River (12-16) and the Reserve Basin consistently showed the highest concentrations of PAH (Figs. 2A-4A). Naphthalene mineralization was consistently highest in Schuylkill River sediments (stations 12-16) and the Reserve Basin. The rates in June 1998 were the slowest for all stations, almost 2 orders of magnitude (Figs 2B-4B). Phenanthrene mineralization was highest in Schuylkill

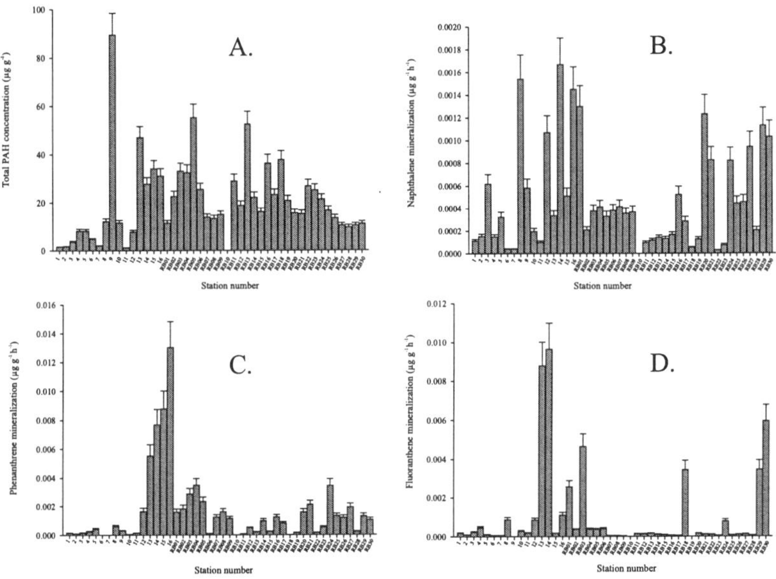

Figure 2. December 1997. A. Total PAH Concentration. B. Naphthalene Mineralization. C. Phenanthrene Mineralization. D. Fluoranthene Mineralization

River (stations 12-16) and Reserve Basin sediments during all seasons sampled (Figs 2C-4C). Phenanthrene mineralization rates were consistently highest in the Schuylkill River sediments. There was no seasonal variation observed in phanthrene mineralization rates. Fluoranthene mineralization rates were highest in Reserve Basin sediments in the June 1998 and September 1998 and highest in the Schuylkill River stations during the December 1997 sampling event (Figs 2D-4D). The high variation found between stations precluded determining an average value for most of the measured parameters (i.e. standard deviation greater than sample mean). Mineralization rates showed no correlation with termperature over the three sampling events. In fact, naphthalene mineralization rates were lowest in the June sampling (Fig. 3B). Laboratory scale studies have found increased mineralzation of naphthalene and other PAHs at higher incubation temperatures (Bauer and Capone, 1985). We did not measure porewater nutrient concentrations, however previous

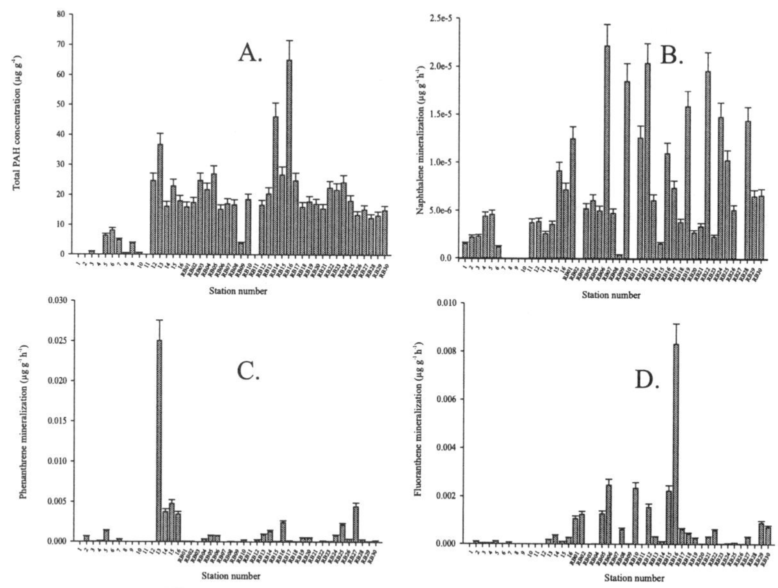

Figure 3. June 1998. A. Total PAH Concentration. B. Naphthalene Mineralization. C. Phenanthrene Mineralization. D. Fluoranthene Mineralization

studies have shown no correlation between biodegradation rates and sediment nutrient concentrations (Bauer and Capone, 1985; Guerin and Jones, 1989).

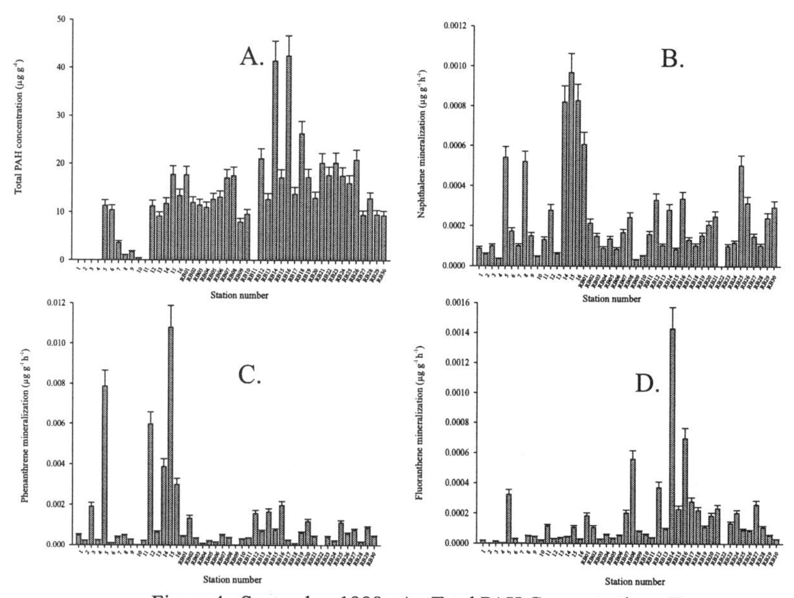

Figure 4. September 1998. A. Total PAH Concentration. B. Naphthalene Mineralization. C. Phenanthrene Mineralization. D. Fluoranthene Mineralization

It is difficult to compare the mineralization rates determined in this project with literature rates because in this work, we added tracer amounts of ^{14}C-labeled substrates and calculated isotope dilution using ambient PAH concentrations. Often mineralization rates are expressed in terms of percent ^{14}C recovery (Boethling and Alexander, 1979; Chung and Alexander, 1998) or with the assumption that ambient substrate concentration is negligible (Bartholomew and Pfaender, 1982). Based on our measured rates, the turnover time for PAHs in the sample area ranges from several days to several years. The shortest turnover times were found in the Schuylkill River which is believed to have the freshest input of PAHs (from numerous refineries along her banks). Fluoranthene had the longest turnover times of the substrates tested. Its rate of biodegradation was fastest in the Reserve Basin.

There was no correlation between oxygen saturation in bottom waters (<1m above sediment surface) and PAH biodegradation, however, the highest mineralization rates always coincided with the highest oxygen saturations (Fig. 5). The lowest dissolved oxygen saturations in bottom waters occured in the warmer months (June and September). Surface dissolved oxygen saturations were high year round (above 85%). Surface waters in the eastern portions of the Reserve Basin had supersaturated oxygen in June and September most likely due to primary productivity and wind forcing.

Sediment trace element concentrations were highest in the Schuylkill River

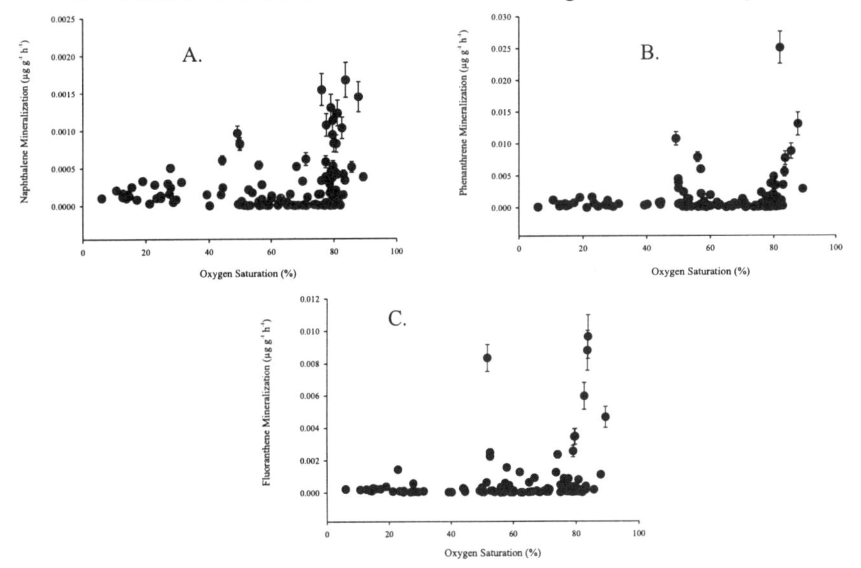

Figure 5. PAH Mineralization Realted to Bottom Water Oxygen Saturation . A. Naphthalene Mineralization. B. Phenanthrene Mineralization. C. Fluoranthene

and the Reserve Basin. Several stations (with ore-related past activities) in the Reserve Basin (RB05, RB06, RB22) had the highest concentrations of lead, cadmium, nickel, chromium, copper and zinc. Concentrations in all other parts of the watershed correlated well with total organic carbon throughout the sampling area indicating a general non-point source for these compounds.

REFERENCES

Bauer J.E., and D.G. Capone. 1985. "Degradation and Mineralization of the Polycyclic Aromatic Hydrocarbons Anthracene and Naphthalene in Intertidal Marine Sediments." *Appl. Environ. Microbiol.* 50:81-90.

Boethling R.S., and M. Alexander. 1979. "Microbial Degradation of Organic Compounds at Trace Levels." *Environ. Sci. Technol.* 13:989-91.

Boyd, T.J., B.J. Spargo, and M.T. Montgomery. 1996. "Improved Method for Measuring Biodegredation Rates of Hydrocarbons in Natural Water Samples." In B.J. Spargo (Ed.) *In Situ Bioremediation and Efficacy Monitoring*, pp. 113-122. US Naval Research Laboratory (NRL/PU/6115-96-317), Washington, DC.

Chung N., and M. Alexander. 1998. "Differences in Sequestration and Bioavailability of Organic Compounds Aged in Dissimilar Soils" *Environ. Sci. Technol.* 32:855-860.

Daskalakis K.D., and T.P. O'Connor. 1995. "Distribution of Chemical Concentrations in US Coastal and Estuarine Sediment." *Mar. Environ. Res.* 40:381-398.

Dickhut R.M., and K.E. Gustafson. 1995. "Atmospheric Washout of Polycyclic Aromatic Hydrocarbons in the Southern Chesapeake Bay Region" *Environ. Sci. Technol.* 29:1518-1525.

Fisher J.A., M.J.Scarlett, A.D. Stott. 1997. "Accelerated Solvent Extraction: An Evaluation for Screening of Soils for Selected U.S. EPA Semivolatile Organic Priority Pollutants." *Environ. Sci. Technol.* 31:1120-1127.

Fox J.L. 1994. "Field Testing for Bioremediation Called Logical Next Step" *ASM News.* 60:319-322.

Mudroch, A., and S.D. MacKnight. 1994. "Bottom Sediment Sampling. In A. Mudroch and S.D. MacKnight (Eds.), *Handbook of Techniques for Aquatic Sediments Sampling*. Lewis Publishers, Boca Raton, FL.

National Research Council. 1993. *In Situ Bioremediation: When Does It Work?* National Academy Press, Washington, DC.

Pfaender F.K., and G.W. Bartholomew. 1982. "Measurement of Aquatic Biodegradation Rates by Determing Heterotrophic Uptake of Radiolabeled Pollutants" *Appl. Environ. Microbiol.* 44:159-164.

Sims R.C., and J.L. Sims. 1995. "Chemical Mass Balance Approach to Field Evaluation of Bioremediation" *Environ. Prog.* 14:F2-F3.

QUANTIFYING FACTORS THAT INFLUENCE PAH BIODEGRADATION RATES IN MARINE SEDIMENTS

Barbara Krieger-Brockett, Jody W. Deming,
Yves-Alain Vetter and Allison Geiselbrecht†
Univ. of Washington and †National Marine Fisheries/NOAA, Seattle, WA USA

ABSTRACT: In continuing and new research efforts, we are examining marine environmental constraints and attempting to quantify the factors that limit or accelerate the rate of natural attenuation of polycyclic aromatic hydrocarbons (PAH) in contaminated marine sediments. Most previous research in bioremediation has been directed toward terrestrial sites. We are fortunate to enjoy collaborations with Hart Crowser, a contractor evaluating alternative remediation designs, and the National Marine Fisheries (NOAA). In these collaborations, we examine some of the research questions raised by the ongoing monitoring of a partially-capped, approximately 50-acre submerged Superfund Site in Puget Sound, marine habitat to fish important to the regional economy. This paper presents some conceptual models and experimental results regarding PAH mineralization, using a sensitive radioisotopic method to monitor the evolution of CO_2 from labelled target compounds, in the undisturbed upper ~10 cm of marine sediment. We are addressing target compound mineralization by natural microbial populations. In addressing contaminant sorption and bioavailability, we focus on the potential for microorganisms to influence bioavailability using extracellular enzymes.

INTRODUCTION

Because of the need to dredge marine sediments to maintain navigability, and also the need to "manage" contaminated or dredged sediments, many of the options for site remediation on land are severely constrained in the marine environment. For example trade-offs, both economic and ecological, exist between immediate removal of the contaminated but submerged sediment and the detrimental dispersal of contaminants to the often economically-important near-shore fishery. Thus, determining natural attenuation or *in situ* rates of microbial degradation of organic contaminants in submerged marine sediments and the long-term viability of sediment capping, remain significant challenges to the assessment of site remediation strategies (Wang, et al. 1991; Truitt, 1986). In a previous article (Krieger-Brockett et al. 1997), we illustrated the cap's ability to prevent hydrophobic contaminant breakthrough for variable periods in part depending on thickness, particle size and organic content of the cap material. In this article, we present information obtained in support of assessing remediation design alternatives for contaminated sites in Puget Sound which have included capping a portion of the submerged sites and monitoring the natural attenuation occurring in other locations. The scope and timetable of the sampling were limited, and consisted of 1 box core at each of three locations in the contaminated region with the approximate characteristics in Table 1. The arbitrary location numbers increase as BEP, BaP and HPAH increase in concentration. Several factors such as contaminant concentration and fine sediment content were similar in the locations; others such as total organic content, TOC, were different at all three locations of interest to our collaborators (Thornburg, et al, 1998). The two recalcitrant organic contaminants under investigation were bis(2-ethylhexyl) phthalate (BEP) and benzo(*a*)pyrene (BaP). To our knowledge, no rate data are available for the biodegradation of these compounds in Puget Sound sediments and little information is available from other marine environments. Investigators examining slurried sediments have indicated biodegradation of BEP is much faster than that of BaP (Johnson and Heitkamp, 1984; Staples et al. 1997).

Table 1 -Attributes of biodegradation sampling locations

Box Core Location	Water Depth (ft)	Percent Fines	Percent TOC	BEP (μg/kg)	BaP (μg/kg)	HPAH (μg/kg)	Temperature (°C)	pH
1	30	58	2.3	710	1100	10460	12-13	7
2	2	56	14.0	8700	1900	19220	12-12.5	6
3	17	97	6.9	8800	2000	26240	13-13.5	7

During the remediation design phase, our goal was to estimate rates of natural attenuation at different locations in the site. We applied a "whole core injection" method (Meyer-Reil, 1986) previously used by one of us to assess rates of natural organic matter degradation in benthic sediments (Relexans et al, 1996). It is a sensitive radioactive tracer method based on labeled CO_2 evolution (with unknown stoichiometry) from mineralization of the radiolabeled BaP or BEP. This method can detect as little as 1-5% degradation (Meyer-Reil, 1986) from undisturbed sediment horizons within closed, sampled sediment cylinders after a known period of incubation. Because the method relies on degradation of the target compound provided in dissolved form, resulting rates probably represent an upper bound for *in situ* degradation of the overall pool of dissolved (freshly available) and sorbed (aged) compound. The method may slightly underestimate degradation of the dissolved pool, since the incubating cores will have lower oxygen and nutrient fluxes compared to field sediments subject to bioturbation and fluid motion which increase the flux of materials to and from the bacteria. On balance, the injection core method should provide estimates of biodegradation rates in minimally disturbed sediments. i.e., estimates that are better suited to the goal of assessing natural attenuation in polluted marine sediments than traditional assays that require sediment slurrying which leads to unrealistic stimulation of bacterial activities.

METHODS AND MATERIALS

The contaminated submerged marine sediment samples were taken using a box corer from three locations with approximate attributes shown in Table 1. Replicate subcores were removed from the box corer, without disturbing the sediment, using polycarbonate cylinders with internal plungers and pre-drilled and sealed injection ports at 0.5-cm intervals down the side. Radioactive substrate diluted in a miscible organic solvent acetone was injected into each sediment horizon at trace levels to avoid altering the natural concentration of dissolved compound. An amount of approximately 0.020 μCi was added to 1 cc of core sediment. Subcores were incubated “upright” in the dark at 13.0 C, the *in situ* temperature of the sediments on the day of sampling.

After a period of time during which detectable mineralization of the BEP and BaP was expected to have occurred, the samples were sacrificed (acidified) and the labeled and unlabeled CO_2 were quantitatively trapped in a basic hydroxide solution. Our project scope limited us to duplicate endpoint analyses only, in this case after an incubation period of 37 days for both substrates. Radioactivity of the evolved CO_2 in the hydroxide solution was measured by liquid scintillation counting. The raw counts (counts per minute or cpm) were corrected to disintegrations per minute (dpm) using a quench correction curve and standard calculations were made using the methods described in Gerhardt, et al, 1994.

Killed controls for some of these studies indicated we were indeed measuring microbial degradation rates.

RESULTS AND DISCUSSION

The variation within the same box core (location 1) in apparent degradation rate for both BEP and BaP are shown in Fig. 1 and 2 respectively. Location 1 exhibited the lowest contaminant concentrations and the lowest total organic carbon content (TOC) of the three sites we examined. The apparent biodegradation rates are simply expressed as %CO_2 evolved per day measured as endpoints after 37 day incubations. We cautiously interpret these apparent CO_2 evolution "rates" since they were obtained using only endpoint analyses after a fixed incubation time. That is, we could have "missed" a more rapid rise to this level by our particular incubation time, chosen *apriori* owing to the lack of guidance from prior studies in undisturbed marine sediments. Both the average of duplicates (bold trendline and filled symbols) and the separate samples are presented to illustrate the variability within this box core. Rather than show error bars at each depth, the error bar was constructed using the pooled standard deviation as an estimate (Box, et al, 1978) to the uncertainty from all depths for a given box core location. The error bar is placed arbitrarily for clarity here at the 5 cm horizon, but applies to all averages (bold trendline) from the particular boxcore. Note from the figure ordinate that at all depths, the BEP biodegradation rate is faster; the depth-averaged apparent BEP biodegradation rate (about 0.006 %/day) appears to be more than one order of magnitude faster than that of BaP (about 0.0004 %/day). When examining the individual injection core rates (plotted with dashed lines), there appears to be a rate variation with depth in the sediment but given the variability between duplicates in this box core, this cannot be ascertained from this small number of determinations.

The apparent biodegradation rate variation between duplicates within the boxcore taken at location 2 is considerably less than that in location 1 as shown in Figs. 3 and 4 for BEP and BaP respectively. Location 2 is a high organic matter (14% TOC) site, with very shallow water. The BEP and BaP concentrations were near the highest sampled, but the total high molecular weight polyaromatic hydrocarbon (HPAH) were mid-range of those sampled. The pooled estimate to the variation (error bar) in Fig. 3 was calculated omitting the single very high (0.08 %/day) apparent BEP degradation rate at 0.5 cm from the sediment-water interface. Judging from all location 2 determinations, however, the apparent BEP degradation rate appears to decline with depth into the sediment. This decline in biodegradation with depth appears to be greater than the variability in the samples, and thus is "real". In contrast, the apparent BaP biodegradation rate within boxcore 2, which also shows good agreement between the duplicates, appears to be very low and comparable at all depths in the sediment. However, we may be near the detection limits for the method since some of the BaP rates were measured as zero. Again, by noting the 25 fold different ordinates of the figures, it can be seen that the BEP biodegradation rates are 1-2 orders of magnitude faster than those for BaP. For a boxcore taken from location 3, a highly contaminated but intermediate total organic matter site with a high fraction of fine particles, the apparent degradation rates (data not shown) were lowest of the three locations for both BEP and BaP, with BEP still exhibiting a 100-fold greater depth-averaged rate than BaP.

Some tentative conclusions emerged from this limited exploration of factors affecting biodegradation rates at each sampled site. First, apparent biodegradation rates at all depths were highest for the lowest contaminant concentrations of BEP, BaP, and HPAH shown in Table 1. The lowest apparent biodegradation rates occurred at sites having the highest target pollutant concentrations, as well as the finest particle sizes, which have been found to affect sorption (Karickhoff and Brown, 1978). This pattern is consistent with the conceptual model of natural

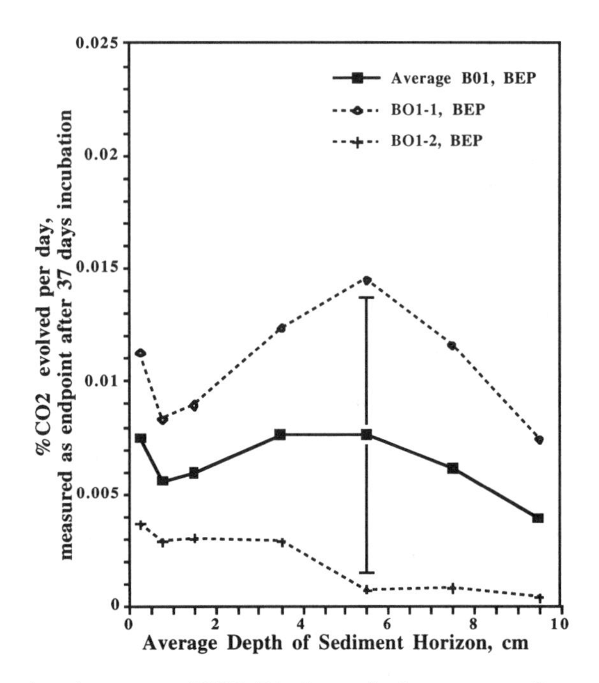

Figure 1 - Apparent BEP Biodegradation rates - Location 1

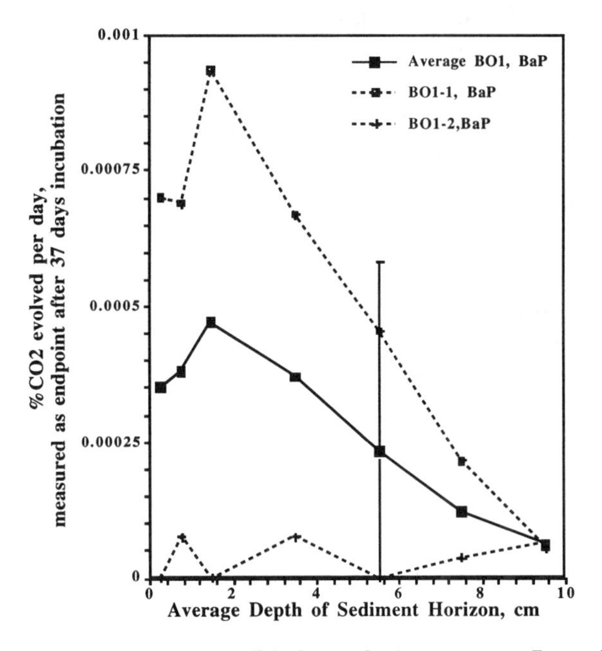

Figure 2 - Apparent BaP Biodegradation rates - Location 1

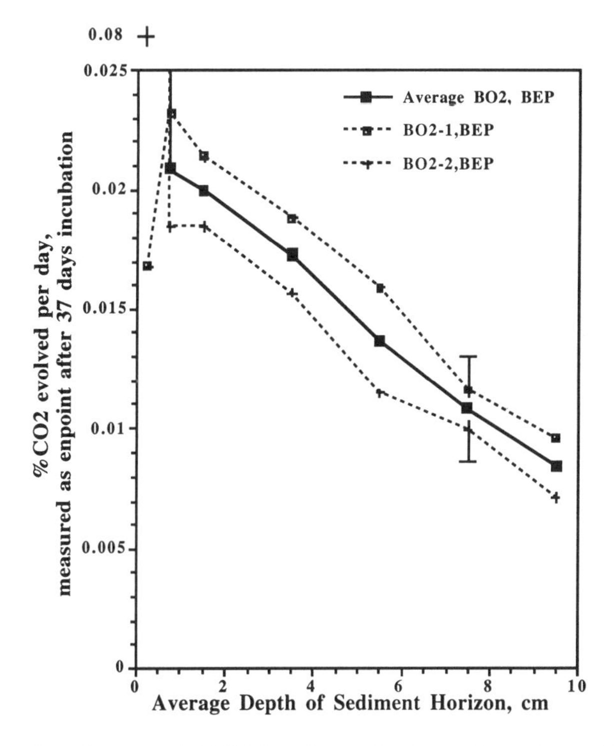

Figure 3 - Apparent BEP Biodegradation rates - Location 2

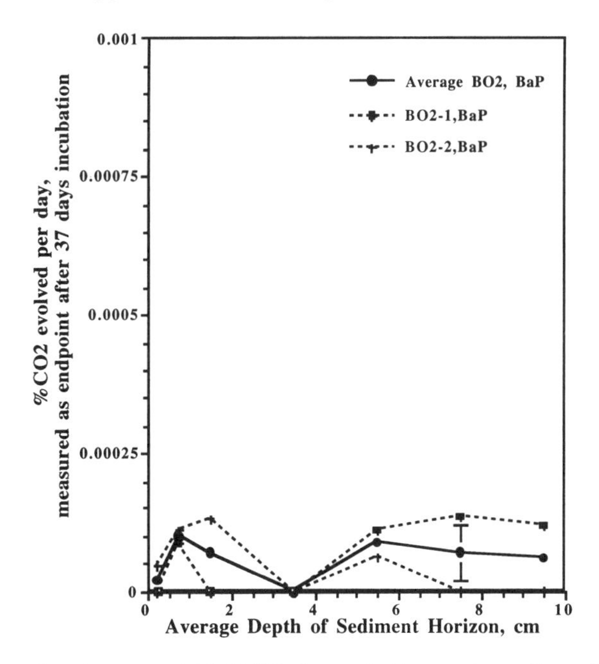

Figure 4 - Apparent BaP Biodegradation rates - Location 2

attenuation occurring maximally at the "edges" of a contaminated site, effectively reducing the overall size of the contaminant "plume" without necessarily reducing a site-specific concentration. This concept, however, remains to be demonstrated in the field with statistical rigor. In our limited study, the observed variability between samples from the same boxcore at a relatively "clean" site (location 1) is consistent with the manifold sources of heterogeneity expected in actual undisturbed, natural marine sediments (for example, Aller, 1980; Vetter et al., 1998). This degree of natural variability was emphasized by Venosa et al. (1996) regarding heterogeneity of biodegradation rates of treated Delaware beaches where 5-9 or more replicates were required to achieve sufficient statistical power to see a difference in treatments. The more limited within-boxcore variability observed at the more contaminated sites of our study (locations 2 and 3) bode well for being able to develop useful degradation rate estimates in submerged sediments without extensive sampling.

The scope of our studies of BEP and BaP degradation did not allow for specific assessments of the size of the responsible bacterial population. However, we were able to approach such an assessment in a somewhat broader study at Eagle Harbor (Puget Sound), where more extensive sediment characterization data, PAH concentrations, and ongoing monitoring provided us with additional quantitative information. The PAH phenanthrene was one of the principal target contaminants driving the remediation design studies at this site (USEPA,1995). We have several lines of evidence that suggest phenanthrene biodegradation rates, measured in the same way, in Eagle Harbor sediments are about the same to approximately 10 times faster than those measured for BEP in the previous figures. Eagle Harbor, a Superfund Site, was partially capped in 1993-4 during our research studies which altered our ability to return to the same sampling sites in some cases. Apparent phenanthrene biodegradation rates were measured for shorter incubation times, 12-14 days, and were accompanied in some cases by bacteria counts using acridine orange stain and epifluorescence microscopy which allows direct counting of intact cells without distinguishing PAH degraders. One of us (Geiselbrecht, 1998) measured actual PAH degraders using most probable number (MPN) dilution methods to estimate culturable PAH-degrading bacteria. Direct counts were found to range from 3.2 to 7.0 x10^8 bacteria/cc wet sediment, but showed no obvious difference in numbers between two harbors: grossly contaminated Eagle Harbor, and nearby Blakely Harbor, selected as a control site for its low background levels of PAHs. In contrast to direct counts, phenanthrene-degrading bacteria were enriched in Eagle Harbor sediments by almost two orders of magnitude over those enumerated in Blakely Harbor (1.2 to 4.5 x10^7 bacteria/cc wet sediment compared to 1.2 to 4.5 x10^5 bacteria/cc wet sediment). These results suggest that, in addition to physical and chemical characterization of target sediments, the size of the bacterial population responsible for degrading the compound of interest is an important parameter to quantify when assessing natural attenuation. We have shown quantitatively (Vetter et al, 1998) how extracellular enzymes may influence the availability of a given compound for bacterial degradation in marine sediments. Limitations on such enzyme activity may provide additional factors accounting for the effect of soil aging on declining pollutant (phenanthrene) bioavailability and degradation rates as measured by Hatzinger and Alexander (1995), while significant or stimulated levels of extracellular enzyme activity may contribute to detectable contaminant degradation rates even in "aged" marine sediments (this study).

ACKNOWLEDGMENTS

The authors would like to acknowledge the financial support of the Office of Naval Research - University Research Initiatives, NOAA/Seagrant, and Hart Crowser, Inc., the collaboration with T. Thornburg, and the able laboratory and field assistance from Shelly Carpenter and Andy Walker.

REFERENCES

Aller, R.C., 1980. "Quantifying Solute Distributions in the Bioturbated Zone of Marine Sediments by Defining an Average Microenvironment", *Geochimica et Cosmochimica Acta, 44*, 1955-1965.

Box, G.E.P., W.G. Hunter, J.S. Hunter, 1978. *Statistics for Experimenters*, Wiley, p 76.

Geiselbrecht, A, 1998. Ph.D. Thesis, Univ. of Washington, Dept of Microbiology.

Gerhardt, P., R.G.E. Murray, W.A. Wood, and N.R. Krieg, 1994. *Methods for General and Molecular Bacteriology,* Amer. Soc. for Microbiology, p 490-498

Hatzinger, P.B., and M. Alexander 1995. "Effect of Aging of Chemicals in Soil on their Biodegradability and Extractability", *Env. Sci. Technol.* 29 537-545.

Johnson, B.T., Heitkamp, M.A., Jones, J.R., 1984. "Environmental and Chemical Factors Influencing the Biodegradation of Phthalic Acid Esters in Freshwater Sediments", *Environ. Pollut.* Ser. B. 8, 101-118.

Krieger-Brockett, B., J.W. Deming, and R.P. Herwig 1997. *In situ and On Site Bioremediation*, (4) p 427-433.

Karickhoff, S.W. and D.S. Brown 1978. "Paraquat sorption as a function of particle size in natural sediments" *J. Environ. Qual* 7, 246.

Relexans, J.C., J.W. Deming, A. Dinet, J.F. Gaillard, and M. Sibuet, 1996. "Sedimentary organic matter and micro-meiobenthos with relations to trophic conditions in the northeast tropical Atlantic", Deep-Sea Res. I 43 (8): 1343-1368.

Staples, CA Peterson, DR Parkerton, T.F. Adams-W-J. 1997. "The environmental fate of phthalate esters: A literature review", *Chemosphere* (35)4, 667-749.

Thornburg, T, B Krieger-Brockett, J Deming, R Herwig, 1998. in preparation.

Truitt, C.L., 1986. The Duwamish Waterway Capping Demonstration Project: Engineering Analysis and Results of Physical Monitoring; Technical Report D-86-2; U.S. Army Engineer Waterways Experiment Station: Vicksburg, MS.

US Environmental Protection Agency and US Army Core of Engineers, 1995. *Operations, Maintenance and Monitoring Plan for the Wyckoff-Eagle Harbor Superfund Site*, Draft proposal, April, 1995.

Venosa, A.D., M.T. Suidan, B.A. Wrenn, K.L. Strohmeier, J.R. Haines, B.L. Eberhart, D. King, and E. Holder, 1996. Bioremediation of an experimental oil spill on the shoreline of Delaware Bay. *Environ. Sci., Technol.* 30:1764-1775.

Vetter, Y.A, J.W. Deming, P.A. Jumars, B. Krieger-Brockett, 1998. "A Predictive Model of Bacterial Foraging by Means of Freely-Released Extracellular Enzymes",*Microbial Ecology, 36*, 75-92.

Wang, X.Q., L.J. Thibodeaux, K.T. Valsaraj, and D.D. Reible 1991. "Efficiency of Capping Contaminated Sediments in Situ. 1. Laboratory Scale Experiments on Diffusion-Adsorption in the Capping Layer", *Env. Sci. Technol. 25* 1578-1584.

COMPARISON OF BENCH-SCALE AND FIELD TREATMENTS FOR SEDIMENT BIOREMEDIATION

T.P. Murphy (Environment Canada, Burlington, Ontario, Canada)
A. Lawson (Environment Canada, Burlington, Ontario, Canada)
J. Babin (Golder Associates, Kowloon Tong, Hong Kong)
M. Kumagai (Lake Biwa Research Institute, Shiga Prefecture, Japan).

ABSTRACT: Sediment samples from several contaminated sites were evaluated for the potential for polycyclic aromatic hydrocarbon (PAH) biodegradation, sulphide oxidation, and phosphorus inactivation. Sediments with a high total petroleum hydrocarbon (TPH) content had rates of PAH biodegradation as high as 1.75% a day. Sites with low TPH concentration had low to negative rates of biodegradation. In two samples, the short term negative rates of biodegradation were related to production of two and three ringed PAHs, presumably from very large PAHs that could not be measured. Only one site appeared to have metal suppression of bio-oxidation/biodegradation. Nutrient inactivation could be achieved but conditions were then not ideal for bioremediation. Laboratory results from glovebox enclosures agreed well with results obtained in pilot-scale treatments. In pilot-scale treatments in Hamilton Harbour for PAHs, Hong Kong Harbour for odour, and Lake Biwa for phosphorus, about half of the PAHs were biodegraded, 99% of sulphides were oxidized, and 80% of phosphorus precipitated, respectively.

INTRODUCTION

In spite of the common demonstration of PAH biodegradation, PAHs can be resistant to biotreatment. There are several reasons for PAHs not to biodegrade but sometimes they can be overcome. Some sediments are relatively easy to bioremediate and some sediments cannot be bioremediated. The difficulty for managers of in situ treatment is to optimize treatment without increasing project costs excessively. Bench-scale and pilot-scale testing are essential to ensure that treatments are optimized and the expectations for full-scale treatment are realistic.

Reasons for poor biodegradation include;

- lack of oxidant
- lack of nutrients including phosphate precipitation
- bacterial community not adapted to contaminants
- poor biodegradability caused by a lack of natural surfactants, bioemusifiers, or natural "solvents"
- toxicity caused by other toxins such as metals or other organic chemicals
- physical limitations such as poor mixing of added materials or cold temperature
- ongoing discharges of contaminants

Objective. In this paper we will review the relative biotreatability of PAHs in sediments in bench-scale treatments conducted over four years from several sites. Biotreatments had first been optimized with sediments from two sites in southern Ontario that are rich in

PAHs and TPHs. Pilot-scale results will be compared to bench-scale testing. Sulphide and phosphate data are used to evaluate the effectiveness of injection of oxidants in situ.

Site Descriptions. Hamilton Harbour is an industrial harbour with Canada's largest steel mills (Murphy et al. 1995). Marine Site B is an urban marine lagoon in southern Europe with rich organic sediments and serious odour problems. Lake Ontario Site A is in the outer harbour of Toronto; it is moderately contaminated, mostly with TPHs. Industrial Site B is an American aluminium manufacturer on a freshwater lake. Lake Ontario Site B is in the inner harbour of Toronto near a storm sewer discharge. Industrial Site C is a waste oil pond of an European car maker. Marine Site A is an aluminium manufacturer on an estuary in Canada. The Hong Kong Site is adjacent to the former airport; it is rich in sewage wastes. Lake Biwa in Shiga prefecture, Japan is eutrophic from municipal and rural waste discharges. Asian Site B is a brackish eutrophic Japanese lake.

MATERIALS AND METHODS

Bench-scale treatments. Amber bottles (1 L) were filled with approximately 600 ml sediment and 100 ml of water plus proprietary amendments. The bottles were purged with helium and the bottles were stored in a sealed glovebox under helium atmosphere and shaken by inversion twice per week. In all treatments, three microcosms were left as controls and calcium nitrate at a concentration of 1 g NO_3-N/L was added to three microcosms. Additional nitrate was added to treated microcosms to bring concentrations back to original values after 10 and 17 days incubation.

Pilot-scale treatments. Two injection procedures were used. The injections in Lake Biwa and Asian Site B used a device 1m by 1m square with 100 17-cm rods with holes to inject at a depth of 15 cm. Pilot-scale sediment injections in Hamilton Harbour and Hong Kong were done with a 8 m wide boom (Babin et al. 1998).

At least three sediment cores were collected from each treatment and control. The sediment was extruded then analyzed for AVS concentrations by ion selective electrodes using a diffusion method (Brouwer and Murphy, 1994). Diffusion chambers were added to each treatment and control in triplicate two weeks after the injection and were left to incubate in the sediment for two weeks. Pore water samples were immediately acidified to a pH of 2 with concentrated nitric acid then analysed for iron and manganese by atomic absorption spectroscopy. The total reactive phosphorus in the acidified pore water samples was measured by a an ascorbic acid method.

Gas Chromatographic/Mass Spectrographic (GC/MS) Procedure. The sediment sample (10 g) was spiked with a known amount of a surrogate mixture of deuterated PAHs, then extracted in a Soxhlet apparatus with an acetone-hexane (59:41) solvent mixture. The organic extract was base-partitioned with 2% potassium bicarbonate solution to separate the acidic compounds from the PAHs and other neutral compounds. The aqueous medium was back-extracted with 50 ml of hexane. The organic fractions were combined, dried through sodium sulphate and concentrated to ca. 3-5 ml. The resulting solutions were analyzed for 16 selected PAHs by GC/MS under the following conditions: GC: Hewlett-Packard model 5890, Split splitless injection; 30 m fused silica capillary column, DB-5; Injection temperature 300°C

Program: 30°C held for 1 min, 30°C to 285°C at 6°C/min, hold 16.5 min. MS; Hewlett-Packard series 5970 mass spectrometer; Source Temperature 200°C; Electron ionization 70 eV; Select ion monitoring (SIM) mode. The PAH results were done in triplicate in a certified laboratory.

RESULTS AND DISCUSSION

PAH Biodegradation. The bench-scale biodegradation of total PAHs was highly variable but significantly correlated to the TPH concentrations (r^2=0.79, n=8). Two sample sites with low TPH appeared to produce PAHs. At least with Marine Site A, this may represent the production of 2- and 3-ringed PAHs from larger unmeasurable PAHs. In three month incubations in samples from Industrial Site B, also an aluminium production site, 2- and 3- ringed PAHs were increasing significantly but they biodegraded within six months. However, the bioproduction of metabolites from Industrial Site C is less likely; this site had 285 mg/L of Zn in the sediment porewater, had only 24% bioremediation of TPHs and unlike samples from all other treated sites, it remained toxic to Microtox™. Changes in sediment matrix during the treatments may have enhanced PAH extractability by up to 10%. Another unusual observation is that in all 6 month biotreatments a similar proportion of the larger PAHS had also biodegraded (Murphy et al. 1995). These results are consistent with observations of Peters and Fan (1997) that there are only small intrinsic differences in PAH biodegradation; the rate limiting step is mainly physical differences in PAH solubility and bioavailability. Similarly Fogel and Findlay (1997) found oil to enhance biodegradation of benzo(a)pyrene. In our sites, the bioavailable PAHs appeared to be associated with TPHs.

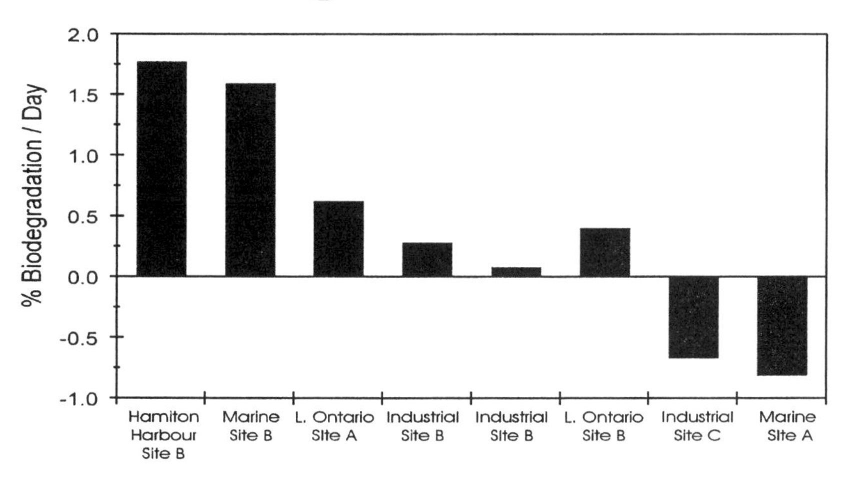

The samples with the highest oil content (Hamilton Harbour) of 2% had much higher rates of PAH biodegradation (1.75%/d). These bench-scale biodegradation rates in Hamilton Harbour were similar but faster than observed in pilot-scale treatments in Hamilton Harbour (Murphy et al. 1995). It has been argued that in situ treatments would be slower due to inadequate mixing (Renholds 1998). This physical aspect could be important but in Hamilton Harbour, the biggest problem was the ongoing discharge of oil sprayed onto coal piles to prevent dust problems (Curran et al. 1999). Ship resuspension of neighbouring untreated sediments were also a problem (Irvine et al. 1997).

The other uncertainties with the Hamilton Harbour project were the end-points of treatment. The bioavailability of the residuals was not assessed but much of the refractory materials was likely coal dust. Although PAHS refractory to biotreatment may still be a serious problem, research has demonstrated that weathered PAHS are less of a risk to the environment (Knaebel et al. 1996, Paine et al. 1996, White and Alexander 1996). The site where Paine et al. (1996) found limited bioavailability of PAHs is the same as our Marine Site A where we found no biodegradation of PAHs. Ideally the end-points should be based upon bioassays, and risk management using ongoing land management to optimize and prioritize remedial efforts. The steel mill study site in Hamilton, is improving its management of coal piles to reduce runoff; however, storm sewers will remain a concern (Irvine et al. 1998).

Sulphide Oxidation. Pilot-scale treatments in Lake Biwa and Asian Site B, Japan oxidized 95-99% of AVS. Pilot-scale treatment of sediments in Hong Kong oxidized 99% of the AVS in the surface sediments (Figure 2, Babin et al. 1998). The bench-scale results for these treatments were similar to the pilot-scale results. Nitrate injections were not compromised by poor mixing. Bioremediation in the bench-scale was minimal (i.e. 15% of PAHs, <15% of TPHs, no change in loss on ignition). The PAH concentrations were very low (3.1 μg/g) but the TPH values were high (~0.5%). The treatment was not optimized for bioremediation of organic contaminants. In this study, oxidation of sulphides and prevention of methane formation were the two main goals. The Hong Kong treatment is scheduled to proceed to full-scale treatment.

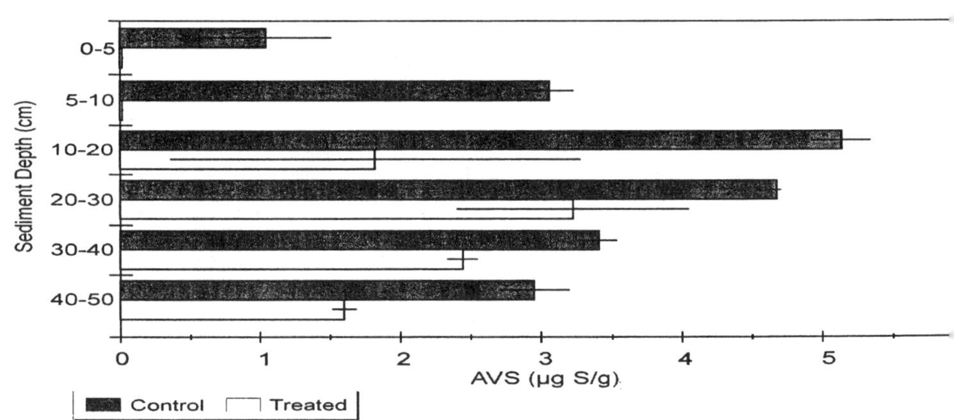

Phosphate Precipitation. Pilot-scale oxidation of sediments of Lake Biwa with calcium nitrate precipitated about 80% of porewater phosphate (Figure 3). This result was less efficient than observed in bench-scale treatments where 90-99% of phosphate precipitated. Since sulphides were oxidized, the weaker response does not likely reflect poor mixing. However, less phosphorus precipitation probably reflects less oxidation. Diffusion of reduced materials from deeper sediments may be a factor resulting in greater sediment oxygen demand than would be measured in bench-scale evaluations. Seasonal changes in AVS measured in these sites indicates that the top 30 cm of sediments oxidizes in winter and goes anoxic in summer. This change must reflect diffusion throughout the sediments. Thus the idea of producing a biotreated in situ "cap" of a few cm is not realistic without ongoing maintenance treatment of the "cap".

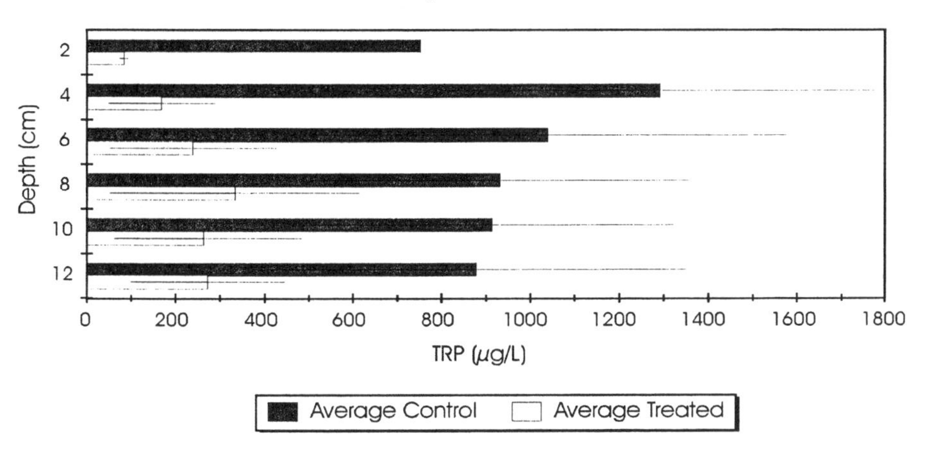

CONCLUSIONS

Bench-scale testing provides an inexpensive method of assessing treatability of PAHs. In some situations, especially when the oil content is high, bioremediation can easily satisfy regulatory guidelines. Pilot-scale testing is very useful to provide an additional check that the site is ready for full-scale treatment. Biodegradation treatments using just calcium nitrate may induce phosphorus deficiency and efforts should be made to provide a form of phosphorus resistant to this oxidant induced precipitation.

ACKNOWLEDGEMENTS

Julie Corsini assisted with many of the sediment measurements. Shinshu Inc. assisted with the sediment treatments at Lake Biwa. Sediment treatments at Asian Site A were done by MIKUNIYA Inc. Water Technology International Corporation did the PAH analysis.

REFERENCES

Babin, J., T. P. Murphy, and J. T. Lynn. 1998. "*In Situ* Sediment Treatment in Kai Tak

Nullah to Control Odours and Methane Production." *Proceedings of the Seventh International Symposium on River Sedimentation, Hong Kong.*

Brouwer, H. and T. P. Murphy. 1994. "Diffusion Method for the Determination of Acid Volatile Sulphide in Sediment." *Envir. Toxicol. Chem. 13*:1273-1275.

Curran, K. J., I. G. Droppo, K. N. Irvine, and T. P. Murphy. 1998. "Suspended Solid, Trace Metal, and PAH Loading from Coal Pile Runoff to Hamilton Harbour." *NWRI contribution No. 98-002, submitted J. Great Lakes Res.*

Fogel, S. and M. Findlay. 1997. "Benzo(a)pyrene Added to Crude Oil:Biodegradation by Natural Microbiota." *In Situ and On-Site Bioremediation, 2*:191. New Orleans Symposium, Battelle Press.

Irvine, K. N., I. G. Droppo, T. P. Murphy, and A. Lawson. 1997. "Sediment Resuspension and Dissolved Oxygen Levels Associated with Ship Traffic: Implications for Habitat Remediation." *Water Poll. Res. J. Canada, 32*(2)421-437.

Irvine, K. N., I. G. Droppo, T. P. Murphy and D. M. Stirrup. 1998. "Annual Loading Estimates of Selected Metals and PAHs in CSOs, Hamilton, Ontario, using a Continuous PCSWMM Approach." In B. James (Ed.), *Advances in Modeling the Management of Stormwater Impacts,* Vol. 6, Ch. 20 , Ann Arbor Science, Ann Arbour, MI.

Knaebel, D. B., T. W. Federle, D. C. McAvoy, and J. R. Vestal. 1996. "Microbial Mineralization of Organic Compounds in an Acidic Agricultural Soil: Effects of Preadsorption of Various Soil Constituents." *Environ. Toxicol Chem. 15*(11)1865-1875.

Murphy, T. P., A. Moller and H. Brouwer, 1995. "In Situ Treatment of Hamilton Harbour Sediment." *J. Aquat. Ecosystem Health 4*: 195-203.

Paine, M. D., P. M. Chapman, P. J. Allard, M. H. Murdoch, and D. Minifie. 1996. "Limited Bioavailability of Sediment PAH Near an Aluminum Smelter: Contamination does not Equal Effects." *Environ. Toxicol. Chem. 15*(11):2003-2018.

Peters, C. A. and J. Fan. 1997. "Multisubstrate Biodegradation and Bioavailability of PAHs." *In Situ and On-Site Bioremediation, 2*:193. New Orleans Symposium, Battelle Press.

Renholds, J. 1998. *In Situ Treatment of Contaminated Sediments.* Http://clu-in.org

White, J. C. and M. Alexander. 1996. "Reduced Biodegradability of Desorption-Resistant Fractions of Polycyclic Aromatic Hydrocarbons in Soil and Aquifer Solids." *Environ. Toxicol. Chem.15*:1973-1978.

MICROBIAL DEGRADATION OF INTERTIDAL CREOSOTE – WYCKOFF/EAGLE HARBOR SUPERFUND SITE

Travis C. Shaw (U.S. Army Corps of Engineers, Seattle, WA)
Russell P. Herwig (University of Washington, Seattle, WA)
John S. Wakeman (U.S. Army Corps of Engineers, Seattle, WA)

ABSTRACT: This study focused on the microbial component of natural recovery at a former wood treatment facility located on Puget Sound in Washington State. Sediment PAH concentrations used in the study ranged between 182-170,000 μg/kg dry weight for low molecular weight PAHs and between 686-24,347 μg/kg dry weight for high molecular weight PAHs. In-situ measures of Eh, NO_3^{-1} and SO_4^{-1} indicated that electron acceptors were not limiting microbial degradation. Mineralization studies by the University of Washington demonstrated the presence of an active PAH-degrading consortium in intertidal sediments. Under aerobic conditions, 40-60% of radio-labeled phenanthrene and pyrene were mineralized to $^{14}CO_2$ within 6 days. Samples incubated under anaerobic conditions did not significantly degrade over the course of the 35 day experiment.

INTRODUCTION

Decades of wood treatment operations at the Wyckoff facility have resulted in widespread soil, groundwater and sediment contamination in and around Eagle Harbor in Washington State. In 1984, the National Oceanic and Atmospheric Administration notified EPA and the Washington State Department of Ecology that samples of sediment, fish and shellfish from Eagle Harbor contained elevated levels of polycyclic aromatic hydrocarbons (PAHs) (Malins, 1984). Since being listed as a Superfund Site in 1987, cleanup activities have proceeded on the upland and subtidal portions of the site. Despite these efforts, intertidal seeps continue to release free oily product and contaminated groundwater into the waters of Eagle Harbor and Puget Sound.

The Record of Decision (ROD) for the East Harbor Operable Unit selected natural recovery as the preferred remedy for intertidal sediments contaminated with PAHs. Cleanup goals for intertidal sediments are to be achieved 10 years after source control has been established on the upland portion of the site in order to allow natural processes adequate time to remediate contaminated intertidal sediments.

Objective. The experimental focus of this study was to estimate the potential contribution of microbial degradation toward achievement of intertidal sediment clean-up goals. Three principle questions were addressed with regard to microbial degradation: (1) Do microbial populations exist in the intertidal sediments adjacent to the Wyckoff facility that are capable of degrading PAHs? (2) At what rate do native microbial populations degrade representative PAHs? (3) Do physical or chemical characteristics of intertidal sediments adjacent to the Wyckoff facility appear to limit microbial degradation of PAHs?

This study was the first step to develop a remedial decision framework for the intertidal area adjacent to the Wyckoff facility. Areas of the intertidal zone that do not recover naturally within a 10 year period may require more active forms of remediation. Future options for remedial action range from enhancement of microbial mediated contaminant degradation to excavation.

Site Description. The Wyckoff / Eagle Harbor Superfund Site is situated on the eastern side of Bainbridge Island in central Puget Sound, approximately 11 kilometers (km) west of Seattle. The bay is approximately 2 km^2 in area, and approximately 3.7 km long; it is widest (0.9 km) at its entrance, becoming progressively narrower to the west. The former Wyckoff facility covers approximately 40 upland acres at the southern entrance of Eagle Harbor. While the entire harbor has been listed as a Superfund Site, only intertidal sediments immediately adjacent to the Wyckoff facility were the focus of this study.

MATERIALS and METHODS

Sediment samples representing a range of contaminant concentrations were collected from the intertidal zone adjacent to the Wyckoff facility. Samples were collected for bulk chemistry analysis, interstitial water analysis and for both aerobic and anaerobic mineralization studies.

Sediment Analytical Procedures. Sediment samples were analyzed for TOC, grain size, Total Volatile Solids, Total Solids, Extractable Petroleum Hydrocarbons and PAHs according to EPA SW-846 methods. Bulk sediment samples were removed from the intertidal area using a stainless steel spoon to a depth of 10 cm, homogenized on site, then shipped to the project laboratory.

Interstitial Water Sampling. Sediment samples for the recovery and analysis of interstitial water were collected concurrently with bulk sediment. In order to preserve the oxidative state of the interstitial water, the samples were collected using dedicated stainless steel sleeves that fit into the center of a soil core sampler. The procedure allowed for the collection of small sediment cores and minimized contact between the sample and atmospheric oxygen.

Sediment Microbiological Sampling. Samples for use in mineralization study were collected using procedures similar to those used to collect sediment cores for interstitial water recovery. Dedicated pre-washed and solvent rinsed stainless steel sleeves inserted into the core sampler were pushed into the sediment to a depth of 10 cm then removed. The stainless steel sleeve was removed from the core sampler and sealed with polyethylene caps or aluminum foil.

Mineralization Study Protocols. Two radioactive substrates were selected to act as representative PAHs in this study, ^{14}C-labeled phenanthrene was used to represent low molecular weight PAHs and ^{14}C –labeled pyrene was used as a representative high molecular weight PAH. Both radiolabeled substrates were added to sediment slurries mixed from each of the ten sampling stations.

Aerobic Mineralization. Slurries used in aerobic incubations were formed by first homogenizing the core samples by hand then mixing 60-100 g of sediment with filter sterilized Puget Sound seawater to form a 1:5 (mass/volume) mixture. A 5 ml aliquot of slurry was added to a 40 ml screw cap vial along with the radiolabeled substrate solution. A 13 x 75 mm disposable glass test tube containing 1 ml of 5N NaOH was placed inside the vial, serving to trap the CO_2 that was released. Three replicate vials were prepared for each sediment sample for each incubation period. Anaerobic and aerobic vials were incubated in the dark in a refrigerated incubator set at 18°C. Aerobic samples were incubated on a shaker and gently agitated at 80 rpm.

Anaerobic Mineralization. Vials used in anaerobic incubations were prepared in a similar manner except they contained a higher ratio of sediment to seawater (3 ml sediment/ 2 ml seawater) and were not vigorously stirred. Anaerobic seawater was used to prepare the slurries. All preparation of the vials for the anaerobic study was performed inside an anaerobic glove box. Anaerobic samples were incubated at 18°C and not agitated.

Measurement of Radioactivity. Periodically, replicate sets of aerobic and anaerobic vials were removed from the incubator. To end the incubation, sediment slurries were acidified with 4 N sulfuric acid. The NaOH solution used to capture the ^{14}C-CO_2 was added to scintillation fluid. Radioactivity was measure using a Tri-Carb 300 Scintillation Counter. The fraction of the labeled PAH substrate that was mineralized to CO_2 could then be calculated.

Sediment Recovery Rate Estimates. The principles of equilibrium partitioning were combined with the site specific information on the rate of microbial degradation to estimate the time required for microbial mediated remediation of Eagle Harbor intertidal sediments.

Calculation of PAH concentrations in each of the three compartments in the phase partitioning model were achieved using the concentration of phenanthrene and pyrene associated with sediment POC and pore water concentrations at time zero (Di Toro et al., 1991). Pore water concentrations are the sum of the dissolved phase and the concentration of PAHs associated with DOC.

A conservative, modified rate of microbial degradation of phenanthrene and pyrene was determined using Michaelis - Menten kinetics for the quantity of contaminant converted by a cell per unit time. The Lineweaver-Burk solution (Morris, 1966) was used to determine mean rates of maximum initial velocity of the conversion reaction (K_{max}) and the mean Michaelis constant (K_s) which estimates the contaminant concentration yielding ½ K_{max} for both phenanthrene and pyrene along each transect. The maximum mineralization rates measured in this study represented the initial velocity (v_o) of the conversion reaction used to calculate these mean degradation rates. The mean degradation rates were then halved to compensate for the higher than ambient incubation temperature.

The change in average sorbed phase concentration of a contaminant with time is related to the change in aqueous phase concentration by microbial degradation and a partitioning coefficient. The equations used to derive the estimated changes in aqueous and sorbed phases concentration changes were modified from Mihelcic and Luthy (1991).

RESULTS AND DISCUSSION

Bulk Sediment. PAH concentrations varied widely between stations. For example, one of the samples from the north transect sediments were non detect (<19 µg/kg, dry wt.) for most LPAHs and contained only slightly elevated concentrations of HPAHs, notably pyrene and chrysene. Conversely, samples collected on the far end of the same transect line exceeded state cleanup goals for all 17 regulated PAHs except acenaphthylene, indeno(1,2,3-cd)pyrene, dibenzo(a,h)anthracene and benzo(g,h,i)perylene. The concentration of total HPAHs at this stations was 87,010 µg/kg (dry wt). Stations on the eastern transect demonstrated a similar pattern. Samples collected from areas with free product on the surface of the sediment exceeded the state cleanup goals for all 17 regulated PAHs. Stations located in areas that appeared to be free of oily product generally had minor exceedance for only one or two compounds.

Interstitial Water Analytical Results. Ammonia concentrations ranged between 2.2 mg/l and 0.67 mg/l. Nitrate was measured above the reporting limit at only one station (0.86 mg/l). Sulfate was detected in samples from all tested stations in uniform concentrations (2,000-2,100 mg/l) and sulfide levels ranged between 0.025-0.10 mg/l. These composite pore water samples are consistent with the Eh measurements that were consistently below –200 mV, ranging to slightly less than –400 mV. Redox measurements in this range are indicative of sediments depleted of oxygen, nitrate and nitrite.

PAH Mineralization Study Results. Maximum rates of mineralization were estimated by calculating the slope of the line that results when the percent mineralization is plotted over incubation time during the period of greatest mineralization activity. Maximum rates of phenanthrene mineralization generally occurred during the first two days of incubation and ranged between 860-56,000 μg/kg/day. Maximum rates of pyrene mineralization generally occurred over the first three days of incubations and ranged between 40-18,000 μg/kg/day.

Estimated Recovery Time. Bacterial communities in Eagle Harbor intertidal sediments are capable of degrading PAHs and may be the predominant natural recovery process. However, the low quantity of PAHs partitioning into assimilable forms appears to be a rate limiting step for microbial degradation of PAHs. Microbial degradation alone will likely not achieve the cleanup goals described in the ROD for the East Harbor across the full range of PAH concentrations measured in all intertidal sediments. Given that the estimated relative contribution of microbial breakdown is the most influential remediation mechanism, it is unlikely that natural recovery will occur throughout the intertidal area within the ten year period designated by the ROD.

REFERENCES

Di Toro, D.M., C.S. Zarba, D.J. Hansen, W.J. Berry, R.C. Swartz, C.E. Cowan, S.P. Pavlou, H.E. Allen, N.A. Thomas and P.R. Paquin. 1991. "Technical Basis for Establishing Sediment Quality Criteria for Nonionic Organic Chemicals Using Equillibrium Partitioning." *Environmental Toxicology and Chemistry*. 10:1541-1583.

Malins, D.C. 1984. *Summary Report on Chemical and Biological Data from Eagle Harbor*. National Marine Fisheries Service, Seattle, Washington.

Mihelcic, J.R and R.G. Luthy. 1991. "Sorption and Microbial Degradation of Naphthalene in Soil-Water Suspensions under Denitrification Conditions." *Environ. Sci. Technol*. 25:169-177.

Morris, J.G. 1966. *A Biologist's Physical Chemistry*. Addison -Wesley Publishing Company, Reading, MA.

AUTHOR INDEX

This index contains names, affiliations, and volume/page citations for all authors who contributed to the eight-volume proceedings of the Fifth International In Situ and On-Site Bioremediation Symposium (San Diego, California, April 19–22, 1999). Ordering information is provided on the back cover of this book.

The citations reference the eight volumes as follows:

5(1): Alleman, B.C., and A. Leeson (Eds.), *Natural Attenuation of Chlorinated Solvents, Petroleum Hydrocarbons, and Other Organic Compounds*. Battelle Press, Columbus, OH, 1999. 402 pp.
5(2): Leeson, A., and B.C. Alleman (Eds.), *Engineered Approaches for In Situ Bioremediation of Chlorinated Solvent Contamination*. Battelle Press, Columbus, OH, 1999. 336 pp.
5(3): Alleman, B.C., and A. Leeson (Eds.), *In Situ Bioremediation of Petroleum Hydrocarbon and Other Organic Compounds*. Battelle Press, Columbus, OH, 1999. 588 pp.
5(4): Leeson, A., and B.C. Alleman (Eds.), *Bioremediation of Metals and Inorganic Compounds*. Battelle Press, Columbus, OH, 1999. 190 pp.
5(5): Alleman, B.C., and A. Leeson (Eds.), *Bioreactor and Ex Situ Biological Treatment Technologies*. Battelle Press, Columbus, OH, 1999. 256 pp.
5(6): Leeson, A., and B.C. Alleman (Eds.), *Phytoremediation and Innovative Strategies for Specialized Remedial Applications*. Battelle Press, Columbus, OH, 1999. 340 pp.
5(7): Alleman, B.C., and A. Leeson (Eds.), *Bioremediation of Nitroaromatic and Haloaromatic Compounds*. Battelle Press, Columbus, OH, 1999. 302 pp.
5(8): Leeson, A., and B.C. Alleman (Eds.), *Bioremediation Technologies for Polycyclic Aromatic Hydrocarbon Compounds*. Battelle Press, Columbus, OH, 1999. 358 pp.

KEYWORD INDEX

This index contains keyword terms assigned to the articles in the eight-volume proceedings of the Fifth International In Situ and On-Site Bioremediation Symposium (San Diego, California, April 19-22, 1999). Ordering information is provided on the back cover of this book.

In assigning the terms that appear in this index, no attempt was made to reference all subjects addressed. Instead, terms were assigned to each article to reflect the primary topics covered by that article. Authors' suggestions were taken into consideration and expanded or revised as necessary. The citations reference the eight volumes as follows:

5(1): Alleman, B.C., and A. Leeson (Eds.), *Natural Attenuation of Chlorinated Solvents, Petroleum Hydrocarbons, and Other Organic Compounds*. Battelle Press, Columbus, OH, 1999. 402 pp.
5(2): Leeson, A., and B.C. Alleman (Eds.), *Engineered Approaches for In Situ Bioremediation of Chlorinated Solvent Contamination*. Battelle Press, Columbus, OH, 1999. 336 pp.
5(3): Alleman, B.C., and A. Leeson (Eds.), *In Situ Bioremediation of Petroleum Hydrocarbon and Other Organic Compounds*. Battelle Press, Columbus, OH, 1999. 588 pp.
5(4): Leeson, A., and B.C. Alleman (Eds.), *Bioremediation of Metals and Inorganic Compounds*. Battelle Press, Columbus, OH, 1999. 190 pp.
5(5): Alleman, B.C., and A. Leeson (Eds.), *Bioreactor and Ex Situ Biological Treatment Technologies*. Battelle Press, Columbus, OH, 1999. 256 pp.
5(6): Leeson, A., and B.C. Alleman (Eds.), *Phytoremediation and Innovative Strategies for Specialized Remedial Applications*. Battelle Press, Columbus, OH, 1999. 340 pp.
5(7): Alleman, B.C., and A. Leeson (Eds.), *Bioremediation of Nitroaromatic and Haloaromatic Compounds*. Battelle Press, Columbus, OH, 1999. 302 pp.
5(8): Leeson, A., and B.C. Alleman (Eds.), *Bioremediation Technologies for Polycyclic Aromatic Hydrocarbon Compounds*. Battelle Press, Columbus, OH, 1999. 358 pp.

A

C

D

E

J

K

L

M

N

O

P

Q

R

S

T

U

V

W

Z